I0640796

Phytochemical Phylogeny

Phytochemical Phylogeny

PROCEEDINGS OF THE
PHYTOCHEMICAL SOCIETY SYMPOSIUM
BRISTOL, APRIL 1969

Edited by

J. B. HARBORNE

Department of Botany
University of Reading, Reading, England

1970

ACADEMIC PRESS
LONDON AND NEW YORK

Library
I.U.P.
Indiana, Pa.

581.38   P569h

C. 1

ACADEMIC PRESS INC. (LONDON) LTD
Berkeley Square House
Berkeley Square,
London, W1X 6BA

U.S. Edition published by
ACADEMIC PRESS INC.
111 Fifth Avenue,
New York, New York 10003

Copyright © 1970 By ACADEMIC PRESS INC. (LONDON) LTD

All Rights Reserved

No part of this book may be reproduced in any form by photostat, microfilm, or any other
means, without written permission from the publishers

Library of Congress Catalog Card Number: 71-117130

SBN: 12-324666-0

PRINTED IN GREAT BRITAIN BY
SPOTTISWOODE, BALLANTYNE & CO. LTD
LONDON AND COLCHESTER

Contributors

K. ALLEN, *Department of Botany, University of Bristol, Bristol, England*

S. BARTNICKI-GARCIA, *Department of Plant Pathology, University of California, Riverside, California, USA*

D. BOULTER, *Department of Botany, University of Durham, Durham, England*

N. G. CARR, *Department of Biochemistry, University of Liverpool, Liverpool, England*

W. G. CHALONER, *Department of Botany, University College, London, England*

I. W. CRAIG, *Department of Biochemistry, University of Liverpool, Liverpool, England*

P. ECHLIN, *Botany Department, University of Cambridge, Cambridge, England*

M. V. LAYCOCK, *Atlantic Regional Laboratory, Halifax, Nova Scotia, Canada*

T. J. MABRY, *The Cell Research Institute and Department of Botany, The University of Texas at Austin, Austin, Texas, USA*

J. W. McCLURE, *Department of Botany, Miami University, Oxford, Ohio, USA*

B. W. NICHOLS, *Unilever Research Laboratory, Colworth House, Sharnbrook, Bedford, England*

J. RAMSHAW, *Department of Botany, University of Durham, Durham, England*

G. SHAW, *School of Chemistry, University of Bradford, Bradford, England*

B. R. THOMAS, *Department of Organic Chemistry, LTH, Lund, Sweden*

E. W. THOMPSON, *Department of Botany, University of Durham, Durham, England*

B. L. TURNER, *The Cell Research Institute and Department of Botany, The University of Texas at Austin, Austin, Texas, USA*

R. L. WATTS, *Department of Biochemistry, Guy's Hospital Medical School, London, England*

H. W. WOOLHOUSE, *Department of Botany, University of Leeds, Leeds, England*

Preface

The inter-related problems of the origin of life and the evolution of life forms, traditionally regarded as the province of the biologist, have recently been tackled with increasing success by phytochemists. Their achievements have been mainly due to the availability of very sophisticated techniques (particularly gas:liquid chromatography–mass spectrometry) for analysing the organic molecules of biological materials. It is the purpose of this book to review the impact of phytochemistry on plant phylogeny and to capture some of the excitement of the new vistas opened up by recent discoveries. This is the common theme running through a book written by a range of experts from many different scientific fields but which should be read with interest by all plant scientists.

Who can fail to be intrigued by evidence for photosynthesis in the isolation of isoprenoid hydrocarbons from fossil specimens 3·1 billion years old and by the presence of chlorophyll-like pigments in fossil plants that have been buried in lignite deposits for 50 million years? Equally exciting are the startling parallel findings of the ammonia-loving bacterium *Kakabekia* as a fossil in Precambrian Gunflint chert of North America and in living form on the walls of Harlech Castle, North Wales. These and other novel reports are reviewed by Echlin and by Chaloner and Allen in the first two of the four chapters devoted to the application of phytochemistry to palaeobotany. In Chapter 3, Shaw describes the first "breakthrough" in elucidating the chemical nature of the intractable organic matters, sporopollenins, present in the outer coating of both present day and fossil pollens. He shows that these sporopollenins arise by polymerization of fatty acid esters of carotenoid pigments produced in the pollen during ripening. The resistance of carotenoid and other isoprenoids to the ravages of time are also clearly brought out in the following chapter by Thomas, who reviews, *inter alia*, the evidence for believing that fossil resins are derived almost entirely from the heartwood terpenoids present in ancient conifer forests of the Cretaceous era.

In the remainder of the book, attention is turned to the relationship between phytochemistry and phylogeny in living plants. In the first of three chapters on lower plants, Bartnicki-Garcia relates cell wall composition and other available biochemical markers to fungal phylogeny. Then, Nichols, Carr and Craig consider the biochemistry of algae with a phyletic viewpoint. The latter two authors also critically examine the evidence for and against the theory of the bacterial origin of higher plant chloroplasts. Higher plants themselves and their small and large molecules are covered in the remaining six chapters. A general account by Rosemary Watts of the evolutionary implications of nucleic acid and protein analysis is followed by a report by Boulter and his colleagues of very recent results of amino acid sequence analyses on three higher plant cytochromes c. The effect of environment on enzyme proteins in plants can be considerable; this is discussed in the next chapter by Woolhouse, who illustrates his theme by reference to the acid phosphatases in the roots and the carboxydismutases in the leaves of plants.

While a study of the macromolecules of higher plants has great potential, the data at present are of limited value to phylogeny because of sampling difficulties. By contrast, low molecular weight constituents can rapidly be screened for in large plant populations and the results have already been used for phylogenetic purposes, as is shown in Chapters 10, 12 and 13. Thus, McClure discusses the relevance of flavonoid data to the evolution of the Lemnaceae, a family of aquatic angiosperms, Turner the application of essential oil analyses to hybridity studies in the North America junipers and Mabry the use of sesquiterpene lactone data in geographical studies of the ragweed genus, *Ambrosia*. While it is shown that phytochemical studies usually support phylogenetic conclusions based on analysis of biological characters, this is not always the case. The chapter by Turner is therefore especially timely since it provides an illustration of the power of phytochemical methods, in a case where the classical morphological approach to an evolutionary problem has been shown to give quite the wrong answer.

The various essays in this book are based on review papers given at a Phytochemical Society Symposium held at the University of Bristol from March 31st to April 2nd 1969. The editor, as Secretary of the Society, would especially like to thank Dr. G. Eglinton, who was responsible for the scientific programme, and Mrs. Wendy Harrison, who dealt with the practical organization of this meeting. The editing of this volume has been made an easy task by the contributors, who have responded quickly to all the demands made upon them. Finally, the editor acknowledges the efficient and expert assistance of Academic Press in preparing the book for publication.

January 1970   J. B. HARBORNE

Contents

CHAPTER 1

The Origins of Plants

Patrick Echlin

CHAPTER 2

Palaeobotany and Phytochemical Phylogeny

W. G. Chaloner and K. Allen

CHAPTER 3

Sporopollenin

G. Shaw

ix

CHAPTER 4

Modern and Fossil Plant Resins

B. R. Thomas

CHAPTER 5

Cell Wall Composition and Other Biochemical Markers in Fungal Phylogeny

S. Bartnicki-Garcia

CHAPTER 6

Comparative Lipid Biochemistry of Photosynthetic Organisms

B. W. Nichols

CHAPTER 7

The Relationship between Bacteria, Blue-green Algae and Chloroplasts

N. G. Carr and I. W. Craig

CHAPTER 8

Proteins and Plant Phylogeny

R. L. Watts

CHAPTER 9

Amino Acid Sequence Studies of Plant Cytochrome *c*, with Particular Reference to Mung Bean Cytochrome *c*

D. Boulter, M. V. Laycock, J. Ramshaw and E. W. Thompson

CHAPTER 10

Molecular Approaches to Population Problems at the Infraspecific Level

B. L. Turner

CHAPTER 11

Environment and Enzyme Evolution in Plants

H. W. Woolhouse

CHAPTER 12

Secondary Constituents of Aquatic Angiosperms

Jerry W. McClure

CHAPTER 13

Infraspecific Variation of Sesquiterpene Lactones in
Ambrosia (Compositae): Applications to Evolutionary Problems
at the Populational Level

Tom J. Mabry

CHAPTER 1

The Origins of Plants

PATRICK ECHLIN

Botany Department, University of Cambridge, Cambridge, England

I. Introduction

The word "plants" is interpreted broadly in this present context and will be considered to include microorganisms as well as the more conventional and readily recognizable plant forms. The studies on early plants and a reasoned explanation of the results embrace a number of very different disciplines including organic and inorganic geochemistry, atmosphere sciences, microbiology and biochemistry, micropalaeontology and cell biology. These and other branches of science provide small pieces of what is little more than circumstantial evidence when seen in isolation; however, when the several lines of evidence come to bear on the same problem, some definitive advances begin to accrue.

Although there still remains some doubt concerning the age of the universe, it is generally agreed that the planet Earth is approximately 5 billion (10^9) years (b.y.) old, and that life originated somewhere between 4·0 and 3·5 b.y. ago. No discussion will be made on the origin of life, since a detailed consideration of the events leading up to the emergence of self-duplicating living systems can be found in the studies of Oparin (1957), Bernal (1967), Wald (1964), Abelson (1966), Calvin (1967) and others.

II. Chemical Analysis and Dating of Precambrian Fossils

It is only within the last decade or so that serious thought has been given to the existence of plants or even life in the geologic period preceding the Cambrian era, and the two reviews by Glaessner (1962, 1966) together with the extensive bibliography of Precambrian organisms compiled by Murray (1965) contain

1

many references to the earlier descriptions of Precambrian palaeontology. The Cambrian period of geological succession existed from approximately 0·5 to 0·6 b.y. ago and contains the fossilized remains of both plants and animals. The Precambrian or Archeozoic period by definition embraces the time between the origin of the earth and the beginning of the Cambrian, and until a few years ago was considered only rarely to contain any fossils of biologic origin. Thus the Precambrian era lasted from approximately 3·8 to 0·6 b.y., a period of time which comprises about 80% of the time of presumed biological activity on this planet.

There has been considerable scepticism in the past concerning the affinities of the Precambrian microfossils and even more doubt as to their metabolism and ecology. However, thanks to the new and sophisticated technology of gas chromatography, mass spectrometry and thin layer chromatography employed in analytical chemistry, it is possible to critically analyse organic matter *in situ* from rocks deposited thousands of million years ago. The techniques of scanning and transmission electron microscopy have revealed the cellular detail of numerous Precambrian microfossils.

An analysis and interpretation of the Precambrian fossil flora is fraught with many difficulties, and not the least of these is the accurate dating of the rocks in which the microfossils are found. Quantitative measurement of geologic time involves radioactive decay, and the estimates of time have been obtained by making radiometric determinations from rocks obtained at critical locations.

The solar system is a closed system and its elemental composition is the same as when it was formed, except in so far as it has been modified by the conversion of hydrogen to helium in the sun and by the decay of radioactive materials. The age of the elements is then calculated from the time when the naturally occurring radioactive series consisted entirely of the parent elements. When an atom breaks down or decays, it is transformed into daughter nuclides. The rate at which any given radioactive nuclide decays does not vary, although different nuclides will decay at different rates. Experiments have shown that the rates of decay of any given nuclide remain virtually constant in the geologic past. Determination of geologic age is thus made possible by the fact that a given radioactive nuclide decays at a uniform rate and thus forms a kind of geologic clock. Radiometric age dates are based on the following assumptions.

1. That the specimen containing radioactive material is representative of the age of the geologic body being dated.
2. That the specimen has neither lost nor gained radioactivity due to external causes such as solution or precipitation.
3. That the rate of decay of the radioactive parent material is known with sufficient accuracy.

Even if these assumptions apply, the radiometric dating methods are difficult. There are sources of error inherent in the analytical process; there exists the possibility of gains or losses in the specimen analysed due to processes other than radioactive decay, and finally there may be difficulty in establishing the stratigraphic position of the geologic body, particularly in the case of igneous rocks.

The principal natural radioactive series of importance in geochronology are given in Table I. There are a number of other long-lived radioactive nuclides, but their rate of decay is so slow (for example ^{174}Hf to ^{170}Yb has a half-life of $4·3 \times 10^{15}$ years) that they are of little use for dating on our planet.

The uranium–lead methods and the thorium–lead method for determining age are accurate, provided corrections have been made for contamination by lead from other sources. Because potassium-bearing minerals are more wide-spread than uranium-containing minerals, the potassium–argon method has certain advantages over the uranium and lead–thorium methods. Both the potassium–argon and the rubidium–strontium methods can be used on sedimentary rocks. This is most important, as it is these rocks which contain the micro objects of biological significance.

TABLE I

Radioactive elements used in geochronology

Parent element	Half-life (b.y.)	End products
^{238}U	4·5	^{206}Pb + ^{4}He
^{235}U	0·71	^{207}Pb + ^{4}He
^{232}Th	14	^{208}Pb + ^{4}He
^{87}Rb	51	^{87}Sr
^{40}K	1·3	^{40}A + ^{40}Ca

Having established the date for the microfossils, an interpretation of the morphological and chemical data must be accepted with caution. It must be quite certain for example that the microfossil and any chemical "fingerprints" it may leave are indigenous to the rock and not a result of contamination. Many of the Precambrian microstructures are morphologically rather simple and must be clearly separable from abiological artefacts. Nagy and Urey (1968) elegantly demonstrated how easy it is to confuse artefacts with microfossils by showing that "double-walled" particles with remarkable cell-like form, were inorganic siliceous spheroidal artefacts from a thin section of basaltic larva. It must be shown that the microfossils which are discovered in a deposit are truly representative of the original fauna and flora of that particular time. This is very difficult to assess accurately, and the probability remains that the Precambrian microflora may represent a highly selected group of organisms. Nearly all the deposits are found in cherts (rocks composed of very fine-grained silica), and the environments which gave rise to these deposits may have favoured initially the growth and later the preservation of the organisms. As will be shown later, the modern counterparts of the Precambrian microflora are invariably invested by a gelatinous sheath. Such a sheath or capsule if present in the Precambrian forms would provide a milieu for the entrapping of silica particles. The organism would become embedded in a sheath of jelly-like silica, which soon after the death of the cell would rapidly harden into the dense chert which would protect the delicate structures from extensive deformation and decay. Organisms which did not live in this siliceous embalming environment would not be preserved, and would leave no morphological trace of their existence. It must also be shown that chemical fossils associated with the cellular debris are not a result of contamination either from material which has leaked in from organic matter deposited later in the geologic record, or from handling or pre-treating the sample prior to its chemical analysis. This aspect of contamination has been highlighted by the preliminary analysis of the meteorite which recently landed in Mexico (Han *et al.* 1969). An analysis which was made within seven days of the meteorite landing revealed that the organic material in the surface layer of the meteorite was of biological origin and a result of terrestrial contamination. There was virtually no organic material in the interior of the meteorite.

It must also be shown that the organic matter has not been abiogenetically produced (McCarthy and Calvin, 1967), for there is now an increasing amount of evidence which suggests that many biologically significant molecules may be produced under abiogenetic conditions. For example, isoprenoids have been recovered from ethylene polymerization plants producing polyethylene for the plastics industry. It is, however, rather unlikely that abiological synthesis would produce the specific isoprenoid patterns found in the products of living cells.

Although geochemists are reasonably certain what happens to large molecules as they are degraded over a long period of time (see Eglinton, 1965; Eglinton and Calvin, 1967; Manskaya and Drozdova, 1968), many gaps still remain in our knowledge concerning the diagenesis of organic material. Nevertheless, the results of organic geochemistry have revealed the existence of a number of chemical fossils, which indicate that biological activity and in particular photosynthesis was occurring throughout most of the Precambrian. There are a large number of chemical fossils which are of interest in a general interpretation of biological activity in the Precambrian, but only three will be referred to in any detail, as they have an immediate bearing on tracing the origins of plants.

An analysis by mass spectrometry of the ratio of the relative amounts of the isotopes ^{12}C and ^{13}C in the organic material associated with a given microfossil compared with the isotope ratio found in the parent rock can provide evidence that photosynthesis was occurring at the time the organism was living. The two stable isotopes of carbon, ^{12}C and ^{13}C, are present in the atmosphere in the ratio of 99:1 and following the original work of Craig (1953) and Wickman (1956) it can be shown that any biological activity, such as photosynthesis, which incorporates inorganic carbon (such as CO_2) into organic carbon, does so with an enrichment effect in the direction of the ^{12}C isotope. This means that organic matter derived by biological activity has a slightly higher $^{12}C/^{13}C$ ratio compared with the $^{12}C/^{13}C$ ratio from the parent rock. Precipitation of calcite

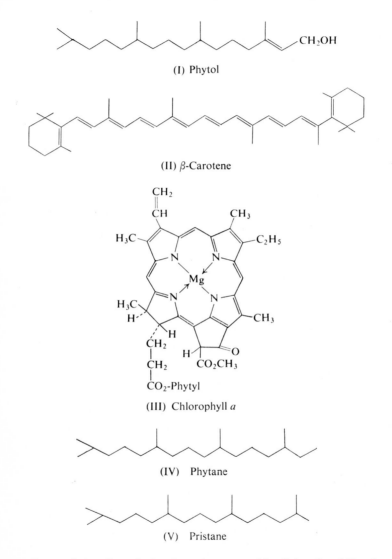

(I) Phytol

(II) β-Carotene

(III) Chlorophyll *a*

(IV) Phytane

(V) Pristane

in some Precambrian deposits has been interpreted by Schopf and Barghoorn (1967) as indicating the presence of free CO_2.

Isoprenoids, which are branched chains of 5-carbon units put together in a regular fashion, are widespread in many biological systems. The 5-carbon units are usually assembled in a head–tail fashion with a methyl group attached to every fifth carbon atom (see formula of phytol, I); but in a few instances such as β-carotene (II) they are linked tail to tail. Chlorophyll (III) is one of the most widely distributed molecules containing an isoprenoid chain, and although its fate under geological sedimentation is not clearly understood, traces of the molecule may be found associated with the organic remains of microfossils. The chlorophyll molecule consists essentially of two parts, the chlorophyllide head, a conjugated cyclotetrapyrrole, and the long chain alcohol phytol (I), which is attached to the tetrapyrrole by an ester linkage. When chlorophyll is decomposed, the phytyl chain is split off and may be converted either to the isoprenoid hydrocarbon phytane (IV) with the same number of carbon atoms or pristane (V) which has one less carbon atom. Both these branched alkanes may be extracted from suitable sediments, and provide reasonably good evidence for the existence of chlorophyll and presumably photosynthesis. The tetrapyrrole probably gives rise to the metallo-porphyrins which can also be isolated from organic material.

Having assembled the various types of evidence presently available from the Precambrian, which indicate that photosynthetic organisms were in existence, and having preached the necessary cautionary tale about the interpretation and validity of the evidence, let us examine the raw data which is presented to us in the Precambrian.

III. Morphology and Geochemistry of Precambrian Fossils

In spite of all the exacting criteria, the pieces of evidence clearly point to the existence of biological activity, three thousand million years ago (see Table II). The oldest known and readily recognizable microfossils have been found in some sedimentary rocks in South Africa which have been dated at about 3·1 b.y. The morphology and geochemistry of the microfossils have been described by Pflug (1966a), Ramsay (1966), Barghoorn and Schopf (1966), Schopf and Barghoorn (1967), Kvenvolden et al. (1969), Schopf et al. (1968) and Oro and Nooner (1967). A number of microfossils have been described which are thought to resemble bacteria and unicellular blue-green algae. The more detailed studies of Schopf and Barghoorn have shown that the bacteria-like organism(s?) is very small and is less than a micrometre in diameter, but in some samples does reveal a distinctly two-layered cell wall. The alga-like organism is much larger, measuring up to 19 μm in diameter with a distinctly reticulate cell wall, and occasionally contains preserved remains of the original cytoplasm. The geochemical findings include $^{12}C/^{13}C$ ratios, phytane and pristane, both of which are indicative of photosynthesis, and a number of optically active amino acids which are good indicators of general biological activity. However, Oro and Nooner (1967) show that the $^{12}C/^{13}C$ ratios in the extractable and non-extractable samples are the same, which would indicate that photosynthesis was unlikely to have occurred at this time. It is well known that data obtained from isotope fractionation are less reliable from older deposits.

Some even older microfossils are thought to be present in the rocks of the Onverwacht series which lie below the Fig Tree deposits. Engel *et al.* (1968) and Nagy and Urey (1968) describe a number of spheroidal and cupshaped carbonaceous alga-like bodies up to 100 μm in diameter together with a number of filamentous structures. These structures occur in carbon-rich cherts which lie 10,000 metres stratigraphically below the Fig Tree series, and have been given a tentative date of 3·2 b.y. The preservation of these microstructures is rather poor, and only minimal geochemical data are available on the organic material associated with the sediments. If these structures are of biological origin, then they represent the oldest recognizable life-forms on our planet. The Onverwacht sediments are considered to be the oldest little-altered sedimentary rocks on Earth, and as there would seem little chance of finding any earlier deposits, the morphological and geochemical remains of cellular life on this planet may never be revealed.

The Bulawayan deposits of Rhodesia dated at 2·7 b.y. contain some structures which in their gross morphology resemble stromatolites (McGregor, 1941). Stromatolites are the most common Precambrian fossil and are calcareous masses of quite distinct form, usually showing in cross section thin curved lamellae which are considered to be due largely to the activity of lithophilic blue-green algae. The stromatolites are thought to be formed by the trapping and binding of pre-existing particulate carbonate sediments by mats of blue-green algae. It is interesting to note here that the principal forms of blue-green algae which form mats are filamentous rather than unicellular. Although the importance of stromatolites decreases from the Cambrian onward, they are found throughout the geologic record up to the present day. They show little microstructure and no geochemical data are presently available, but it is remarkable that when one compares modern and Precambrian forms, there is practically no change in their morphology over a period of nearly 3·0 b.y.

The Soudan Iron Formation, a deposit of carbonaceous rocks from Northeast Minnesota, dated at about 2·7 b.y., has been investigated by Cloud *et al.* (1965) and Cloud and Licari (1968), and the geochemistry described by Meinschein (1965). The morphological investigations of material in pyrite balls associated with iron oxides have revealed microstructures which, although of doubtful affinity, are nevertheless considered to resemble bacteria. The $^{13}C/^{12}C$ ratios, and the finding of phytane and pristane in the organic extracts of these rocks are indicative of a vital origin, though it is not entirely clear whether these materials are endemic to the rock. An analysis of the ratio of the two isotopes of sulphur, $^{34}S/^{32}S$, in the Minnesota deposits indicates that this element may have been metabolized by living organisms. Feely and Kulp (1957) and later other workers (see La Riviere, 1966) have investigated the occurrence of sulphur deposits in sedimentary rocks, and it would appear that an isotope enrichment occurs, similar to that which occurs with carbon. It is known that some anaerobic bacteria are able to reduce sulphates to hydrogen sulphide

PATRICK ECHLIN

Time × 10^6 years	Stratigraphic deposit	Environment	Geochemistry		Metabolism	Organisms of uncertain affinity
500	Sinian China 0·8 b.y.	Terrestrial (?) Aquatic (?)			Oxygen at 10% P.A.L.	
1000	Bitter Springs Aust. 1·0 b.y.	Shallow marine Mats Planktonic (Littoral ?)	Phytane/Pristane, Tetrapyrroles			
	Belt Series Montana 1·1 b.y.	Aquatic				*Fibularix* (bacteria)
	Nonesuch Series 1·1 b.y.	Aquatic (?)	Phytane/Pristane, Tetrapyrroles Optical activity			
	Mount Isa Aust. 1·4 b.y.	Shallow lake bottom		RED BEDS	Aerobic respiration O_2 at 1% P.A.L.	Yes
1500	Huronian Minn. 1·5 b.y.					
	Svecofennian Finland 1·7 b.y.	Aquatic	$^{13}C/^{12}C$		O_2 produced by photosynthesis increases in hydrosphere and atmosphere	*Corycium* (alga ?)
	Ketilidian Greenland 2·0 b.y.	Fresh water	$^{13}C/^{12}C$, Phytane/ Pristane			
						Eosphaera (colonial alga ?)
2000	Gunflint Iron Formation 1·9 b.y.	Shallow water	Phytane/Pristane, $^{13}C/^{12}C$		Aerophilic photoautotrophs	*Kakabekia* (stalked bacteria ?)
	Witwatersrand S. Afr. 2·1 b.y.	Aquatic	Am/Ac Mono- saccharides $^{13}C/^{12}C$			Yes
	Cobalt Series Canada 2·2 b.y.	Lake bottom (Anaerobic)				
2500	Soudan Iron Minn. 2·7 b.y.	Aquatic (?)	$^{13}C/^{12}C$, Phytane/ Pristane	BANDED RED BEDS	Any photosynthetic O_2 used up by reduced acceptors Microaerophilic	
	Bulawayan Africa 2·7 b.y.	Aquatic (10m)	$^{13}C/^{12}C$ ratio (C as CO_2)		Photoautotrophs	
3000	Fig Tree Series S. Afr. 3·1 b.y.	Aquatic (10m)	Phytane/Pristane, $^{13}C/^{12}C$ L Am/Ac (C as CO_2)		Anaerobic photo- autotrophs	
	Onverwacht Series S. Afr. 3·2 b.y.	Aquatic (?)	Carbonaceous filaments		Photophosphoryl- ation (?) Anaerobic hetero- trophs Chemoorganic Chemolithic	Yes
3500						

II

the Precambrian period

| | | Procaryotes | | | Eucaryotes | |
| | | Blue-green algae | | Complex | | |
Bacteria	Unicells	Colonial	Filamentous	filamentous	Algae	Fungi
	Yes	Yes	Yes	(?)		
Yes	Yes	Yes	Yes	Yes	Caryosphaeroides (?)	Eumycetopsis (?)
	Yes	Yes				Yes (?)
		Yes	Nostoc-like	Yes		
				Vallenia (?)		
Eoastrion (iron oxidizing bacteria?)	Yes	Yes	Animikiea	Gunflintia		Huronospora (?)
Actinomycetes (branched)						
Yes ?	Yes (?)					
Eobacterium	Archeosphaeroides					

STROMATOLITES

which may react with more sulphate to give sulphur deposits such as those in the salt domes in Louisiana and Texas. Bacterial reduction of sulphate to sulphide is considered by some workers to result in a higher $^{32}S/^{34}S$ ratio in the sulphide and sulphur than in the sulphate from which it originated, indicating enrichment of ^{32}S during their formation.

The Cobalt series of Canada, dated at 2·2 b.y., contain microstructures which are similar in form to modern Actinomycetes associated with sulphide crystals (Jackson, 1967). No geochemical analysis is presently available, although the authors consider that the deposit represents the remains of anaerobic bacterial forms which metabolized by reducing sulphate. Prashnowsky and Schidlowski (1967) have identified amino acids and monosaccharides in carbonaceous materials from rocks in the South African Witwatersrand dated at 2·1 b.y. The carbon isotope enrichment indicates photosynthesis was occurring, and replicas of freshly cleaved fracture planes revealed single and aggregated microstructures resembling cell colonies. Some Precambrian organisms together with an analysis of the organic material from the Ketilidian series of south-west Greenland have been described by Bondesen et al. (1967) and Pedersen and Jorgen (1968). The organic remains have been described in low metamorphic rocks which have been given an age of approximately 2·0 b.y. Most of the organic remains are microscopic globules and cell-like fragments, although there are a number of structures which are similar in morphology to some modern bacteria. The best preserved structure is a large (up to $\frac{1}{2}$ mm) complex globular structure which has been assigned to a new genus *Vallenia*. The carbon isotope composition gives a ^{13}C figure which is within the range for biogenic carbon produced by photosynthesis and is very close to the figure for fresh water algae. Phytane and pristane have been found in small amounts together with a number of other organic compounds, which the authors interpret as indicating advanced biological activity.

Probably the best known of the Precambrian deposits is the Gunflint iron formation which was described fifteen years ago by Tyler and Barghoorn (1954). Since that time there have been a number of papers describing both the microflora and the chemical analysis of the organic remains (Moorhouse and Beales, 1962; Barghoorn and Tyler, 1965; Oro et al. 1965; Cloud, 1965; Cloud and Hagen, 1965; Schopf et al. 1965; Schopf, 1967; Licari and Cloud, 1968). The Gunflint iron formation, which is thought to be just under 2 b.y. old, is composed of shales, limestones and cherts and the best preserved microstructures are seen in the unmetamorphosed carbon-rich cherts usually associated with oxidized iron deposits, which in their gross morphology resemble stromatolites. The most frequently found microfossils in the Gunflint material are a series of unbranched filaments and various spheroidal bodies, and Schopf and Barghoorn are able to distinguish 8 genera and 12 species of morphologically distinct organisms. Although some of the filamentous forms are thought to resemble fungi, the majority bear a close similarity to modern blue-green algae. The unicellular organisms are similar in some respects to extant unicellular blue-green algae and bacteria, particularly living manganese and iron oxidizing bacteria, while also showing some common characters with some of the early Precambrian microfossils. Some organisms such as the tripartite umbrella like *Kakabekia*, and the multi-compartmented spheroidal *Eosphaera*, are of uncertain relationship. Some recent work by Siegel and Giumarro (1966) and Siegel and Siegel (1968) on the viability of modern microorganism to presumptive primitive atmospheres revealed some rather startling results. These investigators isolated and described what appears to be a living form of an organism (*Kakabekia barghoorniana*) which last appeared in the Gunflint chert (*Kakabekia umbellata*) (Fig. 1). The living organism (Fig. 2) was isolated from soil at the base of a wall of Harlech Castle. The soil was found to contain relatively high amounts of ammoniacal compounds and subsequent experiments by Siegel showed that the newly discovered organism was able to live and grow in low concentrations of oxygen and high concentrations of ammonia and methane.

Chemical analysis of the organic remains from the Gunflint provides further evidence for the existence of biochemical activity during the Gunflint era. The finding of a substantial enrichment of the ^{12}C isotope and the presence of phytane and pristane, together with the geochemical remains of the chlorophyllide head of tetrapyrroles, all suggest that photosynthesis was occurring. In the ensuing billion years, a number of different microfossils have been described from various middle and late Precambrian deposits, all over the world. Curious carbonaceous sac-like structures called *Corycium* have been described from the 1·7 b.y. old Svecofennian rocks of Finland, and isotope and geochemical analysis suggests these may be of algal origin (Marmo, 1959; Rankama, 1954; Ohlson, 1962). Spheres, colonies and complex filaments, some of which have a distinct affinity with the blue-green algae, have been described from the Upper Huronian of Michigan by Tyler *et al.* (1957), and organisms of rather uncertain affinity, possibly bacteria, have been described from the pyritic shale of Mount Isa, Queensland, dated at 1·5 b.y. by Love and Zimmerman (1961).

Recent papers by Cloud *et al.* (1969) and Gutstadt and Schopf (1969) describe some interesting microfossils from the Beck Spring Dolomite from southern California dated at somewhere between 1·2 and 1·4 b.y., Gutstadt and Schopf describe microfossils which resemble blue-green algae and sulphur bacteria. Cloud *et al.* consider that some of the microfossils which contain cellular contents are Eucaryotes and members of the Chlorophycean or Chrysophycean algae. If this interpretation is correct, these microfossils would represent the oldest eucaryotic cells.

Well-preserved fossils, which appear to be the remains of blue-green algae and fungi, have been described by Pflug (1964, 1965a, 1965b and 1966b) from the 1·1 b.y. old Belt series of Montana. Pflug describes aggregated spheroids which may represent the reproductive structures of primitive fungi, although the absence of a recognizable nucleus casts some doubt on this interpretation.

The Nonesuch shale of northern Michigan which was deposited at approximately the same time, although containing only poorly preserved microfossils, contains a wealth of geochemical data. The presence of, amongst other things, phytane, pristane, sterane-type hydrocarbons related to chlorophyll and ^{12}C

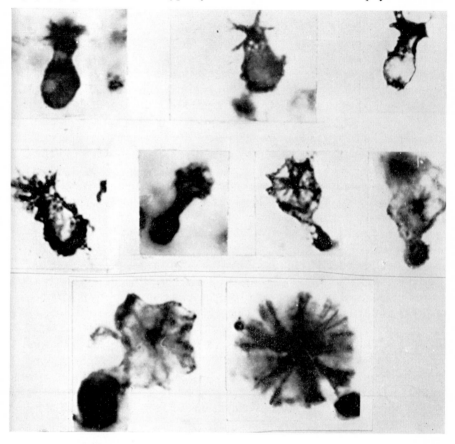

FIG. 1. Photographs of *Kakabekia umbellata* Barghoorn (×1800) obtained from Gunflint chert, Ontario, Canada. The various pictures probably represent different ontogenetic stages in the growth of the organism (Reproduced by kind permission of Professor E. S. Barghoorn, Harvard University)

isotope enrichment all indicate that the metabolic pathways at the time of the Precambrian were essentially the same as in modern plants (Meinschein *et al.* 1964; Barghoorn *et al.* 1965).

The best preserved collection of late Precambrian microfossils is undoubtedly the billion year old material from the Bitter Springs formation in Central Australia. Barghoorn and Schopf (1965) and Schopf (1968) have described a number of structurally and organically well preserved blue-green and green algae from these laminated cherts. Schopf (1968) has made a detailed study of

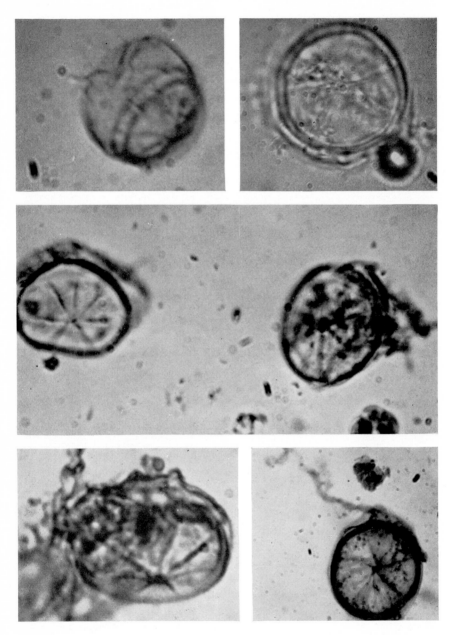

FIG. 2. Morphological variations in *Kakabekia barghoorniana* Siegel (Reproduced by kind permission of Professor S. M. Siegel)

these microfossils which are so well preserved that they retain their three dimensional configuration in which it is possible to make out internal structures within the cells such as pyrenoids and nuclei. Twenty-four new genera of blue-green algae are recognized, as well as a number of algae which may be referred to modern blue-greens, including some morphologically advanced forms. Schopf considers that the presence of green algae clearly establishes for the first time the existence of eucaryotic cells in the Precambrian.

IV. BEARING OF THE FOSSIL EVIDENCE ON PLANT EVOLUTION

There is thus good evidence that by about a billion years ago eucaryotic cells were established on our planet. The palaeobotany of the billion years from the late Precambrian until the present day is relatively well known although there are many gaps. By the beginning of the Cambrian era nearly 600 m.y. ago many invertebrate animals are considered to have been in existence. By late Ordovician about 430 m.y. ago all the major invertebrate group were well established, and the first vertebrates have been common in the sea since Devonian times 400–350 m.y. ago. The search for the earliest vascular plants has occupied the attention of many palaeobotanists for a number of years. A number of different criteria have been suggested to adequately define land plants, but the presence in the fossil of isolated xylem elements is considered to be of prime importance (Chaloner, 1964). There seems to be general agreement that the first truly land plants are to be found in the lower Devonian or possibly the upper Silurian about 400 m.y. ago. Once established, however, the land plants rapidly evolved by Carboniferous times (350–300 m.y. ago).

There is general agreement among botanists that the green algae are the most probable ancestors of vascular plants because of the close biochemical similarity between the two groups, and the fact that the green algae are one of the few algal groups who have successfully made the transition from marine to fresh water. There appears to be little trace of the transition between an algal ancestor and first land dwelling vascular plants.

Calcium-carbonate depositing green and red algae have been reasonably common fossils since the Ordovician between 450 and 500 m.y. ago, and some impressions from Silurian rocks have been interpreted as brown algae. It is not the intention of this present paper to discuss either the origin of vascular plants, or the transition forms which may have existed from the late Precambrian to late Silurian times. For a more detailed account of the origin of land plants, the reader is referred to the following references: Andrew, 1961; Banks, 1966, 1968a, 1968b; Chaloner, 1967; Harland et al. 1967; Cloud, 1968a.

By the beginning of the Cambrian era, plants had become established, at least in the oceans and estuaries, if not on the land itself, and a number of conclusions may be drawn from the Precambrian flora.

An examination of the bacteria and blue-green algae-like organisms reveals an apparent progression from unicellular forms through filamentous forms, to forms which may best be described as "complex filamentous". There is a gradual increase in morphological complexity.

The microfossils in the Precambrian up to the Bitter Springs material appear to have been derived from procaryotic organisms. All living organisms can be divided quite simply into two main groups, the Procaryota and the Eucaryota. The Procaryota which are comprised principally of the bacteria and blue-green algae, lack a discrete membrane bound nucleus and mitosis and meiosis. There are no known multicellular or distinctly differentiated procaryotic organisms and they are characterized by a lack of internal compartmentalization (Echlin and Morris, 1965). The Eucaryota, which comprise all other living organisms from Amoeba to Man, have the complex compartments of mitochondria and chloroplasts (where appropriate), have a discrete membrane bound nucleus and variously exhibit mitosis and meiosis. It has been generally assumed that procaryotic cells were likely to have been the first cell forms and the evidence from Precambrian microfossils gives further support to this contention. Bacteria and blue-green algae constitute a small but important part of the flora of the extant biosphere, and while the bacteria are generally widespread, the blue-green algae tend to occupy specialized, and quite frequently unfavourable, niches. Both groups appear to show little change from their Precambrian ancestors. The first eucaryotic organisms probably first appeared about 1·5 b.y. ago, although the first record of them appears to be the Bitter Springs material. As to how the Eucaryota may have evolved is at the moment an open question, although there is an increasing amount of evidence which suggests that a process of endosymbiosis of small procaryotic cells with larger procaryotic cells may have been involved (see Echlin, 1970 and Carr and Craig, Chapter 7).

It is now clear that the first plants on our planet gave rise to the 20% oxygen presently found in our atmosphere. Urey (1957) had shown some years ago that the short wave ultraviolet photo-dissociation of the small amount of water vapour present in the primordial atmosphere would soon become self-limiting and would have only contributed a small amount of oxygen to the atmosphere. Berkner and Marshall (1964, 1965) and later Cloud (1968b) were able to show that the oxygen arose by green plant photosynthesis. The bacteria-like organisms in the Fig Tree deposits were probably similar in metabolism to modern photosynthetic bacteria, and as it is generally agreed that our planet was initially anoxic and mildly reducing, these conditions would have favoured organisms like photosynthetic bacteria. Such organisms although they are able to reduce carbon compounds to carbohydrate, do not use water as a hydrogen donor, and consequently do not give off oxygen in photosynthesis. In fact, they are strict anaerobes when photosynthesizing and are inhibited by oxygen. The blue-green algae although morphologically quite distinct from higher plants, share a common characteristic in that they both evolve O_2 during photosynthesis and it is thought that the blue-green algae were probably the first true plants (Echlin, 1966a, 1966b, 1969). These algae were probably the only plants on our planet for approximately 1·5 to 2·0 b.y., giving way to more complex eucaryotic cells, which in turn evolved to the higher algae and eventually to land plants.

Initially the appearance of the photosynthetic and oxygenic blue-green algae would have been an embarrassment to anaerobic organisms, and it is likely that the oxygen would have been quickly reduced, either by reactions coupled enzymatically with the oxidation of reduced organic material, or by the oxidation of reduced iron compounds to oxides. As stated earlier, many of the micro-fossil remains in the Precambrian are associated with extensive deposits of oxided iron compounds (Licari and Cloud, 1968). Gradually oxygen became a more tolerable part of the atmosphere and as well as acting as an absorbent for the intensive ultraviolet flux which bathes our planet, would permit aerobic respiration to occur with its twenty-fold increase in available energy.

Surprising as it may seem, it is possible to make some reasonably accurate speculations concerning the habitats of these early plants. Initially the organisms would have been in small geothermally heated pools and then in the primitive seas and in at least thirty feet of water, which would prevent ultra-violet radiation damage. As the oxygen concentration built up, the organisms would have inhabited shallow water or the sea as in the Gunflint deposits, or shallow freshwater as in deposits recently found in Greenland and Australia. By the time of the Bitter Springs deposits, Schopf has good evidence that the plants were in mats, some of which may have been littoral, although appreciable invasion of the land would be unlikely to have occurred much before 500 b.y. ago.

V. CONCLUSION

Some considerable problems still remain. The problem concerning the origin of eucaryotic cells has already been alluded to, and still remains un-resolved. An even greater mystery is the apparent lateness in appearance in the geologic record of multicellular plants and animals. The period of the Pre-cambrian procaryotic microflora lasted for about half the total biological history of the earth, and it is only in the last 400–500 million years that there has been a great outburst in evolutionary diversity. The answer to this may lie in three factors. In order that evolution could proceed to the level of multi-cellular and differentiated plant and animal organisms, cells had to become eucaryotic, aerobic and terrestrial. For there are, as far as I am aware, no anaerobic multicellular organisms, and no fully differentiated procaryotic cells present on this planet, and the conditions for this to occur did not occur until about 500 million years ago.

ACKNOWLEDGEMENT

The author is grateful to Miss Ruth Braverman for her patient secretarial assistance and for help in compiling the bibliography.

REFERENCES

Abelson, P. H. (1966). Chemical events on the primitive earth. *Proc. natn. Acad. Sci.* **55**, 1365.

Andrew, H. J. (1961). "Studies in Paleobotany." Wiley, New York.

Banks, H. P. (1966). Early land plants and some of their relatives. *Bioscience*, **16**, 422.

Banks, H. P. (1968a). The stratigraphic occurrence of early land plants and its bearing on their origin. "Proc. Int. Sym. on Devonian System" (D. H. Oswald, ed.). Calgary, Canada.

Banks, H. P. (1968b). The early history of land plants. *In* "Evolution and Environment" (E. T. Drake, ed.). Yale U.P., New Haven.

Barghoorn, E. S., Meinschein, W. G., and Schopf, J. W. (1965). Palaeobiology of a Precambrian shale. *Science*, **148**, 461.

Barghoorn, E. S., and Schopf, J. W. (1965). Microorganisms from the late Precambrian of Central Australia. *Science*, **150**, 337.

Barghoorn, E. S., and Schopf, J. W. (1966). Microorganisms three billion years ago from the Precambrian of S. Africa. *Science*, **152**, 758.

Barghoorn, E. S., and Tyler, S. A. (1965). Microorganisms from the Gunflint chert. *Science*, **147**, 563.

Berkener, L. V., and Marshall, L. C. (1964). The history of oxygenic concentration in the Earth's atmosphere. *Discuss. Faraday Soc.* **37**, 122.

Berkener, L. V., and Marshall, L. C. (1965). On the origin and rise of oxygen concentration in the Earth's atmosphere. *J. Atmos. Sci.*, **22**, 225.

Bernal, J. D. (1967). "The origin of life." Weidenfeld and Nicolson, London.

Bondesen, E., Pedersen, K. R., and Jorgensen, O. (1967). Precambrian organisms and the isotopic composition of organic remains in the Ketilidian of south-west Greenland. *Meddr om Grønland*, **164**, 1.

Calvin, M. (1967). Chemical evolution. *In* "Progress in Theoretical Biology" (F. M. Snell, ed.). Academic, New York and London.

Chaloner, W. G. (1964). An outline of Precambrian and pre-Devonian microfossil records: evidence of early land plants from microfossils. *Tenth Int. Bot. Cong. Abstracts*, 16–17.

Chaloner, W. G. (1967). Spores and land plant evolution. *Rev. Palaeobotany Palynology*, **1**, 83.

Cloud, P. E. (1965). Significance of the Gunflint (Precambrian) microflora. *Science*, **148**, 27.

Cloud, P. E. (1968a). The pre-metazoan evolution and the origins of the Metazoa. *In* "Evolution and Environment" (E. T. Drake, ed.). Yale U.P., New Haven.

Cloud, P. E. (1968b). Atmosphere and hydrosphere evolution on the primitive Earth. *Science*, **160**, 729.

Cloud, P. E., Gruner, J. W., and Hagen, H. (1965). Carbonaceous rocks of the Soudan Iron Formation (early Precambrian). *Science*, **148**, 1713.

Cloud, P. E., and Hagen, H. (1965). Electron microscopy of the Gunflint microflora: preliminary results. *Proc. natn. Acad. Sci.* **54**, 1.

Cloud, P. E., and Licari, G. R. (1968). Microbiotas of the banded iron formations. *Proc. natn. Acad. Sci.* **61**, 779.

Cloud, P. E., Licari, G. R., Wright, L. A., and Troxel, B. W. (1969). Proterozoic eucaryotes from Eastern California. *Proc. natn. Acad. Sci.* **62**, 623–630.

Craig, H. (1953). The geochemistry of the stable carbon isotopes. *Geochim. cosmochim. Acta*, **3**, 53.

Echlin, P. (1966a). Origins of photosynthesis. *Sci. J.* April 1966.

Echlin, P. (1966b). The blue-green algae. *Scient. Am.* June 1966.

Echlin, P. (1969). Primitive photosynthetic organisms. "Proceedings 2nd International Conference of Organic Geochemistry." Pergamon, Oxford (In press).

Echlin, P. (1970). Photosynthetic apparatus in Prokaryotes and Eukaryotes. *In* "Organization and Control in Prokaryotic and Eukaryotic cells" (H. P. Charles and B. C. J. G. Knight, eds.). Cambridge U.P., London.

Echlin, P., and Morris, I. (1965). The relationship between bacteria and blue-green algae. *Biol. Rev.* **40**, 143.

Eglinton, G. (1965). Recent advances in organic geochemistry. *Sonderdr. Geol. Rdsch.* **55**, 551.

Eglinton, G., and Calvin, M. (1967). Chemical fossils. *Scient. Am.* **216**, 32.

Engel, A. E. J., Nagy, B., Nagy, L. A., Engel, C. G., Kremp, G. O. W., and Dreco, C. M. (1968). Alga-like forms in Onverwacht Series, South Africa. Oldest recognized life-like forms on earth. *Science*, **161**, 1005.

Feely, H. W., and Kulp, J. L. (1957). Origin of Gulf Coast salt-dome sulphur deposits. *Bull. Am. Ass. Petrol. Geol.* **41**, 1802.

Glaessner, M. F. (1962). Precambrian fossils. *Biol. Rev. Cambridge Philo. Soc.* **37**, 467.

Glaessner, M. F. (1966). Precambrian palaeontology. *Earth Sci. Rev.* **1**, 29.

Gutstadt, A. M., and Schopf, J. W. (1969). Possible algae microfossils from the late Pre-Cambrian of California. *Nature*, **223**, 165–167.

Han, J., Simoneit, B. R., Burlingame, A. L., and Calvin, M. (1969). Organic analysis of the Pueblito de Allende meteorite. *Nature*, **222**, 364.

Harland, W. B., *et al.* (1967). "The Fossil Record." Geological Society, London.

Jackson, T. A. (1967). Fossil Actinomycetes in middle Precambrian Varves. *Science*, **155**, 1003.

Kvenvolden, K. A., Peterson, E., and Pollock, G. E. (1969). Optical configuration of amino acids in Precambrian Fig Tree chert. *Nature*, **221**, 141.

Licari, G. R., and Cloud, P. E. (1968). Reproductive structures and taxonomic affinities of some nanofossils from the Gunflint iron formation. *Proc. natn. Acad. Sci.* **59**, 1053.

Love, L. G., and Zimmerman, D. O. (1961). Bedded pyrite and microorganisms from the Mount Isa shale. *Econ. Geol.* **56**, 873.

Manskaya, S. M., and Drozdova, T. V. (1968). "Geochemistry of Organic Substances." Pergamon, Oxford.

Marmo, V. (1959). Elaman meikeista Prekambrisissa kivissa. *Geogr. Sallskap Finl. Tidsskr.* **71**, 150.

McCarthy, D., and Calvin, M. (1967). Organic geochemical studies. I. Molecular criteria for hydrocarbon genesis. *Nature*, **216**, 642.

McGregor, A. M. (1941). A Precambrian algal limestone in Southern Rhodesia. *Trans. Geol. Soc. S. Africa*, **43**, 9.

Meinschein, W. G. (1965). Soudan Iron Formation: organic extracts of early Precambrian rocks. *Science*, **150**, 601.

Meinschein, W. G., Barghoorn, E. S., and Schopf, J. W. (1964). Biological remnants in a Precambrian sediment. *Science*, **145**, 262.

Moorhouse, W. W., and Beales, F. W. (1962). Fossils from the Animikie Port Arthur, Ontario. *Trans. Roy. Soc. Canada*, **56**, 99.

Murray, G. E. (1965). Indigenous Precambrian petroleum. *Bull. Am. Ass. Petrol. Geol.* **49**, 3.

Nagy, B., and Urey, H. C. (1968). Organic geochemical investigations in relation to the analyses of returned lunar rock samples. *8th Meeting of Committee for Space Research.* Pre-print No L.2.6, 9.

Ohlson, B. (1962). Observations on recent lake balls and ancient Corycium inclusions in Finland. *Bull. Commn. géol. Finl.* **196**, 377.

Oparin, A. I. (1957). "The Origin of Life on the Earth," 3rd edition. Oliver and Boyd, Edinburgh and London.

Oro, J., Nooner, D. W., Zlatkis, A., Wikstrom, S. A., and Barghoorn, E. S. (1965). Hydrocarbons of biological origin in sediments about two billion years ago. *Science*, **148**, 77.

Oro, J., and Nooner, D. W. (1967). Aliphatic hydrocarbons in Precambrian rocks. *Nature*, **213**, 1082.

Pedersen, K. R., and Jorgen, L. (1968). Precambrian organic compounds from the Ketilidian of south-west Greenland. *Meddr om Grønland*, **185**, 1.

Pflug, H. D. (1964). Niedere Algen und ähnliche Kleinformen aus dem Algonkium der Belt-Serie. *Ber. oberhess. Ges. Nat.- u. Heilk.* **33**, 403.

Pflug, H. D. (1965a). Organische Reste aus der Belt serie (Algonkium) von Nordamerika. *Paläont. Z.* **39**, 10.

Pflug, H. D. (1965b). Foraminiferen und ähnliche Fossilreste aus dem Kambrium und Algonkium. *Palaeontographica*, **125**, 46,

Pflug, H. D. (1966a). University of Witwatersrand, Johannesburg, S.A. *Econ. Geol. Res. Unit Circ.* 28.

Pflug, H. D. (1966b). Einige Reste niederer Pflanzen aus dem Algonkium. *Palaeontographica*, **117**, 59.

Prashnowsky, A. A., and Schidlowski, M. (1967). Investigation of Precambrian Thucholite. *Nature*, **216**, 560.

Ramsay, J. G. (1966). Structural investigations in the Barberton Mountain land, Eastern Transvaal. *Trans. Proc. geol. Soc. S. Afr.* 66.

Rankama, K. (1954). The isotope composition of carbon in ancient rocks as an indicator of its biogenic or non-biogenic origin. *Geochim. cosmochim. Acta*, **5**, 142.

Riviere la, J. W. M. (1966). The microbial sulphur cycle and some of its implications for the geochemistry of sulphur deposits. *Geol. Rdsch.* **55**, 568.

Schopf, J. W. (1967). "Antiquity and Evolution of Precambrian Life." McGraw-Hill, New York.

Schopf, J. W. (1968). Microflora of the Bitter Springs formation late Precambrian, Central Australia. *J. Paleont.* **42**, 651.

Schopf, J. W., and Barghoorn, E. S. (1967). Alga-like fossils from the early Precambrian of South Africa. *Science*, **156**, 508.

Schopf, J. W., Barghoorn, E. S., Maser, M. D., and Gordon, R. O. (1965). Electron microscopy of fossil bacteria two billion years old. *Science*, **149**, 1365.

Schopf, J. W., Kvenvolden, K. A., and Barghoorn, E. S. (1968). Amino acids in Precambrian sediments: an assay. *Proc. natn. Acad. Sci.* **59**, 639.

Siegel, S. M., and Giumarro, C. (1966). On the culture of a microorganism similar to the Precambrian microfossil Kakabekia umbellata Barghoorn in NH_3-rich atmospheres. *Proc. natn. Acad. Sci.* **55**, 349.

Siegel, S. M., and Siegel, B. Z. (1968). A living organism morphologically comparable to the Precambrian genus Kakabekia. *Am. J. Bot.* **55**, 684.

Tyler, S. A., and Barghoorn, E. S. (1954). Occurrence of structurally preserved plants in Precambrian plants of the Canadian Shield. *Science*, **119**, 606.

Tyler, S. A., Barghoorn, E. S., and Barrett, L. P. (1957). Anthracite coal from Precambrian upper Huronian black shale of the Iron river district, Northern Michigan. *Bull. Geol. Soc. Am.* **68**, 1293.

Urey, H. C. (1957). Primitive planetary atmospheres and the origin of life. *From* "The Origin of Life on the Earth." Symposium of International Union of Biochemistry Moscow, **1**, 16. Macmillan, New York.

Wald, J. D. (1964). The origins of life. *Proc. natn. Acad. Sci.* **52**, 595.

Wickman, F. E. (1956). The cycles of carbon and the stable carbon isotopes. *Geochim. cosmochim. Acta*, **9**, 136.

CHAPTER 2

Palaeobotany and Phytochemical Phylogeny

W. G. CHALONER

Department of Botany, University College, London, England

and

K. ALLEN

Department of Botany, University of Bristol, Bristol, England

I. INTRODUCTION

This article is an attempt to review the potential of a biochemical approach to fossil plants as a means of elucidating their evolutionary history. While phytochemistry has made great contributions to our understanding of relationship among living plants (particularly, for example, in the algae), an extension of this method to fossil plants is still only in its very early stages. Our aim is to consider, from the standpoint of palaeobotanists, some aspects of "palaeophytochemistry" which may extend the body of evidence from which we may draw in postulating interrelationships among fossil and living plants.

II. FOSSIL PLANTS IN THE TIME SCALE OF EARTH HISTORY

If the age of the Earth is taken as approximately 5 b.y., then we have some evidence of plant life in the form of structurally preserved bacteria-like bodies, extending back for about 3 b.y. (see Echlin, Chapter 1). The evidence afforded by the structure of these very simple fossils is inevitably somewhat equivocal, and our acceptance that they demonstrate the existence of life when these early rocks were formed is reinforced by the presence of associated phytane and pristane (possible chlorophyll residues), ^{13}C enrichment (evidence of photosynthesis) and the presence of optically active amino acids. While the record of simple aquatic plants (bacteria, and later, algae) goes back for about three-fifths of the Earth's known history, the record of plants adapted to life on land is very much shorter. Such land plants, possessing the water-conducting tissue which characterizes the group (the tracheophyta) which now dominates the Earth's land flora, have a record going back only some 400 m.y. It seems, rather surprisingly, that algae existed in the seas (or possibly in fresh or fresher water than that of the present oceans) for at least 1 b.y. before accomplishing the adaptations which made possible sustained life on land. The major part of palaeobotanical research has been concentrated on interpreting this last phase of the history of plant life, the rise of the vascular land plants. Inevitably, like the fossil record of any group of organisms, this is extremely incomplete. A whole series of fortuitous events have to coincide for a plant living on land to have become fossilized, and subsequently to be subjected to scientific investigation. These circumstances have operated so rarely that only a minute proportion of the sequence of plant life can have been preserved, and of this only an infinitesimal fraction can have come to our knowledge. It is appropriate to review here briefly the different ways in which plants may become fossilized. This of course affects both the kind of information that they can yield as to the structure of the plant, and the extent of the biochemical change affecting the composition of the plant material.

III. The Formation of Fossil Plants

Aquatic plants, particularly algae, generally decompose soon after death, and unless they are actually involved in mineral deposition (as in the case of calcareous algae) are very rarely preserved as fossils. Only in exceptional circumstances do algae lacking their own mineralization of the cell walls become fossilized. We do not deal with these further, as our main interest is in the phylogeny of vascular plants. These will normally only become fossilized if they are transported after death into an environment in which they can become incorporated into some sedimentary rock before they are broken down by microbial action. If in such an environment of sedimentation there is an

---

Fig. 1 (top left). Peel section of petrified fossil spores inside a sporangium of the Upper Carboniferous plant *Peltastrobus reedae*, showing what appear to be nuclei. ×1000. From a preparation by Prof. R. W. Baxter

Fig. 2 (top right). Section of a petrified fossil vascular plant, cf. *Rhynia* sp., showing fungal spores within the vascular plant tissue (crumpled remains of the epidermis are at left). Lower Devonian, Rhynie, Scotland. ×80

Figs. 3–5 (bottom left, centre, bottom right). A Devonian vascular plant (Fig. 3) and two animal fossils of comparable age (Figs. 4 and 5), showing their similar morphology.

Fig. 3. *Psilophyton princeps*, a spiny vascular land plant preserved as a compression fossil; Lower Devonian (ca. 400 m.y. old) from Canada. ×2. Fig. 4. The arms of a crinoid (a sedentary marine animal) from the Upper Devonian of Delabole, Cornwall. ×1. (Specimen no. KN 967, Institute of Geological Sciences.) Fig. 5. A fossil acanthograptid, a colonial marine organism, from the Upper Silurian of the Welsh border. ×2. (Specimen no. C309, University Museum, Oxford.)

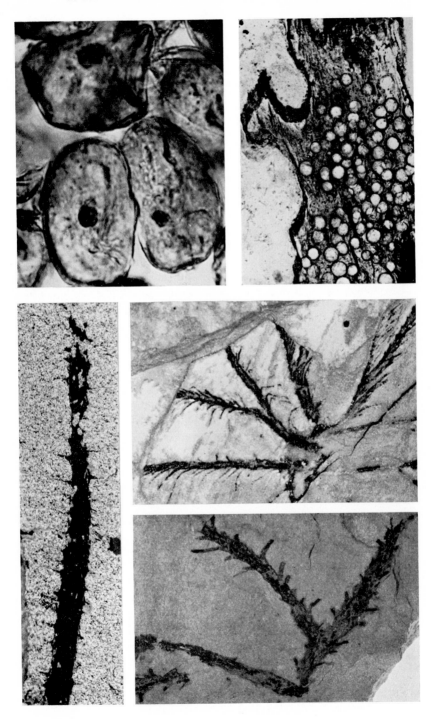

abundance of mineral matter in solution, it has sometimes happened that inorganic ions have penetrated relatively rapidly into the plant tissue, and actually into the cells, before the weight of accruing sediments causes distortion and collapse. This mineral infiltration results in the type of fossil called a "petrifaction," in which the internal structure of the fossil is preserved, at least in the form of the remains of the cell walls, so that the general features of the plant's anatomy may be seen. Usually each cell cavity in such a petrifaction has been infilled with mineral matter, the protoplast having been entirely destroyed by microbial activity. The cell wall has normally undergone drastic change in composition, to what may in a broad sense be termed coal. We still know relatively little about the chemistry of petrifaction, and the process has not been reproduced in the laboratory. But a number of quite different types of mineral may be involved in the petrifaction process, the commoner ones including silica, calcium carbonate and iron sulphide.

Petrifactions may be investigated either by cutting and grinding a thin section of the enclosing rock until it is transparent enough for microscopic examination, or by a simple and ingenious method called "peel sectioning." This involves etching away a thin layer of the petrifying mineral, leaving the coalified cell walls protruding from the rock, and then incorporating them in a softened thin film of a cellulose ester, which is subsequently stripped off and may be examined microscopically. (For a fuller description of the technique, and for references, see Andrews, 1961.)

Very exceptionally, even cell contents may be preserved in a petrifaction— organelles such as nuclei survive as discernible entities under favourable circumstances. Figure 1 shows what may be nuclei inside spores enclosed in a sporangium of a 300 m.y. old plant, petrified in calcium carbonate and sectioned by the peel technique. In one case even chromosomes have been reported and figured (Darrah, 1938). Occasionally other cytoplasmic inclusions may be recognizable, if they have a sufficiently distinctive morphology, as for example the starch grains reported by Baxter (1964) in a Carboniferous fossil seed.

Of much commoner occurrence than petrifactions are fossils in which the accruing and compaction of the sediment were accompanied by more or less extensive microbial breakdown of the plant tissue. If the sedimentary environment becomes sufficiently anaerobic, the more chemically inert constituents of the plant material may survive with relatively little alteration—particularly the cuticle and the exine of spores and pollen. In a fossil leaf, for instance, the mesophyll cells may become converted to a very thin film of coal, sandwiched between the upper and lower cuticle. Controlled oxidation of material removed from such a "compression fossil", with nitric acid and potassium chlorate, can dissolve the coaly layer, freeing the two cuticles, which reveal much of the cellular architecture of the leaf epidermis. (Further details of this technique are given in Andrews, 1961). Such compression fossils with cuticles are less informative botanically than a petrifaction, but will at least reveal cellular detail of the outer surface of the plant, as well as its original external form. Of course in

Library
I.U.P.
Indiana, Pa.

581. 38   P569h
C. 1

many fossil plant compressions the cuticle has undergone drastic alteration in composition in the process of coalification, and in such cases the cuticle cannot be separated from the other coaly residue by this differential oxidation. The irony of any chemical study of fossil plant compression remains is that the very quality favouring survival is likely to make them intractable on analysis. Fossil cuticles and exines, which by their inertness are highly resistant to biological attack, present a considerable challenge to any investigation of their composition. However, Brooks and Shaw (1968; see also Chapter 3) have already assailed the problem in the analysis of fossil sporopollenin and analogous substances which are similar to cutin at least in so far as their behaviour on fossilization is concerned.

A final category of fossil plants must be mentioned, in which virtually all the original substance of the plant has been destroyed. In such a case a copy of some cavity within the plant—for example, that of a hollow pith—may be preserved by becoming infilled with sediment which later becomes indurated to form a natural cast of the shape of the original cavity. Latin names have been given to such fossils (e.g. many of the *Calamites* species of the Carboniferous period). Such casts reveal something about the form of the plant, but of course the composition of the substance of the cast is simply that of the enclosing detrital rock, and offers little biochemical interest.

Before leaving the process of plant fossilization, it is particularly important to acknowledge in any palaeobiochemical work the extensive role of microorganisms whose activity must accompany the formation of almost any kind of fossil. Even in petrifaction, the most rapid and "efficient" form of fossilization, higher plant tissue has presumably been extensively invaded by microorganisms soon after death in most cases. In the Lower Devonian Rhynie chert, one of our earliest records of petrified vascular plants, we see plenty of evidence of the activity of fungi in the higher plant tissue. In this instance, plants growing in a peat bog were killed and fixed, apparently very rapidly by boiling water from a volcanic fumarole, and incorporated in a petrified state into a bed of colloidal silica derived from the same source. Figure 2, illustrates a condition which is quite common in this material in which a mass of fungal spores are seen occupying the site of the original tissue within the epidermis of the plant that they have invaded. Where fossilization has been slower, and the preservation less complete, as in a compression fossil, the resulting coaly material is an aggregate of original plant substance plus a whole microcosm of invading organisms. Because interest has usually been directed towards the higher plant fossil itself, palaeobotanists have generally tended to ignore this post-mortem microflora (which is, in any case, usually poorly preserved in terms of recognizable structures). A useful review of this important but hitherto rather neglected study of fossil microorganisms and their activity ("geomicrobiology") is given by Moore (1966). For this reason, in compression fossils especially, any biochemically active constituents extracted from the coaly residue cannot be automatically regarded as original attributes of the source plant.

IV. PHYTOCHEMICAL WORK ON FOSSIL PLANTS

A. PREVIOUS WORK

For the purpose of this brief review, it may be said that biochemical investigation of fossil plants has fallen into two main categories. The first has been the biochemical study of fossil plant material for the intrinsic interest of the organic residues present; while the second, a much more recent development, has been the attempt to use such residues as a contribution to taxonomic or phylogenetic study. As a recent convergence of these two aspects, we have also the current work on compounds associated with early Precambrian microfossils, dealt with in Chapter 1 by Echlin. There has of course been an enormous amount of work done on coal and petroleum chemistry, but here the orientation has generally been towards the genesis, composition and exploitation of the end product rather than the nature of the starting point *per se*.

A good illustration of the first type of study may be taken from recent work on the remarkable survival of plant products in the Geiseltal lignite, near Halle, Germany. This 50 m.y. old (Eocene) lignite deposit has long been noted for the quality of the preservation of both plant and animal substances which became incorporated in the highly anaerobic environment of the accruing plant debris. Latex residues for example, still in a rubbery, elastic state, are preserved as fibrous looking masses (Kautschukfäden), the so-called "apes hair" of the early German workers. These are believed to have been derived from the latex-bearing bark of members of the Moraceae, Sapotaceae or Apocynaceae, families represented by systematically identifiable fossil material in the lignite (Mägdefrau, 1968). It has also long been noted that some of the leaf compressions from the Geiseltal show a green coloration, and Noak's early spectroscopic investigation of the pigment suggested that a chlorophyll residue was responsible (see Mägdefrau, 1968 for references). This study has been refined by the recent work of Dilcher (1967) who used a soxhlet extraction to obtain a green pigment from the Geiseltal leaves, to which he applied paper chromatography. He was able to separate three constituent pigments with the characteristic absorption bands of phorbides (phaeophytin *a*, chlorophyllide *a* and phaeophorbide *a*) all containing the cyclopentane ring, and known to be simple derivatives of the chlorophyll molecule. The survival of such relatively complex and reactive substances over this period of time is remarkable; but the leaves were of course manifestly those of flowering plants on their morphology alone, and for this reason chlorophyll residues might be considered the most likely cause of a green coloration. To this extent there is no element of phyto-taxonomic significance in such material. But Dilcher's work none the less suggests exciting prospects within the field of palaeophytochemical taxonomy, by demonstrating the potential of the method. If phorbides can survive in a suitable geological environment for 50 m.y., then why not for 500 m.y.? Comparable evidence of affinity from plants of more problematical status than that of the Geiseltal leaves could be of great value.

Two examples may be taken to illustrate recent attempts to use biochemical studies of fossil plants for directly phytotaxonomic ends. One of these is based on residual carbohydrates, and the other on hydrocarbons. Swain *et al.* (1967, 1968) have studied a considerable number of Palaeozoic plants (250–400 million years old), extracting soluble carbohydrates by acid hydrolysis of coalified compression fossils and petrifactions. They have then assayed the monosaccharides obtained by paper chromatography, enzymatic methods, and by gas chromatography of their trimethylsilyl ethers (Swain *et al.* 1967, 1968 and references there cited). One of the most significant pieces of quantitative data that they derive from these studies is the D-galactose/D-glucose ratio in the extracts from the fossils. They suggest that the wood of the Devonian plant *Callixylon* (a Progymnosperm) has a higher value for this ratio than that of *Cordaites* (a true Gymnosperm) of the succeeding Carboniferous period. Swain *et al.* suggest that this might be correlated with a change from algal-like galactan-rich structural polysaccharides to cellulose, in the evolution of *Cordaites* from the *Callixylon* ancestral group. It would certainly be rather remarkable if *Callixylon* had retained such a fundamental biochemical difference from living vascular plants, while resembling them in so many structural aspects; so that while this possibility must be accepted, the several other alternatives must also be considered. For example, the extent to which various organisms (including algae) may exist as epiphytes or saprophytes on or in dead plant material lying in water will obviously affect the ratio of polysaccharide residues in the final fossil. But whatever the explanation, the immediate significance of the work of Swain and his colleagues is in establishing that biochemical criteria can be obtained from such fossil material, and that these constitute an additional set of characters against which hypotheses of phyletic relationship based on purely morphological observations can be tested. More of this type of data must be sought, so that monosaccharide ratios (and other biochemical features of the fossils) can be shown to be genuine attributes of particular species in a number of specimens, and if possible from different environments and states of fossilization.

An analogous study of the Triassic (200 m.y. old) fossil *Equisetum brongniarti* was conducted by Knoche and Ourisson (1967). The coaly material of this compression fossil was extracted with a benzene–methanol mixture, and the extract subjected to gas chromatography. A series of prominent peaks on their chromatogram correspond with *n*-hydrocarbon chains of 23, 25, 27 and 29 carbon atoms, and these same four *n*-hydrocarbons show a similar prominence in the hydrocarbon fraction extracted from living *Equisetum* plants. Here, as in the work of Swain *et al.*, there is a significant new source of data for testing hypotheses of phyletic relationship. But much still remains to be done in order to establish the relationship between observations on the fossil, and the plant from which it was derived. It is known that in the coalification process which is involved in the preservation of any organic matter in a compression fossil, polymerization occurs in addition to simple breakdown of original plant constituents (as for example in the coalification of lignin—see Flaig, 1968). It is obviously desirable that further work on such hydrocarbons should seek to confirm that they are original constituents of, and peculiar to, that particular species, and are absent from, or occur in significantly different ratios in, other plants occurring as fossils in the same sediments.

B. FUTURE APPLICATIONS OF PALAEOPHYTOCHEMISTRY

If we review the history of the evolution of land plants it is evident that broadly speaking the further back in time we go the less complete and secure our knowledge of plants becomes; and the greater the scope and need for phytochemical evidence to support our hypotheses of evolutionary relationship. One possible future application of a phytochemical approach to palaeobotany lies in the recognition of the earliest land plants. One of the innate problems in palaeobotany is that the "first record" of many land-plant groups is based on a relatively poorly preserved fossil, possibly lacking many of the features by which the group is characterized in its later representatives. This has resulted in a considerable controversy concerning land-plant origins, and the occurrence of vascular land plants much older than about 400 m.y. Two opposed schools of thought exist; one, the views of which are well set out by Axelrod (1959), favours an early, possibly Precambrian (600 m.y.) origin of land plants. Some proponents of this thesis further believe that vascular plants were probably polyphyletic, arising as several separate lines from the algae. Support for both these ideas is sought in the several reports of putative land plants from various pre-Devonian sources (see Chaloner, 1960). The opposite view, favoured by Stewart (1960) and Banks (1968) accept a late Silurian (ca. 400 m.y.) appearance of the first vascular plants, and these authors dispute the validity of any older fossil land plants. One of the problems at the core of this controversy is the fact that the putative earliest land plants are mostly compression fossils lacking secure criteria (presence of xylem, a cuticle or stomata) which would help to corroborate their identity as a terrestrial plant. One of the earliest groups of generally accepted vascular land plants, exemplified by *Psilophyton* from the Lower Devonian of North America (Fig. 3) is characterized by having spiny processes on the stem surface. Now several of the putative pre-Devonian land plants (e.g. *Aldanophyton, Boiophyton, Saxonia* and *Psilophyton hedei*) are fragments of axes with features reminiscent of *Psilophyton*, but lacking the microscopic detail which could confirm their land-plant status. There are unfortunately at least two groups of marine animals which in their gross structure are similar to the Psilophytes. One of these is the dendroid graptolites, floating colonial organisms of which fossils are common in marine rocks from Cambrian to Silurian in age (Fig. 5); the other is the crinoids ("sea lilies"), a group of sedentary marine animals of which the arms may show a rather Psilophyte-like appearance (Fig. 4). We are therefore faced with the absurdity that several putative pre-Devonian fossil land plants (a number of which, including the four cited above, are reported from marine sediments) are under suspicion of being marine animals! If an analysis of the composition of such fossils could confirm or deny this possibility it would be of tremendous value in clarifying the record of the earliest land plants. The separation of the organic residue of chitin or protein-based animal remains from the largely carbohydrate-based residues of plant material (even allowing for considerable post-depositional change) should be relatively straightforward.

There are a number of other early land-plant fossils which, while undoubted plants, are of uncertain status in other respects. There are several Devonian fossils which seem to have algal attributes, but also show features of adaptation to life on land (e.g. the large petrified tree trunks of *Prototaxites*, the cutinized thallus and resistant spores of *Protosalvinia* and *Parka*). In fossils such as these, still totally enigmatic in their systematic position, any biochemical evidence of affinity would be of value. The exines of triradiate spores occurring as dispersed fossils in Silurian and Devonian sedimentary rocks are widely held to be those of vascular plants, although it is known that by the end of the Devonian, bryophytes had already evolved. It may be argued that the perishable structure of bryophytes militates against their survival as fossils, and that they may actually have preceded vascular plants but are represented only by their spore exines as the most inert part of the plant structure. Unfortunately, there is no morphological basis for separating the exines of vascular plant spores from those of all bryophytes. Here again, if the composition of their constituent sporopollenin offered a basis for separation, it would be of enormous value in interpreting the spore record of early land plants.

These are only a few of the many controversies on which biochemical evidence would offer a valuable second opinion in phylogenetic problems. The study of plant phylogeny, as it is seen in the fossil record, has in the past been based almost solely on the comparative morphology and anatomy of the fossils and their living representatives. Hypotheses based on such structural features may now be put to the test offered by a study of the composition of the fossil material.

REFERENCES

Andrews, H. N. (1961). "Studies in Paleobotany." Wiley, New York.

Axelrod, D. I. (1959). Evolution of the Psilophyte Palaeoflora. *Evolution*, **13**, 264–275.

Banks, H. P. (1968). The Early History of Land Plants. *In* "Evolution and Environment" (E. T. Drake, ed.), pp. 73–107. Yale U.P. New Haven.

Baxter, R. W. (1964). Paleozoic Starch in Fossil Seeds from Kansas Coal Balls. *Trans. Kansas Acad. Sci.* **67** (3), 418–422.

Brooks, J., and Shaw, G. (1968). Identity of Sporopollenin with Older Kerogen and New Evidence for the Possible Biological Source of Chemicals in Sedimentary Rocks. *Nature*, **220**, 678–679.

Chaloner, W. G. (1960). The Origin of Vascular Plants. *Sci. Progr.* **191**, 524–534.

Darrah, W. C. (1938). A remarkable fossil *Selaginella* with preserved female gametophytes. *Harvard Bot. Mus. Leafl.* **6** (6), 113–136.

Dilcher, D. (1967). Chlorophyll in der Braunkohle des Geiseltales. *Natur und Museum*, **97** (4), 124–130.

Flaig, W. (1968). Biochemical Factors in Coal Formation. *In* "Coal and Coal-bearing Strata" (D. G. Murchison and T. S. Westoll, eds.), pp. 197–232. Oliver and Boyd, Edinburgh.

Knoche, H., and Ourisson, G. (1967). Organic compounds in fossil plants (Equisetum: Horsetails). *Angew. Chem. Internat. Edit.* **6** (12), 1085.

Mägdefrau, K. (1968). "Paläobiologie der Pflanzen," 4th edition. Fischer, Jena.

Moore, L. R. (1966). Frontiers in Geology–Geomicrobiology. *Adv. Sci.* Oct. 1966, 313–330.

Stewart, W. N. (1960). More about the Evolution of Vascular Plants. *Plant Sci. Bull.* **6** (5), 1–5.

Swain, F. M., Bratt, J. M., and Kirkwood, S. (1967). Carbohydrate components of some Paleozoic plant fossils. *J. Paleo.* **41** (6), 1949–1954.

Swain, F. M., Bratt, J. M., and Kirkwood, S. (1968). Possible biochemical evolution of carbohydrates of some Paleozoic plants. *J. Paleo.* **42** (4), 1078–1082.

CHAPTER 3

Sporopollenin

G. SHAW

School of Chemistry, University of Bradford, Bradford, England

I. INTRODUCTION

The male genetic material of flowering plants is contained in the pollen grains which are produced in the anthers from a sporogenous tissue. The production of pollen and the associated female ovule which it fertilizes is therefore the most important function of the plant and one which we can assume has played a major part in its evolution.

In the evolution of higher plants pollen grains are derived from asexual structures called spores, and in certain groups of plants, both living and fossil, the distinction between pollen and spores is arbitrary since they perform a similar function in reproduction. The term pollen is however usually restricted to the male reproductive structure of seed bearing plants including the gymnosperms (conifers, cycads, etc.) and all angiosperms (flowering plants). The ovule is contained in an ovary at the base of the pistil. At fertilization a pollen grain adhering to the sticky stigma at the tip of the pistil develops a long tube which reaches into the ovary. The genetic material of the pollen grain travels down the tube and fertilizes the ovule. Cryptogam spores are very similar to pollen grains although the fertilization process may take place in a slightly different manner.

Pollen grains vary in size from about 5 microns (*Myosotis*) to 250 microns (pumpkin) but the size may depend on the state of nutrition of the plant and the degree of hydration so that the living pollen grains may expand or contract like a balloon. The shape of the pollen grain is generally an ellipsoid or spheroid with a characteristically sculptured reticulate pattern on its surface which varies diagnostically from pollen grain to pollen grain. Enormous amounts of pollen are produced. A hemp shoot produces 500 million grains (a single anther of hemp contains about 70,000 grains), a male plant of *Rumex acetosa* produces 400 million whereas the flax plant (*Linum catharticum*) can only manage 20,000. Estimates have been made that the spruce forests alone of southern and central Sweden produce 75,000 tons of pollen per annum (Faegri and Iversen, 1964).

Architecturally the pollen grain consists of a sac which contains in addition to the cytoplasmic and genetic material large quantities of fat (50–60% or more). The pollen wall consists of at least two basic layers. The inner layer consists of an almost pure cellulose sac which retains much of the surface feature of the whole grain associated with some other polysaccharide material (possibly of a hemicellulose or pectin nature) and an outer layer, the exine, formed largely of substances known as sporopollenins. This article will be concerned with these latter substances.

II. POLLEN AND SPORE EXINES AND POLLEN ANALYSIS

"The exine is formed by one of the most extraordinarily resistant materials known in the organic world" (Faegri and Iversen, 1964).

The feature of pollen and related spore grains which has unquestionably attracted the most attention is their remarkable resistance to both chemical and biological decay. This has resulted in their survival for at least 500 m.y. in the case of pollen and into the early Precambrian in the case of microspores. Pollen grains and microspores constitute in fact the most universal plant fossils, and it is largely because of this that the subject of pollen analysis has become possible.

Fundamentally this technique depends for its existence on the fact that pollen exines and spore walls are generally preserved in peat and sediments as morphologically intact bodies even when most other organic materials have been reduced to morphologically amorphous structures, or completely degraded by biological or chemical reactions to simple soluble chemicals. The characteristic sculpturing and reticulation of the pollen wall exine provide an excellent means of recognition and classification; these and other related surface features (furrows, grooves, etc.) are associated with chemical stability. In the living pollen the external wall invariably possesses pores and these appear to allow of a certain degree of diffusion of water, and the living pollen grain may vary considerably in size according to the degree of hydration or nutrition, where in the latter case the fat content especially may vary in quantity. Emergence of the pollen tube during the early stages of the fertilization process also appears to involve penetration of one of the pores or fissures in the surface. In fossil material however the empty sac retains a constant size which is of particular diagnostic value during analysis (Faegri and Iversen, 1964).

The chemical stability of the pollen exine is further underlined when one considers the manner of isolation of fossil material from sedimentary rocks. The basic technique involves crushing the rock and treatment of the powder with hot potassium hydroxide solution, hydrochloric acid, hydrofluoric acid, then acetolysis with a mixture of acetic anhydride and concentrated sulphuric acid to remove cellulose and other polysaccharides, and finally, in many cases, an oxidative treatment with either fuming nitric acid, chloric acid or Schulze's reagent (nitric–chloric acid) to remove lignins, humic acids and other similar compounds. Although pollen exines are quite stable to non-oxidative reagents, oxidizing agents must be used with care since prolonged action results in decomposition of the exine. This susceptibility to oxidation is emphasized when one considers the conditions under which pollen is preserved in sediments. Thus, corrosion is most prevalent in aerated peats especially at high pH values near alkaline springs. In acid conditions with minimal oxygen, maximum preservation can be expected.

Botanists have frequently suggested that some corrosion of pollen grains may be due to bacterial or fungal action and some evidence for slight etching of pollen in the presence of fungi has been presented (Goldstein, 1960; Elsik, 1966). However it must be emphasized that, at the time of writing, there is no evidence known to the author for the existence of any microorganism capable of affecting in any decisive manner pollen exines. We have ourselves searched for such organisms in soil cultures without success to date. All the evidence points to corrosion of the exine as being due to chemical (generally aerial) oxidation which is catalysed by basic conditions. Many biological observations on pollen exine behaviour are best explained by this hypothesis. Thus, the claim that certain exines are soluble in ethanolamine (Bailey, 1960) we believe is almost certainly due to aerial oxidation of the exine in the presence of the base. Many of the observations which have been made by biologists on the reactivity, solubility, etc. of pollen and spore exines are carried out on microscope slides so that with minute amounts of pollen and maximum availability of oxygen the conditions are invariably ideal for oxidative degradation. Recorded cases of fungal attack on living pollen certainly prove that several fungi are capable of sending their hyphae into the interior of the pollen grain and presumably metabolizing the contents, but during this process there is little suggestion of attack of the exine (Goldstein, 1960). The reports of slight etching could equally be due to either aerial oxidation or even expansion of areas originally occupied by pores or furrows and hence most susceptible via the inner membranes to fungal penetration. Microphotographs showing fungal attachment to fossil pollen or spore grains in a Palaeozoic sediment equally reveal general retention of the spheroidal structure of the grain (Moore, 1963). Partial breakage of the spore wall shown in some of these samples could equally be due to physical rather than microbiological abrasion. It would be

2

reasonable to expect that if there had been any major fungal attack on the exine in these examples then it would have resulted in their collapse to amorphous structures.

III. CHEMISTRY AND BIOCHEMISTRY OF SPORE AND POLLEN EXINE COMPONENTS

A. THE EARLY WORK

The earliest chemical study of the exine of pollen grains appears to have been by John (1814) and Braconnot (1829) who introduced the term "pollenin" to describe the resistant wall material. In the first of a series of papers devoted to a study of the walls of spores and pollen, Zetzsche and Huggler (1928) coined "sporonin" to describe the related material from spores with special reference to the asexual spores of *Lycopodium clavatum* which because of their ready availability have been a major source of sporopollenin. Since the resistant material obtained from either pollen or spore exines appeared to be of the same general chemical character, Zetzsche subsequently introduced the collective name "sporopollenin" to describe either substance. Until our recent work, the only chemistry of any significance on sporopollenin was carried out by Zetzsche and his co-workers and published in a series of papers, mainly in the 1930s.

We will use the term sporopollenin to describe that particular type of highly resistant (to non-oxidative chemical and to biological attack) chemical substance which appears to be a constant and characteristic component of the exine of the large number of pollen and spore walls, including some microfossils, which have been studied to date. To date we know of no pollen or spore wall which does not contain sporopollenin and although its content in different grains varies considerably we cannot of course be certain that it is always present in all grains. Indeed the pollen of a few aquatic plants where pollination occurs under water (*Zostera, Ceratophyllum, Zannichellia*) are reported to have as far as is known, no exine (Faegri and Iversen, 1964).

Zetzsche and Huggler (1928) first showed a pollen or spore wall preparation morphologically intact yet free from nitrogen could be obtained by extraction of the material with organic solvents followed by a treatment with boiling sodium hydroxide solution, which served to release the pollen contents, presumably by prior breakage of the membrane at the base of the pores. Subsequent treatment with acid gave a wall material generally free from nitrogen and varying in yield from 5–25% of the original pollen weight. A major part of the material extracted under these conditions was fat.

They discovered that the pollen wall consisted of at least two layers, an inner (intine) which was stained blue with iodine in sulphuric acid and could be removed with cuprammonium hydroxide solution or better by contact with 85% phosphoric acid for about a week, and the outer exinous layer of sporopollenin. That the intine was partly cellulose (comprising some 10–15% of the wall) was confirmed by examination of the insoluble material which remained after treatment of the walls with diacetylnitrate in acetic acid which removed the sporopollenin. This residue showed the staining properties of the intine and hydrolysis with acid and acetylation of the product gave octa-acetyl cellobiose. Zetzsche *et al*. (1937) also prepared several other pollen membranes including microfossil material especially from brown coal and from tasmanin spore shale (white spore coal). Spores from the latter material have since been classified as planktonic algae (Muir and Sargeant, 1968). From the results of elemental analyses they concluded all sporopollenins could be represented by a general empirical formula which they chose to formulate arbitrarily on a C_{90} basis and showed that they varied from $C_{90}H_{134}O_{31}$ in *Secale cereale* pollen to $C_{90}H_{150}O_{33}$ for *Phoenix dactylifera*. In the case of fossil material there is a tendency for a lower hydrogen and oxygen content, and often sulphur and possibly nitrogen is also present.

TABLE I

Early analytical data for sporopollenins

Material	Formula	C-Me/mole	OH/mole
Sporonins			
Lycopodium clavatum	$C_{90}H_{144}O_{27}$	2·04	15
Equisetum arvense	$C_{90}H_{144}O_{31}$		13
Ceratozamia mexicana	$C_{90}H_{148}O_{31}$	4·08	
Pollenins			
Corylus avellana	$C_{90}H_{138}O_{22}$		11
Picea excelsa	$C_{90}H_{144}O_{26}$		
Picea orientalis	$C_{90}H_{144}O_{25}$		14
Phoenix dactylifera	$C_{90}H_{150}O_{33}$	3·45	
Pinus sylvestris	$C_{90}H_{144}O_{24}$	1·70	13
Taxus baccata	$C_{90}H_{138}O_{26}$	2·74	
Fossil Pollenins			
Tasmanin	$C_{90}H_{136}O_{17}$	3·00	
Geiseltalpollenin	$C_{90}H_{129}O_{19}S_7N$	3·40	
Lange sporonin	$C_{90}H_{82}O_{17}N$	4·50	

Data from Zetzsche *et al*. (1937). All membranes had their cellulosic intine removed with phosphoric acid.

Zetzsche *et al*. (1937) also carried out several experiments with sporopollenin especially from *L. clavatum* in an attempt to arrive at a chemical structure but most of this work was unsuccessful largely because of the great resistance of the material to non-oxidative chemical or biological degradation. They showed however from Kuhn–Roth oxidations that the sporopollenin typically contained C-Me groups (Table I). Zetzsche and Huggler (1928) also found that the material was highly unsaturated and readily absorbed large quantities of bromine to give a bromo-sporopollenin. At the same time they showed the material contained hydroxyl groups since it could be acetylated with acetic anhydride and the resulting acetylsporopollenin could then be saponified with alkali, and figures obtained for the number of hydroxyl groups per C_{90} unit.

Attempts by Zetzsche and Huggler (1928) to degrade sporopollenin with reagents, e.g. long boiling with hydriodic acid or phosphoryl chloride, merely resulted in introduction of iodine or chlorine into the (unsaturated) polymer with no apparent change in the morphology of the grain and no perceptible breakdown into small fragments. Zetzsche *et al.* (1937) did however find sporopollenin was quite susceptible to chemical oxidation and could be converted into soluble compounds by a wide variety of oxidizing agents, which again underlined the polyunsaturated nature of the material. The reagent which appeared to produce the most useful results was ozone in acetic acid followed by hydrogen peroxide, when much of the wall dissolved and relatively large amounts of a mixture of simple dicarboxylic acids especially oxalic, malonic, succinic, glutaric and adipic acid were obtained. The inherent difficulties in work of this kind during the 1930s is illustrated by Zetzsche's separation and identification of these acids. The mixture was converted with methanolic hydrogen chloride into esters which were fractionally distilled. The purified esters with ammonia gave crystalline amides which were isolated and their structure established by comparison with pure compounds. The same mixture of dicarboxylic acids was also obtained in a very similar manner by an analogous ozonization of the microfossil material tasmanin. Zetzsche made (necessarily rough) quantitative measurements of the amounts of the foregoing acids and according to these the yield of succinic acid amounted to 15–16% by weight of the sporopollenin (in *L. clavatum*) and for every two molecules of succinic acid there was formed 1 mole each of malonic, glutaric and adipic acid. Together with C-methyl analyses (2 mole) these figures accounted for 40–50% of the weight of the original walls.

B. LATER WORK

Following the early work of Zetzsche the subject remained dormant until the 1960s when it was re-examined by the author and his co-workers (Shaw and Yeadon, 1964, 1966; Brooks and Shaw, 1968a). Zetzsche's work had clearly established that a quite unique group of substances called sporopollenins form part of the exines of many if not all pollen grains or spore walls. The sporopollenins had similar empirical formulae and were highly unsaturated polymers which contained C-methyl, and hydroxyl groups, and on oxidative degradation gave a mixture of simple dicarboxylic acids (C_2–C_6) as the main products. However none of these results gave any indication of the type of monomeric unit or units (if any should exist) from which the polymeric material might have arisen.

Because of the great advances and sophistication of techniques for the separation and identification of organic compounds which have occurred since the 1930s, it seemed to us that a re-investigation of the problem was propitious. It soon became clear however from initial work that the difficulties experienced by Zetzsche were not easily overcome. After a survey of the effect of various oxidizing agents on sporopollenins especially from *L. clavatum* spores or *Pinus sylvestris* pollen (chosen largely because of their availability in quantity and their high sporopollenin content), the use of ozone was found to give the most useful results (Shaw and Yeadon, 1964, 1966). Zetzsche had used mainly ozone followed by hydrogen peroxide whereas we found some advantage in using ozone alone in acetic acid when in the case of *L. clavatum* spores a solid (80–90 % by weight of the original membrane wall) and a solution from which syrup (30–35 %) could be isolated were obtained. The solid was largely soluble in warm sodium hydroxide solution to leave an insoluble residue of almost pure cellulose. This cellulose sac fully retained the original shape of the spore with all its surface reticulation, and its chemical structure was confirmed by infrared spectral comparison with an authentic specimen of cellulose, and by its hydrolysis to cellobiose and glucose. In addition, extraction of the solid from the ozone reaction, with aqueous ammonium acetate, gave after precipitation with ethanol a polysaccharide fraction with all the properties of a hemicellulose, which on hydrolysis gave a mixture of simple sugars including glucose, xylose and galactose.

From the acidified alkaline solution was obtained a mixture of acids which were largely the simple dicarboxylic acids (oxalic, malonic, succinic, glutaric and adipic acids) isolated by Zetzsche but in addition small amounts of long-chain dicarboxylic acids including 7-hydroxyhexadecandioic acid, 6,11-dioxo-hexadecandioic acid, traces of other related compounds, together with some monocarboxylic acids were isolated. The compounds were examined by thin layer chromatography and also, after conversion to methyl esters with methanol/boron trifluoride, by gas–liquid chromatography. The syrup obtained during the ozonization was found after a preliminary extraction and esterification to consist largely of monocarboxylic acids both straight chain and branched, but also including smaller amounts of simpler dicarboxylic acids together with a small quantity of neutral material which contained ketone groups. Zetzsche's ozonization of *L. clavatum* sporopollenin in acetic acid, followed by reaction of the products with hydrogen peroxide, was repeated. In this case the majority of the wall entered solution leaving a residue (35 % by weight of original wall). The solution contained the same sort of mixture of mono- and dicarboxylic acids as before. The insoluble material after hydrolysis with dilute sulphuric acid left a residue of cellulose and a solution which contained a mixture of sugars, chiefly xylose, glucose and galactose.

Ozonization of *P. sylvestris* sporopollenin gave similar results. Ozone in acetic acid gave a residue (ca. 80 % of the weight of the starting material), which was largely soluble in warm sodium hydroxide solution to leave an insoluble residue of cellulose amounting to some 15 % of the wall. The alkaline solution after acidification gave a mixture of ether soluble mono- and dicarboxylic acids similar to those obtained from *L. clavatum*. In addition, a water and ether-insoluble hemicellulose-like product was precipitated and this on acid hydrolysis gave mainly glucose. The acetic acid solution from the ozonization contained mainly monocarboxylic acids (C_4–C_{18}).

These general results suggested that the wall material of the sporopollenins examined consisted of (i) a cellulose intine (10–15% in these examples but clearly capable of considerable variation in other species) which may remain (e.g. with *L. clavatum*) after ozonization as a clearly defined morphological structure retaining the shape and character of the grain, (ii) some polysaccharide material, possibly a hemicellulose or pectin type, which appears as a water–ether insoluble fraction more readily hydrolysed than cellulose. The amount of this also appears to be of the order of 10–15% and (iii) sporopollenin outer exinous material, a tough poly-unsaturated substance resistant to non-oxidative chemical attack but fairly susceptible to oxidation. In most of the pollen examined this would constitute a major part of the exine but again considerable variation appears to be possible (Tables II and III). Sporopol-

TABLE II

Membrane percentage and molecular formula of
certain pollens and spores

Material	% Membrane[a]	Molecular formula
Lycopodium clavatum	29·0	$C_{90}H_{149}O_{38}$
Pinus silvestris	24·0	$C_{90}H_{157}O_{37}$
Alnus glutinosa	15·8	$C_{90}H_{148}O_{37}$
Betula verrucosa	15·3	$C_{90}H_{156}O_{44}$
Ulmus procera	12·0	$C_{90}H_{156}O_{37}$
Holcus lanatus	10·4	$C_{90}H_{153}O_{40}$
Chenopodium album	9·8	$C_{90}H_{133}O_{30}$
Fagus sylvatica	9·4	$C_{90}H_{145}O_{35}$
Dactylis glomerata	8·5	$C_{90}H_{151}O_{35}$
Poa annua	7·4	$C_{90}H_{159}O_{41}$
Quercus robur	7·1	$C_{90}H_{144}O_{33}$
Phleum pratense	6·1	$C_{90}H_{152}O_{33}$
Salix fragilis	6·8	$C_{90}H_{152}O_{40}$
Scilla non-scripta	4·4	$C_{90}H_{138}O_{38}$
Narcissus pseudonarcissus	2·4	$C_{90}H_{146}O_{30}$

Data from Shaw and Yeadon (1966).
[a] All membranes probably contain cellulose.

lenin after ozonization gives a bewildering mixture of mono- and di- straight and branched chain carboxylic acids together with keto acids, hydroxy acids and ketones and smaller amounts of other miscellaneous substances including aldehydes. There is a remarkably high concentration of carboxylic acids following ozonization without further oxidation, and unless ozone is behaving

TABLE III

Membrane compositions of plant pollens

| Material | Composition of the membrane | |
	% Cellulose	% Sporopollenin
Polytrichum commune	2·2	
Lycopodium clavatum	2·7	23·4
Equisetum arvense	14·1	1·8
Onoclea struthiopteris	1·2	8·2
Ceratozamia mexicana		20·1
Pinus silvestris	3·6	19·6
Pinus uliginosa	8·5	18·1
Pinus mughus	9·7	17·3
Typha angustifolia	5·3	11·3
Corylus avellana	2·7	8·5
Alnus incana	1·8	8·8
Betula verrucosa	3·2	8·2
Carpinus betulus	3·5	8·2
Ulmus sp.	5·8	7·5
Quercus rubra	2·6	5·9
Quercus sessiliflora	1·9	5·9
Populus alba	4·2	5·05
Acer negundo	2·4	7·4
Sambucus nigra	1·7	12·2
Tilia sp.	8·1	14·9

Data from Kwiatkowski and Lubliner-Mianowska (1957).

unusually in these particular circumstances it would appear that much of the unsaturation in the polymer is of a type which would lead, after ozonization, directly to acids. Such groups are of the type

$$C:C.O.R. \rightarrow OC.OR \quad \text{and} \quad C:C.O.C.:C \rightarrow OC.O.CO$$
$$\quad\quad | \quad\quad\quad\quad | \quad\quad\quad\quad\quad\quad | \quad\quad | \quad\quad\quad\quad\quad | \quad\quad |$$
$$\quad\quad R' \quad\quad\quad\quad R' \quad\quad\quad\quad\quad R \quad\quad R' \quad\quad\quad R \quad\quad R'$$

giving rise to esters or anhydrides respectively. The latter and perhaps some of the former compounds would readily hydrolyse probably in aqueous acetic acid whilst the esters would require more vigorous hydrolysis as with sodium hydroxide solution. This is in agreement with the sort of behaviour observed where following ozonization a proportion of the acidic substances enters the acetic acid solution but the majority is only released from the insoluble product after warming with sodium hydroxide solution. Fusion of the sporopollenin of both *L. clavatum* and *P. silvestris* with potassium hydroxide gave in addition to carboxylic acids, the phenolic acids, *p*- and *m*-hydroxybenzoic acids and protocatechuic acid from *L. clavatum* and mainly *p*-hydroxybenzoic acid from *P. silvestris*. Originally we had thought that the formation of phenolic acids might suggest a lignin component in the structure but we have subsequently withdrawn this suggestion (Brooks and Shaw, 1968).

C. BIOCHEMICAL EXPERIMENTS WITH *Lilium henryii* AND THE CHEMICAL STRUCTURE OF SPOROPOLLENIN AS AN OXIDATIVE POLYMER OF CAROTENOIDS

The bewildering number of different products isolated by the necessarily limited degradation techniques might lead one to think that sporopollenin was evolved from a large mixture of different types of compounds all linked together to form some vast heterogeneous polymeric structure. However there were a number of reasons for thinking otherwise. Firstly it seemed clear that the sporopollenin basic structure although varying somewhat from plant to plant was in the main, essentially of a similar character throughout, and this implied the existence of a similar type of monomeric unit or units. Secondly, for a material as important as the covering for the carrier of genetic material in the plant, that nature should use a haphazard mixture of compounds seemed most unlikely. It appeared clear that if further useful information about the structure was to be unearthed some completely new approach had to be found. We decided to seek out directly possible monomeric precursors of the sporopollenin structure, by following the course of formation of pollen exine in a particular plant and correlating this development with parallel development of chemical substances in the anthers (Brooks and Shaw, 1968a, 1968b). A valuable clue here was that it seemed certain that the precursors would have to be essentially lipid in nature and to provide a source of unsaturation.

The concept of the development of the exine of pollen grains in the tapetal cells of the anthers during the early stages of meiosis is well known (Heslop-Harrison and MacKenzie, 1967), and correlation of development of the pollen wall with anther or bud size is roughly possible in certain *Lilium* species. For our initial experiments we chose therefore *L. henryii* and 100 plants were grown under greenhouse conditions. At varying stages of bud growth, anthers were removed, their weights and lengths recorded and portions examined microscopically. Remaining material was extracted with aqueous and organic solvents and the extracts examined chemically. About 10–15% of the plants were allowed to develop to maturity and the pollen collected and the nitrogen and cellulose walls isolated by the general methods. They were very similar in all respects to related sporopollenin from other plants.

It rapidly became apparent during these experiments that the formation of sporopollenin and concomitant development of the pollen grain was accompanied in a quite dramatic manner by a parallel formation of carotenoids (see Fig. 1). Our quantitative measurements of the extracts revealed that, apart from very small quantities of saturated hydrocarbons, the great majority of the lipid material being formed at quite early stages of growth (95%) consisted of a mixture of carotenoids and carotenoid esters. The carotenoids were found to consist of a mixture of free carotenoids and carotenoid esters (ratio 2·2:1). Saponification of the esters and examination of the resultant fatty acids by gas–liquid chromatography showed that the mixture contained approximately 90% straight-chain fatty acids: C_{16} (80%), C_{18} (6·6%), C_{7-13} (1·5%) and 10%

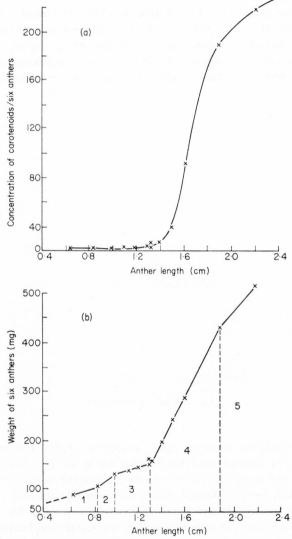

FIG. 1. Relation between carotenoid formation, anther length and pollen grain development in *L. henryii*.
(a) Variation of the concentration of carotenoids in the anther with anther length.
(b) Relation of the anther length of *L. henryii* with the ontogeny of the pollen grain

    1. Sporogenous tissue   4. Exine deposition
    2. Dyad stage   5. Pollen grains
    3. Tetrad stage

branched chain acids: C_{16} (4·6%), C_{14} (2·7%), C_{11} (1·2%), C_{7-13} (1·5%). The major part (90%) of the carotenoids appeared to consist of two compounds in roughly equal proportions one of which was identified (thin layer chromatography and UV absorption spectroscopy) as antheroxanthin (Fig. 2).

Examination of the basic skeletal structure of the carotenoid and carotenoid ester molecules (Fig. 2) suggested that they could provide just those features required to provide by some partial oxidative process a macromolecular structure analogous to sporopollenin, i.e. a highly insoluble lipid-like macromolecule retaining a substantial degree of unsaturation and yet able to yield on

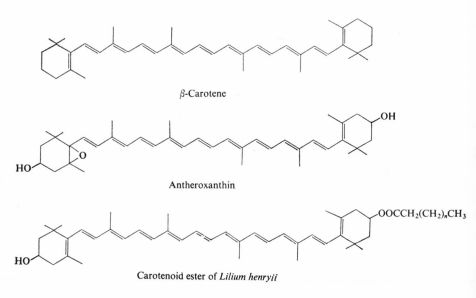

β-Carotene

Antheroxanthin

Carotenoid ester of *Lilium henryii*

FIG. 2. Carotenoid formulae

oxidative degradation saturated chains of varying lengths. The apparent lack of function of plant carotenoids has long puzzled botanists and biochemists for many years and has given rise to suggestion about the inefficiency of natural selection in this context (Burnett, 1965). Their possible function as precursors of sporopollenin has hitherto been completely unsuspected. The presence of carotenoids in quantity in anthers is well known (Karrer and Toappi, 1948) and equally of interest is their presence in spore-forming algae, related planktonic material and in fungi suggesting that sporopollenin could equally be involved in the formation of spore walls from these plant species.

To further establish a relationship between carotenoids and sporopollenin we examined (Brooks and Shaw, 1968a, 1968b) the polymerization of a number of carotenoids using both free radical and ionic catalysts in the presence and

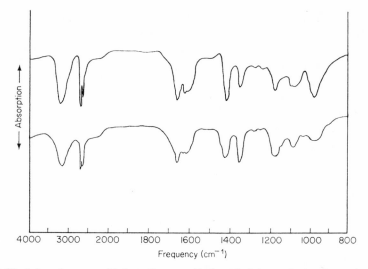

FIG. 3. The infrared spectra of *L. henryii* sporopollenin and of the synthetic polymer from the carotenoids of *L. henryii*

absence of an oxygen atmosphere. In methylene chloride solution and with a trace of an ionic catalyst and in the presence of oxygen, the carotenoid fraction from *L. henryii* anthers smoothly gave in excellent yield, an insoluble oxygen-containing unsaturated polymer with properties virtually identical with those

TABLE IV

Fatty acids produced by ozonization of natural and "synthetic" sporopollenins

Material	Molecular formula	% acids produced by ozonization		
		Branched chain mono-acids	Straight chain mono-acids	Dicarboxylic acids
Lilium henryii pollen wall	$C_{90}H_{142}O_{36}$	59·3	8·4	32·9
L. henryii carotenoid/ carotenoid ester polymer	$C_{90}H_{148}O_{38}$	57·3	6·5	36·4
L. henryii carotenoid polymer	$C_{90}H_{110}O_{33}$	64·1	3·4	33·4
β-Carotene polymer	$C_{90}H_{130}O_{30}$	61·3	1·9	35·0
Vitamin A palmitate polymer	$C_{90}H_{150}O_{13}$	28·3	47·6	25·8
Lycopodium clavatum spore wall	$C_{90}H_{144}O_{27}$	38·8	21·4	37·7

of the related sporopollenin. The comparisons used included resistance to acetolysis, elemental analyses, infrared spectra (Fig. 3), examination of the oxidation (ozone) products and comparison of pyrolysis gas–liquid chromatograms (Fig. 8) (Brooks and Shaw, 1969). Similar polymerizations of pure model carotenoids including β-carotene and vitamin A palmitate gave similar insoluble unsaturated oxygen-containing polymers (Table IV) with the characteristic properties of sporopollenin.

Potash fusion of both the *L. henryii* sporopollenin or the synthetic polymer derived from its anther carotenoids gave in each case an almost identical

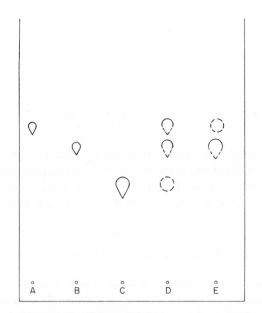

FIG. 4. Thin-layer chromatography of the potash fusion products of *L. henryii* sporopollenin and the synthetic polymer from *L. henryii* carotenoids

    A. *m*-Hydroxybenzoic acid
    B. *p*-Hydroxybenzoic acid
    C. Protocatechuic acid
    D. *L. henryii* sporopollenin
    E. Synthetic polymer from *L. henryii* carotenoids

"spectrum" on thin layer chromatograms of phenolic acids with *p*-hydroxybenzoic acid predominating (Fig. 4). Similar fusion of the polymerized β-carotene also readily gave *p*-hydroxybenzoic acid and these results clearly eliminated the need to postulate lignin precursors in the sporopollenin structure.

Our results suggested therefore that sporopollenin is an oxidative polymer of carotenoids and carotenoid esters formed in the anthers and the polymerization process occurs some time early in the growth.

IV. The Post-tetrad Ontogeny of the Pollen Wall

The exine of pollen is known to develop in the tapetal cells during the early stages of meiosis (Heslop-Harrison, 1968b). Recent observations (Godwin and Echlin, 1968) have shown that spheroidal bodies (orbicules or Ubisch bodies) appear at about the same time as the exine, and these bodies have been reported to be coated with sporopollenin. We believe that these bodies, which from our photographs are some 2–5 microns in diameter, consist of solution of carotenoids and carotenoid esters in a fat solvent undergoing what appears to be a typical emulsion polymerization process with formation of sporopollenin.

The post-tetrad ontogeny of the pollen grain can then be interpreted as involving (1) release of the microspore from the tetrad; the microspore consists of a sac with the surface features related to those in the mature pollen grain (Heslop-Harrison, 1968a). (2) At this or perhaps slightly later time a layer of hemicellulose glue is laid down on the sac. (3) This is followed by adherence of the partially polymerized sporopollenin globules on the surface of the microspore in alignment with the preformed sculpture initially in a reasonably mobile manner. (4) The sporopollenin ultimately becomes fully polymerized, the fat solvent runs off (this can be observed) and the sporopollenin coat is formed. Unused orbicules appear to harden into discrete solid balls of sporopollenin. In the early stages of these reactions the amount of free carotenoid extractable is quite small and presumably what is available is being used in the above process. As the process nears completion the carotenoid concentration rapidly increases (Fig. 1).

In the case of *L. henryii* little or no carotenoids were detected in anthers less than 1·1 cm long and this corresponded to the presence of sporogenous tissue alone (anthers 0·9 cm long) or to tetrads only without exine (anthers 1·1 cm long); when anther lengths were greater than 1·1 cm, exine was being increasingly deposited and corresponded to increasing concentrations of carotenoids. The results suggest that there are at least two sets of reactions in progress, both presumably controlled by appropriate enzyme systems. One of these involves the production of carotenoids possibly at a constant rate and the other their polymerization to sporopollenin as an emulsion or fine suspension process. When polymerization is at a maximum, carotenoid concentrations are low and as utilization of carotenoids nears completion the carotenoid concentration increases before finally becoming more or less constant (see Fig. 1). The effect of two such rate-controlled processes would be expected to result in the type of curve (Fig. 1) observed. The final process to this sequence of reactions may be seen as an expansion of the developing microspore either by absorption of water or fat or both into the completed spheroid or ellipsoid. During this process the final external architecture of the wall will be laid down both by expansion of the tacky but hardening sporopollenin polymer layer and also by further invasion of the template intine into this layer. The process could be likened to the deposition of soft sticky spheroids on to an adhesive surface when there is a tendency for a rough crystalline structure to be laid down. This will, however, be considerably modified by the stickiness of the balls which tends to favour formation of high points and these will be augmented or depressed by the shape and subsequent expansion of the microspore intine.

V. Geochemistry of Sporopollenin

A. KEROGEN

Organic chemists are frequently surprised to learn that many (especially sedimentary) rocks contain quite large amounts of organic substances, and indeed it has been estimated that more than 90% of the organic matter on the planet is present in these forms (Meinschein, 1963). The organic matter occurs in three major forms: (a) morphologically intact insoluble material of known biological origin, (b) amorphous insoluble material and (c) simple and relatively soluble (in aqueous or non-aqueous solvents) organic chemicals; they occur in the combination (a + b + c) and (b + c). The nature of the morphologically intact material which includes fossils and microfossils varies considerably as one examines rocks of increasing age. As one progresses back in time the fossil material becomes increasingly less differentiated until in older Precambrian sediments virtually the only recognizable bodies are intact spore walls, which have been clearly found in, for example, sediments of the Bitter Springs formation of Australia (1 b.y.) (Schopf, 1968) and there is good evidence for their presence in the fig-tree system of rocks (at least 3·2 b.y.) (Schopf and Barghoorn, 1967; Pflug, 1967). These results reflect (1) the greater stability (i.e. to the type of environment imposed during the sedimentation processes) of the sporopollenin of spore walls; (2) the lesser stability of material (celluloses, lignins and cutins, etc.) which make up the grosser parts of both flowering plants and lower forms, especially to hydrolysis which must be a major process during sedimentation (Breger, 1963); (3) the impact of evolutionary processes.

The detailed concepts of biological evolution have been based to a large extent on the examination of fossil and microfossil remains but because of the considerable variation in stability of the organic chemicals involved, little reliance can be placed on the more quantitative aspects of this type of work. This concept of differential chemical stability of fossils is a major concern in palynology (Faegri and Iversen, 1964) and is to some extent overcome by only accepting analyses as reasonably reliable when there is no evidence in the environment studied of any pollen or spore decomposition. Results from areas where there is any suspicion of partial degradation are always treated with great reserve.

The organic insoluble amorphous material which is known to occur in some of the oldest sediments including the fig-tree system (3·2 b.y.) is generally called "kerogen" (Forsman, 1963). However there are different types of kerogen and their nature depends in part on the age of the rock under investigation and the degree of metamorphosis to which it has been subjected. The youngest identifiable kerogen will include the immediate decomposition products of much of the recent plant and possibly the animal kingdom. As one goes back in time the kerogen tends to become more readily definable in terms of simple organic chemistry when one reaches areas of moderate stability as evidenced by the coaly kerogens. These appear to be mainly of two types: (1) the most important include the ordinary black coals (carboniferous era, 250 m.y.) which appear to be largely the metamorphosed product of decayed vegetable matter especially

TABLE V

Molecular formulae of some sporopollenins

Material	Age	Molecular formulae[a]
Lycopodium clavatum spore exine with cellulose intine removed	Present	$[C_{90}H_{144}O_{27}]_n$
Lycopodium clavatum spore exine[b]	Present	$[C_{90}H_{156}O_{35}]_n$
Lycopodium exine after heat treatment, at 275°C for 6 hours	—	$[C_{90}H_{95}O_{26}]_n$
Lilium henryii pollen exine, with cellulose intine removed	Present	$[C_{90}H_{142}O_{36}]_n$
Synthetic polymer from *L. henryii* carotenoids and carotenoid esters	—	$[C_{90}H_{148}O_{38}]_n$
Polymerized β-carotene	—	$[C_{90}H_{130}O_{30}]_n$
Green river shale kerogen	60 m.y.	$[C_{90}H_{134}O_{25}]_n{}^d$
Kukersite shale kerogen	400 m.y.	$[C_{90}H_{136}O_{10}]_n$
Carboniferous megaspore	250 m.y.	$[C_{90}H_{102}O_{16}]_n$
Tasmanites huronensis (Dawson)	350 m.y.	$[C_{90}H_{134}O_{17}]_n$
Tasmanites punctatus	250 m.y.	$[C_{90}H_{132}O_{16}]_n{}^d$
Tasmanites erratius	350 m.y.	$[C_{90}H_{133}O_{11}]_n{}^d$
Gleocapsamorpha prisca	350 m.y.	$[C_{90}H_{131}O_{17}]_n{}^d$
Geiseltal pollenin	250 m.y.	$[C_{90}H_{129}O_{19}]_n{}^d$
Lange-sporonin	250 m.y.	$[C_{90}H_{82}O_{17}]_n$
Orgueil Meteorite, insoluble organic matter	4·5 b.y.	$[C_{90}H_{65}O_9]_n{}^d$
Murray Meteorite, insoluble organic matter	4·5 b.y.	see note[c]
Pinus montana pollen exine[b]	Present	$[C_{90}H_{150}O_{40}]_n$
Pinus radiata pollen exine[b]	Present	$[C_{90}H_{148}O_{35}]_n$
Pinus silvestris pollen exine[b]	Present	$[C_{90}H_{154}O_{36}]_n$
Alder pollen exine[b]	Present	$[C_{90}H_{122}O_{30}]_n$
Beech pollen exine[b]	Present	$[C_{90}H_{144}O_{35}]_n$

[a] Molecular formulae are arbitrarily recorded on a C_{90} basis to facilitate comparison with earlier references (Zetzsche *et al.* 1937).

[b] Cellulose has not been removed from these exines.

[c] Insufficient material was available for detailed elementary analysis, but the %C: %H ratio was 8·80 (cf. *Tasmanites punctatus* = 8·15; *Tasmanites erratius* = 8·03; Green River Shale Kerogen = 8·03; *Gleocapsamorpha prisca* = 8·25; *Lycopodium clavatum* before heating = 7·25; *Lycopodium clavatum* after heating = 11·50; and *Lilium henryii* exine = 7·65).

[d] These materials all contain small amounts of nitrogen and sulphur.

cellulose and lignin and (2) the spore-algal or "white coal" (actually generally grey) which is largely derived from algal or other spores and is essentially metamorphosed sporopollenin (Table V) and may contain 80% exinite (spore exines) (Smith, 1969).

As rocks become older and as the more readily degraded organic material (e.g. cellulose) is lost, the nature of the kerogen changes until it assumes a

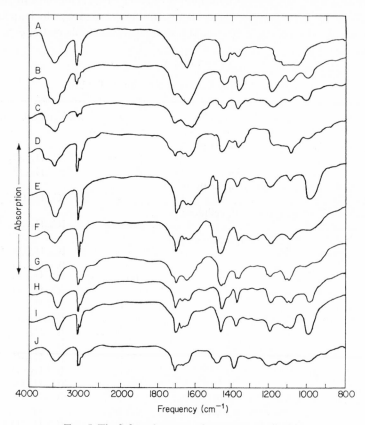

FIG. 5. The infrared spectra of some sporopollenins

A. *Tasmanites huronensis* (Dawson)
B. Insoluble organic matter from the Orgueil Meteorite
C. Insoluble organic matter from the Murray Meteorite
D. Carboniferous megaspore (*Valvisisporites auritus*)
E. *Lycopodium clavatum* spore exine
F. *Lycopodium clavatum* spore exine after heat treatment
G. Polymerized β-carotene
H. *Pinus montana* pollen exine
I. *Lilium henryii* pollen exine
J. Synthetic polymer from *Lilium henryii* carotenoids and carotenoid esters

Each material was intimately ground with dry KBr, the mixture compressed *in vacuo* to a disc at pressures up to 10 lb per square inch for several minutes, and infrared spectra measured on the Perkin–Elmer Infrared Spectrometer Model 157

constancy of structure which one would expect to be representative of the most stable (under the geochemical processes considered) material present in living matter. By comparison of analytical data (Table V) from older (especially Precambrian) kerogens and their oxidative breakdown products, with analogous results from pure sporopollenin and from younger spore kerogens we have been able to confidently suggest that this older constant Precambrian kerogen fraction is in fact an amorphous sporopollenin derived from many spore wall exines (Brooks and Shaw, 1968c). We have confirmed this suggestion by comparison of infrared spectra (Figs. 5, 6) and pyrolysis gas–liquid chromatograms (Fig. 7) of a number of kerogens with those of typical sporopollenins both natural and synthetic. It will be realized of course that in many

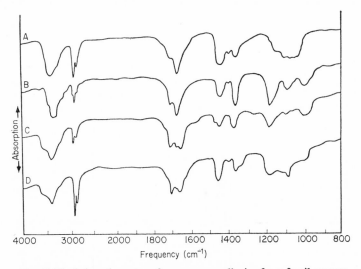

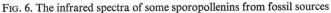

FIG. 6. The infrared spectra of some sporopollenins from fossil sources

A. *Tasmanites huronensis*, a planktonic alga
B. Insoluble organic matter from the Orgueil Meteorite
C. Insoluble organic matter from the Murray Meteorite
D. *Valvisisporites auritus*, a megaspore

cases the amorphous kerogen is accompanied by morphologically intact material (Schopf, 1968) and this adds further weight to the suggestions; it has not always been realized that many pollen and spore grains after treatment to remove contents and intine yield sporopollenin with all the typical and characteristic chemical and physicochemical properties but structurally amorphous and bearing no obvious relationship to the pollen source so that mixtures of intact pollen or spore coats associated with amorphous material of the same general chemical composition is entirely to be expected in sediments (Brooks and Shaw, 1968c). There have also been reports (Barghoorn and Schopf, 1966) of the presence in the nonesuch shale of small spherical (ca. 5 μm diameter) solid bodies (pseudospores) and although at the moment there is

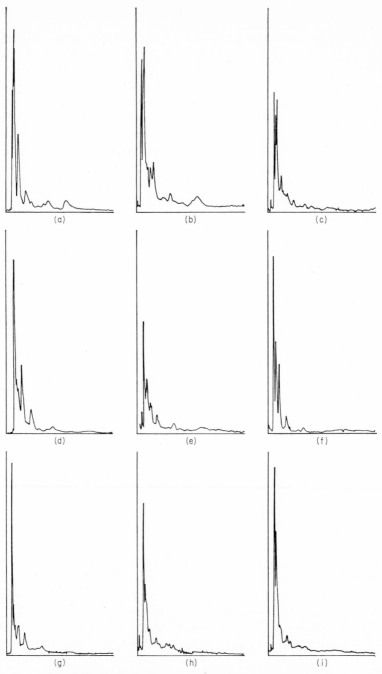

FIG. 7. Pyrolysis gas chromatograms of some sporopollenins

(a) *Lycopodium clavatum* spore exine including cellulose intine
(b) *Lilium henryii* pollen exine including cellulose intine
(c) *Pinus montana* pollen exine including cellulose intine

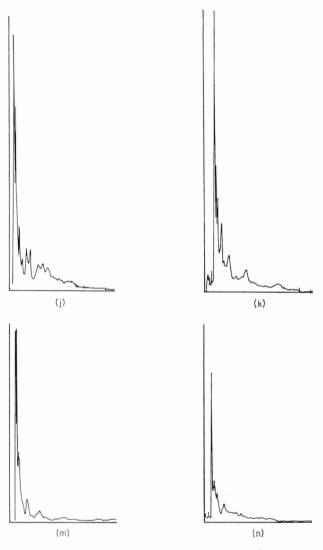

(j)

(k)

(m)

(n)

(d) *Lycopodium clavatum* spore exine with cellulose intine removed
(e) *Lilium henryii* pollen exine with cellulose intine removed
(f) *Pinus montana* pollen exine with cellulose intine removed
(g) *Lycopodium clavatum* spore exine after heat treatment
(h) Insoluble organic matter from the Orgueil meteorite
(i) Insoluble organic matter from the Murray meteorite
(j) *Valvisisporites auritus*, a carboniferous megaspore
(k) *Tasmanites huronensis* (Dawson), a planktonic-alga
(m) Polymerized β-carotene
(n) Synthetic polymer from *Lilium henryii* carotenoids and carotenoid ester

Column 1/8 inch × 5 feet stainless steel packed with 5 % SE 52 stationary phase on Chromo-sorb W (DMCS treated, acid washed) support. Varian-Aerograph 1520 gas chromatograph with flame ionization detectors. N₂ flow rate = 20 ml/min. Temperature = 87°.
Philips chromatography (P:V:4000 series) pyrolysis head and control unit was used to give pyrolysis temperatures of the materials at 770°. This pyrolysis temperature was attained using a 0·55 mm Fe filament (PV 4198) wire which reaches its Curie point (770°) in 0·6 seconds.

nothing known about their chemistry, their size and appearance is remarkably reminiscent of the orbicule or Ubisch bodies (see section on pollen ontogeny).

The nature of the relatively simple soluble chemicals which are found in sediments have been extensively reviewed on several occasions (Calvin, 1969) and will not be dealt with here.

B. SPOROPOLLENIN AND CARBONACEOUS CHONDRITES

Published records suggest that there is a close parallel between the pattern and type of organic matter present in terrestrial (especially older Precambrian) sediments (Nagy et al. 1961; Meinschein, 1963) and that in typical carbonaceous chondrites (Nagy, 1968). In each case organic matter is present as relatively simple and solvent-extractable organic compounds and as amorphous insoluble polymeric material (Eglinton and Calvin, 1967; Hayes, 1967) which may include (in the case of terrestrial sediments) morphologically intact microfossils of undisputed biological origin (Schopf, 1968). There have been strong claims for the similar presence of related microfossils in carbonaceous chondrites (Claus and Nagy, 1961; Staplin, 1962; Timofejer, 1963; Nagy et al. 1963) but the results have come under heavy criticism (Mueller, 1964), not least because of the possibility of contamination with present day pollen or spore material (Chaloner, 1968). At the same time, Urey has stated that in his opinion the evidence for the former presence of living matter in these meteorites would have been readily accepted if recorded from experiments with terrestrial sediments (Urey, 1966). It would appear however that evidence for life in a chondrite which is based solely on the histological evidence of a very few possible microfossils is always going to be open to these criticisms. Although it might be thought that histological evidence of this type would provide the best test for life this is not necessarily so since such evidence is subjective, frequently not repeatable, and subject to the presence of artefacts which even to the most careful observer can appear incredibly like remains of plant (filaments) or spore (spheres, ellipsoids) remains. Unless chondrites should turn up with massive amounts of microfossil contents the main hope of reaching a decision on the question of extraterrestrial life has rested with the unstructured organic matter which they contain.

Apart from some preliminary analytical work (Bitz and Nagy, 1966) the insoluble carbonaceous matter in the chondrites has been completely neglected and virtually the whole of the geochemical studies concentrated on the readily solvent extractable material (Hayes, 1967; Nooner and Oro, 1967), and have attempted to establish whether such compounds which are largely of a hydrocarbon nature have had a biogenic or abiogenic origin. This is to some extent understandable since such compounds are easier to deal with and when obtainable in a pure state can be given precise molecular formulae. However, the emphasis that has been made here is in some ways unfortunate for a variety of reasons. Firstly the soluble matter in chondrites, and for that matter Precambrian sediments is generally of a minor nature, the insoluble substance providing the greater proportion (up to 90 % in Precambrian terrestrial sediments) (Meinschein, 1963) and up to 70 % in carbonaceous chondrites (Hayes, 1967). Secondly the very mobility of the soluble materials makes it more likely that they may have undergone movement in total or in part from their points of

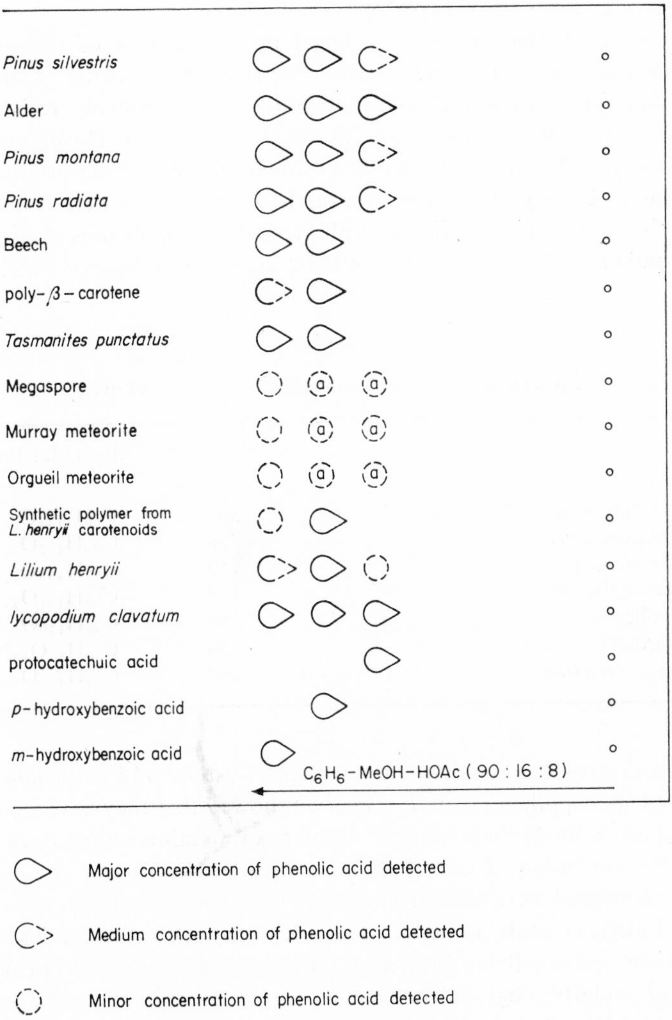

Pinus silvestris	◯ ◯ ◌>	o
Alder	◯ ◯ ◯	o
Pinus montana	◯ ◯ ◌>	o
Pinus radiata	◯ ◯ ◌>	o
Beech	◯ ◯	o
poly-β−carotene	◌> ◯	o
Tasmanites punctatus	◯ ◯	o
Megaspore	◌ ⊙ ⊙	o
Murray meteorite	◌ ⊙ ⊙	o
Orgueil meteorite	◌ ⊙ ⊙	o
Synthetic polymer from L. henryii carotenoids	◌ ◯	o
Lilium henryii	◌> ◯ ◌	o
lycopodium clavatum	◯ ◯ ◯	o
protocatechuic acid	◯	o
p−hydroxybenzoic acid	◯	o
m−hydroxybenzoic acid	◯	o

$C_6H_6 - MeOH - HOAc \ (90 : 16 : 8)$

◯   Major concentration of phenolic acid detected

◌>   Medium concentration of phenolic acid detected

◌   Minor concentration of phenolic acid detected

FIG. 8. Thin layer chromatography of the phenolic acids produced from some sporopollenins by potash fusion

Adsorbent: Microgranular cellulose powder (Whatman thin layer chromedia CC41)
Detection: Spraying with diazotized p-nitroaniline produces coloured spots corresponding to the phenolic acids
a. There was insufficient material to allow a complete study of the potash fusion products. The absence of these spots from the thin layer plates does not necessarily mean that these phenolic acids are absent

origin in aqueous or non-aqueous phases so that it may not be possible to align precisely the chemical with a particular era of time. Yet another criticism which is rarely aired is that if one is to provide acceptable organo-chemical evidence of former life then the more complex that chemical is, the more likely it will be possible to identify it with a larger part of a differentiated biological unit, and hence increase enormously the value of the evidence.

Comparison of the insoluble material in two species of carbonaceous chondrite, the Orgueil (Type I) (Wiik, 1956) and the Murray (Type II) by similar techniques used for sporopollenin analysis, namely resistance to acetolysis, infrared spectroscopy, pyrolysis gas–liquid chromatography, potash fusion followed by thin layer chromatography of the products, and some elemental analyses (results from the last experiments were restricted by the scarcity of material available) with typical sporopollenins derived from modern pollen and spore exines, synthetic analogues prepared by oxidative

TABLE VI

Molecular formula of some fossil sporopollenins

Material	Age (m.y.)	Molecular formula
Tasmanites huronensis	350	$C_{90}H_{134}O_{17}$
Tasmanites punctatus	350	$C_{90}H_{132}O_{16}$
Tasmanites erratius	350	$C_{90}H_{133}O_{11}$
Gleocapsamorpha prisca	350	$C_{90}H_{131}O_{17}$
Geiseltalpollenin	250	$C_{90}H_{129}O_{19}S_7N$
Lange-sporonin	250	$C_{90}H_{82}O_{17}N$
Valvisisporites auritus	250	$C_{90}H_{102}O_{16}$

polymerization of carotenoids and carotenoid esters, and artificially meta-morphosed sporopollenin from *L. clavatum* showed that they have a common identity (Figs. 5–8). At the same time, similar examination was made of several microfossils, including *Tasmanites huronensis* (Dawson) and *T. punctatus* (fossil planktonic algae) and a carboniferous megaspore (*Valvisisporites auritus* from the Lawrence shale, Lone Star Lake, Kansas) and these also had similar characteristic sporopollenin properties (Brooks and Shaw, 1969) (Table VI). Additional oxidative degradative evidence for the identity of *T. punctatus* (the tasmanin of Zetzsche) with sporopollenin had earlier been carried out by Zetzsche *et al.* (1937).

These results which have the advantage of being completely objective, readily repeatable on quite small amounts of material and not subject to trace con-tamination by modern pollen or spore material since detection of such minute amounts would be outside the detectable limits of the techniques used, provide powerful evidence for the existence of extraterrestrial life.

C. SPOROPOLLENIN AND THE ORIGIN OF LIFE ON EARTH

In recent years there has been a considerable increase in speculation about the origin of life on earth. This stimulation of interest is largely a result of greatly increased chemical and biochemical knowledge of the simpler life processes, especially the role of nucleic acids in the genetic process, and protein biosynthesis and the structure and nature of simple viruses, many of which appear to consist of nucleic acid and protein alone. It is also due to the experiments of Miller (Miller, 1953, 1955, 1967) in the 1950s which were based on earlier theories by Urey and Oparin (Urey, 1952) on the origin of planets and of organic chemicals and indicated that when energy in the form of sparks or silent electric discharges is imposed on a hypothetical reducing atmosphere demanded by these theories, containing hydrogen in addition to common chemicals such as carbon dioxide, methane and nitrogen, then a mixture of simple organic compounds was formed which included a number of substances of biochemical interest especially several α-amino acids, urea and sugars. Many experiments of a similar type have been repeated and there is little doubt that almost any simple organic molecule may be elaborated by these types of abiogenic processes.

The combination of the two concepts of a virus having life-like properties yet being only a combination in the simple form of two organic molecules and of the discovery that simple organic molecules can be obtained under "primordial" conditions, which are possible precursors of the nucleic acids and proteins required for the theoretical elaboration of simple forms of life, has proved irresistible to many biochemists and biologists. These factors have led to a current and widely accepted theory about the origin of life on earth. The theory is as follows: simple chemicals were first formed by processes analogous to those described originally by Miller (1957) and later expanded by many others and these collected in the sterile (life-free) oceans to form a "soup". By various processes and given lots of time the simpler smaller molecules come together to form macromolecules, especially proteins and nucleic acids. These join together in a mutually acceptable and profitable partnership and by further addition of proteins which might function as enzymes and hence by metabolizing simpler substances provide sources of energy, and of some membrane to tie the whole thing together into a state of steady equilibrium, one eventually arrives at a primitive unicellular organism which may then evolve by the more or less well known pathways of biological evolution. The development of the pre-organism stage has equally been likened to a type of "chemical evolution". Now whether or not the sort of life with which we are familiar originated in this way is one thing, whether or not it originated this way on this planet is capable of investigation, and there seems to be no reason why evidence pointing in this direction should not be readily forthcoming. I refer of course to the chemical evidence to be found in the oldest of terrestrial rocks which were around at about the time when all this is assumed to be proceeding.

Now the oldest rocks on earth are about 3·5 b.y. old (Sutton, 1967) although the earth is generally considered to be about the same age as the meteorites which fall on it and most of these are approximately 4·3–4·5 b.y. in age (Anders, 1968). However, Sutton has recently pointed out that there is very good geological evidence that the whole surface of the earth was reworked at about 4·0 b.y. ago. Now it is reasonable to assume that since this reworking process would require quite high temperature no life would be possible and for that matter even simple macromolecular structures such as nucleic acids and proteins would be subject to denaturing and hydrolytic process under quite mild (at this level) conditions of heat. This would suggest therefore that if life began on this planet it began not earlier than 4·0 b.y. ago and probably (since the reworking process may have taken many millions of years) a good deal later than this. Now the oldest known Precambrian sedimentary rocks are at least 3·2 b.y. old (fig-tree and Onverwacht systems of Swaziland) and there is good evidence for quite advanced types of life in these materials (Schopf and Barghoorn, 1967; Pflug, 1967; Nagy and Nagy, 1969). In that case life would have had to form and evolve to a very advanced state in a maximum of something like 0·8 b.y. years. It may well be possible yet to cut this figure down considerably as older rocks are found. There is a possibility that older rocks may soon become available and their organic contents should prove most interesting. Now the figure of 0·8 b.y. years might appear to provide too little time for the sort of chemico- cum biological evolutionary processes to have reached what appears to be a very advanced state of biological differentiation. However one can approach the problem in a somewhat different manner. If one did have large quantities of amino acid, protein and nucleic acid soups accumulating presumably for millions of years then this accumulated material would be expected to sediment by adsorption on depositing particles of rocks and clays to which in fact proteins and nucleic acids do very readily adhere most tenaciously, and one would expect to find therefore some evidence for these accretions in the oldest of sediments. In fact, examination of increasingly old sedimentary rocks reveals that they contain very largely hydrocarbon type compounds associated with insoluble polymeric material much of which is unquestionably of the sporopollenin type. It is especially interesting that the amount of nitrogenous matter found in these sediments is extremely small and much evidence even for nanomolecular quantities of simple amino acids is very limited indeed (Schopf *et al.* 1968). One thing is quite clear; there is no evidence in any of the oldest of rocks which would suggest the former presence of vast quantities of nitrogenous material. One might expect to find products of metamorphosis of amino acids and nucleic acids, say various nitrogenous cokes, complexes of amino acids or nucleic acid bases in quantity associated with various mineral reserves. In fact this is not so. What does the current evidence from the rocks suggest? It suggests that advanced life was present possibly in far greater quantity than generally accepted (because of the principle of differential degradation of, for example, exinous material) at a very early stage of the Earth's formation and that this life form evolved into the typical series of forms which we know today.

The most plausible explanation of all the facts is that life on Earth had an extraterrestrial origin.

An important prerequisite of such a theory requires evidence that *any* life has existed extraterrestrially. With our own and other workers' evidence for the pre-existence of life in bodies from which the carbonaceous chondrites have come, these requirements seem to have been met.

REFERENCES

Anders, E. (1968). *Accounts of Chemical Research*, **1**, 289.

Bailey, I. W. (1960). *J. Arnold arb.*, **41**, 141.

Barghoorn, E. S., and Schopf, J. W. (1966). *Science*, **152**, 758.

Braconnot, H. (1829). *Ann. chim. phys.* **2**, 42.

Brooks, J., and Shaw, G. (1968a). *Nature*, **219**, 532.

Brooks, J., and Shaw, G. (1968b). *Grana Palynologica*, **8** (2–3), 227.

Brooks, J., and Shaw, G. (1968c). *Nature*, **220**, 678.

Brooks, J., and Shaw, G. (1969). *Nature*, **223**, 754.

Burnett, J. H. (1965). *In* "Chemistry and Biochemistry of Plant Pigments" (T. W. Goodwin, ed.). Academic, London and New York.

Breger, I. A. (1963). *In* "Organic Geochemistry" (I. A. Breger, ed.), p. 50. Pergamon, Oxford.

Bitz, Mary C., and Nagy, B. (1966). *Proc. natn. Acad. Sci.* **56**, 1323.

Calvin, M. (1969). "Chemical Evolution." Oxford U.P. London.

Claus, G., and Nagy, B. (1961). *Nature*, **192**, 594.

Chaloner, W. G. (1968). *The Listener*, **79**, 529.

Elsik, W. C. (1966). *Micropaleontology*, **12** (4), 515.

Eglinton, G., and Calvin, M. (1967). *Scient. Am.* **216** (1), 32.

Faegri, I., and Iversen, J. (1964a). *In* "Textbook of Pollen Analysis," p. 15. Blackwell, Oxford.

Faegri, I., and Iversen, J. (1964b). *In* "Textbook of Pollen Analysis," p. 120. Blackwell, Oxford.

Forsman, J. P. (1963). *In* "Organic Geochemistry" (I. A. Breger, ed.), p. 148. Pergamon, Oxford.

Godwin, H., and Echlin, P. (1968). *J. cell. Sci.* **3**, 161.

Goldstein, S. (1960). *Ecology*, **41** (3), 543.

Heslop-Harrison, J., and Mackenzie, A. (1967). *J. cell. Sci.* **2**, 387.

Heslop-Harrison, J. (1968a). *Canad. J. Bot.* **46**, 1185.

Heslop-Harrison, J. (1968b). *Science*, **161**, 230.

Hayes, J. M. (1967). *Geochim. cosmochim. Acta*, **31**, 1395.

John (1814). *Schweigg. Journ.* **12** (cf. Zetzsche and Huggler, 1928).

Karrer, P., and Toappi, G. (1948). *Helv. chim. Acta*, **32**, 50.

Kwiatkowski, A., and Lubliner-Mianowska, K. (1957). *Acta Societatis Botanicorum Poloniae*, **26**, 5.

Miller, S. L. (1953). *Science*, **117**, 528.

Miller, S. L. (1955). *J. Am. Chem. Soc.* **77**, 2351.

Miller, S. L. (1957). *Biochim. biophys. Acta*, **23**, 480.

Meinschein, W. G. (1963). *Space Science Reviews*, **2**, 653.

Moore, L. R. (1963). *Palaeontology*, **6** (2), 349.

Muir, Marjorie D., and Sargeant, W. S. (Private Communication).

Mueller, G. (1964). *In* "Advances in Organic Geochemistry" (U. Colombo and G. D. Hobson, eds), p. 109. Pergamon, Oxford.

Nagy, B., Meinschein, W. G., and Hennessy, D. J. (1961). *Ann. N. Y. Acad. Sci.*, **93**, 25.

Nagy, B., Fredriksson, K., Urey, H. C., Claus, G., Andersen, E., and Percy, J. (1963). *Nature*, **198**, 121.

Nagy, B. (1968). *Endeavour*, **27**, 81.

Nagy, B., and Nagy, L. A. (1969). *Nature*, **223**, 1226.

Nooner, D. W., and Oro, J. (1967). *Geochim. cosmochim. Acta*, **31**, 1359.

Pflug, H. D. (1967). *Rev. Palaeobotan. Palynol.* **5**, 9.

Schopf, J. W., and Barghoorn, E. S. (1967). *Science*, **156**, 508.

Schopf, J. W., Kvenvolden, K. A., and Barghoorn, E. S. (1968). *Proc. natn. Acad. Sci.* **59**, 639.

Schopf, J. W. (1968). *J. Paleontology*, **42**, 651.

Shaw, G., and Yeadon, A. (1964). *Grana Palynologica*, **5** (2), 247.

Shaw, G., and Yeadon, A. (1966). *J. Chem. Soc.* (C), 16.

Smith, A. (1969). (Private Communication).

Staplin, F. L. (1962). *Micropaleontology*, **8**, 343; *J. Alberta Soc. Petrol. Geol.* **10**, 575.

Sutton, J. (1967). *Proc. Geol. Ass.* **78**, 498.

Timofejer, B. (1963). *Grana Palynologica*, **4**, 92.

Urey, H. C. (1952). *Proc. natn. Acad. Sci.* **38**, 351.

Urey, H. C. (1966). *Science*, **151**, 157.

Wiik, H. B. (1956). *Geochim. cosmochim. Acta*, **31**, 1625.

Zetzsche, F., and Huggler, K. (1928). *Annalen*, **461**, 89.

Zetzsche, F., and Kälin, O. (1931). *Helv. chim. Acta*, **14**, 517.

Zetzsche, F., Kalt, P., Liechti, J., and Ziegler, E. (1937). *J. Prakt. Chem.* **148**, 267.

CHAPTER 4

Modern and Fossil Plant Resins

B. R. THOMAS

Department of Organic Chemistry, LTH, Lund, Sweden

I. INTRODUCTION

From a phytochemical point of view the interest of plant resins lies in their origin, development and function and how they reflect the evolutionary history of the plants that produce them. Primarily this review will therefore be concerned with the biochemistry of plant resins and particularly with the more conspicuous bled resins.

Investigation of the chemical phylogeny of resin-producing systems must depend largely on modern material. However there are numerous fossil resins that offer the tantalizing possibility of direct evidence of the past history of plant resins. Fossil resins also offer potential information of biological and geological interest on the past occurrence of resin-producing species.

II. ORIGIN AND FUNCTION OF RESINS

The basic function of plant resins is protection by physical means. Thus they are produced in very much larger quantities than compounds with more specific toxic properties. As the early land plants expanded into their new environment they had to meet new physical problems such as dehydration and mechanical damage. Interaction with the bacteria and insects that accompanied them required a steady modification and improvement of defensive means, and the development of long-lived woody plants brought a need for coating and plugging materials.

Some of the secondary metabolites from plants may represent biological variation, history or even waste, but the value of resins is shown by their very wide occurrence, by the durability of many heartwoods that contain them and by the success of resinous plants such as the pines. Resins are formed in a number of different structures: in canals and glands in wood, bark and leaves, in canals formed in response to injury, in special cells, or generally diffused through heartwood. Cracks produced by mechanical damage in the trunks of trees are sometimes filled with resin; *Dacrydium cupressinum* heart-shakes for instance fill with a resin that is mainly podocarpic acid.

Bled resins have evolved to provide a seal over damaged areas of bark or wood. They harden rapidly on exposure to air by loss of water or a solvent such as pinene and sometimes harden further by gradual polymerization. Wood resins serve to protect dead areas of wood within a tree from internal attack by fungi and bacteria.

Epidermal layers often contain large amounts of resins or oils, for instance the epidermal layers of some conifer leaves or the glandular hairs of *Nicotiana.* There are several possible functions for this material. It may provide surface coating or the solvent to carry the coating to the surface of the leaf; it probably functions as an insect repellant and it might be related to systems for the control of gibberellin production.

Presumably resins have developed from protective compounds in the plant epidermis as plants became bigger and longer lived. The occurrence of triterpene mixtures in angiosperm leaf and bark and also in many angiosperm resins is in accord with this. The origin of resins might therefore be sought in the epidermal cells of primitive plants, in the coatings that protected the first land plants from dehydration and in the substances present in primitive cell membranes.

III. RESIN CONSTITUENTS

A. INTRODUCTION

A number of materials that appear to have a protective function have evolved from the biosynthetic resources available to plants: tannins and polyesters occur in large amount in barks; polyesters and waxes on leaf surfaces (Eglinton and Hamilton, 1967); wood resins often contain aromatic substances diverted from lignin synthesis; some angiosperm latexes contain large amounts of linear polyterpenoids (e.g. *Hevea*). In both bled resins and wood resins the main components are usually diterpenes and triterpenes. Although they are exuded as comparatively freely-flowing emulsions or oils, the terpenoid resins are often well adapted to the formation of tough, impermeable coatings.

Terpenoid compounds were incorporated into living organisms at a very primitive stage (Margulis, 1968). Mevalonic acid is essential for the growth of some lactobacilli and is apparently involved in cell membrane formation (Kodicek, 1954). Phytol and carotenoids appear in the photosynthetic bacteria and steroids and gibberellins seem to occur in the algae.

The softer resins, sometimes known as balsams, may contain substantial amounts of monoterpenes and sesquiterpenes. The pleasant smelling angiosperm balsams may contain compounds such as benzoic, cinnamic and phenylpropionic acids together with corresponding alcohols, esters, aldehydes and other more highly oxidized derivatives such as coumarin, vanillin and ferulic acid. The phenols that often occur in wood resins and in wound exudates and kinos are sometimes of considerable taxonomic interest but for brevity this discussion must be largely confined to diterpenoids and triterpenoids (see however Hillis, 1962; Erdtman, 1969).

B. CONIFER RESINS

The conifers appear towards the end of the Carboniferous era as the older groups of plants dwindled. They are very diverse and show a marked division into northern and southern groups that seems to have been already present in the first known conifers (Florin, 1963). This is presumably associated with the barrier to north–south movement formed by the Tethys Sea from the end of the Carboniferous to the Cretaceous. This would imply that all the main conifer groups and probably even the minor groups of southern Cupressaceae (*Callitris*) and Taxodiaceae (*Arthrotaxus*) had already separated by the end of the Carboniferous. Thus after the first rapid diversification of the group further change seems to have been slow. *Agathis* species that have probably been isolated since the Cretaceous or earlier, still show a close similarity.

Bled resins and wood resins occur in all the conifer families but particularly in the Pinaceae and Araucariaceae. As shown in Table I, all conifer families have much the same range of diterpenoid constituents. The genera within each family can sometimes show considerably greater variation. The Pinaceae and especially *Pinus* usually give resins with a high content of abietic acid (II) and its isomers, but the bled resins from some *Abies* (Gray and Mills, 1964) and *Picea* species can contain large amounts of labdanes such as abienol (I).

In the Araucariaceae, *Agathis* usually gives standard labdane–pimarane– abietane resins with agathic acid and communic acid derivatives often prominent (see below) but *Araucaria* gives labdanes in which the 13:14 double bond is reduced, such as imbricatolic acid (VI) (Bruns, 1968). This may be connected with the exposed situations in which *Araucaria* is frequently found. It often projects above the forest canopy and occurs in hurricane areas. Bark hardened by a communic acid polymer, for example, would be quite unsuited to the flexibility required in the long slender trunk.

The Cupressaceae frequently give phenols and these have attracted most

TABLE I

Some gymnosperm resin constituents

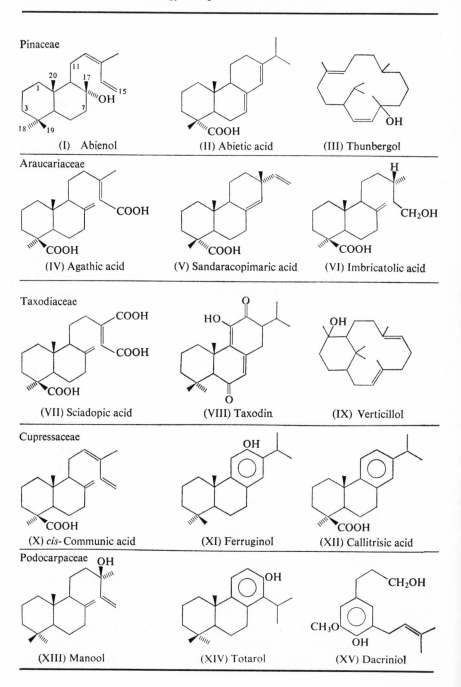

Pinaceae

(I)   Abienol   (II) Abietic acid   (III) Thunbergol

Araucariaceae

(IV) Agathic acid   (V) Sandaracopimaric acid   (VI) Imbricatolic acid

Taxodiaceae

(VII) Sciadopic acid   (VIII) Taxodin   (IX) Verticillol

Cupressaceae

(X) cis-Communic acid   (XI) Ferruginol   (XII) Callitrisic acid

Podocarpaceae

(XIII) Manool   (XIV) Totarol   (XV) Dacriniol

attention here (Erdtman and Norin, 1966). However a recent very comprehensive survey of the diterpenes of the resins and leaf extracts has shown that they characteristically contain labdatrienes, pimaradienes, both usually as acids, and phenolic diterpenes such as ferruginol (XI) though a variety of other compounds including dihydroagathic acid derivatives may occur (Gough, 1969). The Taxodiaceae show a similar range of compounds including phenolic diterpenes.

Of the two principal genera of the Podocarpaceae (Cambie and Weston, 1968), *Podocarpus* gives characteristically phenolic diterpenes while *Dacrydium* gives manool and related compounds. However *D. cupressinum* is somewhat unusual. It is a member of a group of *Dacrydium* that is closely related to *Podocarpus* and although botanically it appears to be a *Dacrydium* it gives a *Podocarpus* resin. In two species, diterpene synthesis appears to be blocked; *D. franklinii* gives instead eugenol and dacriniol (Baggaley *et al.* 1967) and *P. spicatus* lignans.

The only marked separation among the diterpene constituents of the conifer resins is the presence in the Taxodiaceae, Cupressaceae and Podocarpaceae of the oxidation-rearrangement system leading to the aromatic diterpenes such as ferruginol and totarol. Although the conifer resins thus yield little phylogenetic evidence they can however provide valuable fingerprint characteristics that might be useful in tracing fossil resins of individual genera.

C. ANGIOSPERM RESINS

Resins can apparently occur almost throughout the angiosperms but the most prolific producers are tropical forest trees belonging to some of the earlier angiosperm orders (Engler, 1964). Some typical compounds from the more important resins are shown in Table II (see also Ponsinet *et al.* 1968).

Some of the angiosperm bled resins are diterpenoid. The most important of these are the copals from the tribe Amherstiae of the Leguminosae such as Congo copal, which occurs in very large quantity as semifossil resin, and Madagascar copal (*Trachylobium*). They contain mixtures of labdanes, sometimes with normal (*anti-*) configuration, sometimes with *enantio* configuration. In the angiosperms, the tri- and tetracyclic diterpenes are largely represented by *ent*-kaurene derivatives together with *ent*-beyerenes and *ent*-pimarenes, which are related to *ent*-kaurene synthesis. Anti-pimarenes are also produced but seem to have a strong tendency to undergo a friedo shift to cassanes.

Most frequently the angiosperms give triterpenoid bled resins with complex mixtures of penta- and tetracyclic triterpenes. Some components of these mixtures such as β-amyrin and the related oleanolic acid (XVII) are of rather common occurrence. However, they often also contain components with a skeleton or oxidation pattern that can be more or less specific for particular groups of plants. Thus the dammars obtained from a very large number of the Dipterocarpaceae give compounds related to dammarenediol (XVIII) (Ponsinet *et al.*

TABLE II

Some angiosperm resin constituents

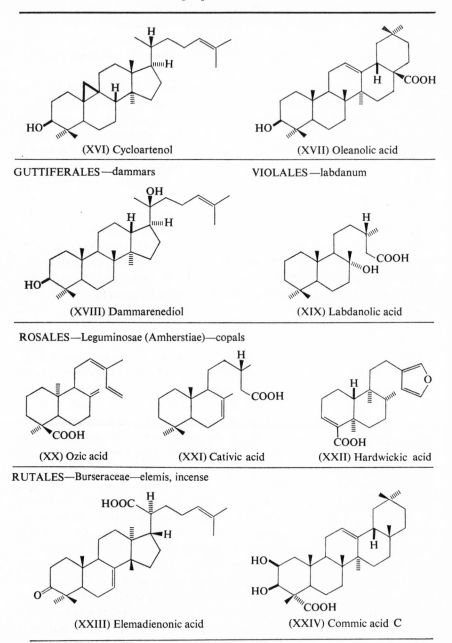

(XVI) Cycloartenol

GUTTIFERALES—dammars

(XVII) Oleanolic acid

VIOLALES—labdanum

(XVIII) Dammarenediol

(XIX) Labdanolic acid

ROSALES—Leguminosae (Amherstiae)—copals

(XX) Ozic acid

(XXI) Cativic acid

(XXII) Hardwickic acid

RUTALES—Burseraceae—elemis, incense

(XXIII) Elemadienonic acid

(XXIV) Commic acid C

TABLE II—*continued*

GERANIALES—Euphorbia

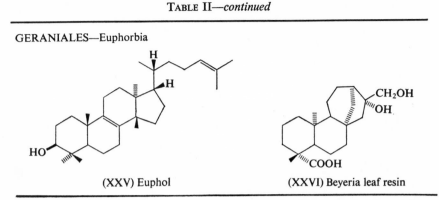

(XXV) Euphol   (XXVI) Beyeria leaf resin

1968; Boiteau *et al.* 1964) while groups of the *Euphorbia* are characterized by compounds such as euphol (XXV) and cycloartenol (XVI). As a phytosterol precursor, cycloartenol is also found in small amount in non-resinous tissue in many plants.

The triterpene resins also contain a number of minor components such as monoterpenes, sesquiterpenes, linear polyterpenes and aromatic ester type compounds. Sometimes these can form a large or even, as in *Hevea*, a predominant component of the latex.

Since some of the triterpenes are good taxonomic indicators and are reasonably stable, they should be quite useful in identifying resins and plant tissues. They should, for instance, be able to offer valuable information on the origin and development of the angiosperms which apparently spring into sudden prominence with well developed diversity during the Cretaceous period.

IV. THE CHEMICAL PHYLOGENY OF RESINS

A. RESIN PRODUCING TISSUES

A wide variety of resins and resin-producing systems can apparently occur in almost any group of the higher plants. The constituents of the resin from any particular tissue will reflect both present requirements and the origin of the tissue.

Epidermal tissues tend to produce *ent*-kaurene derivatives, possibly by a diversion from gibberellin biosynthesis. Some conifer leaves contain *ent*-kaurene (Aplin *et al.* 1963), *Euphorbia* leaf resins give a number of *ent*-kaurene derivatives (Henrick and Jefferies, 1965) and the bitter or toxic components of leaves can be kaurenes such as enmein (XXVII) (Kubota *et al.* 1964).

Conifer resin canals are presumably related to epidermal tissue but the compounds of normal configuration that can accompany *enantio* compounds in the leaf surface are here almost the sole product. Diterpene resin canals in the angiosperms give compounds of both configurations. More usually however angiosperm resin canals give mixtures of triterpenes often together with aromatic ester type compounds. The same type of product is also found in the

3

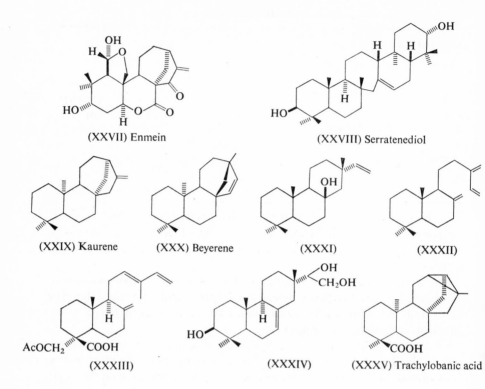

(XXVII) Enmein

(XXVIII) Serratenediol

(XXIX) Kaurene   (XXX) Beyerene   (XXXI)   (XXXII)

(XXXIII)

(XXXIV)

(XXXV) Trachylobanic acid

"barks" of both angiosperms and conifers. Thus serratenediol (XXVIII) occurs together with ferulic esters in the bark of *Pinus banksiana* (Rowe *et al.* 1969).

In both conifers and angiosperms the exudates from exposed xylem surfaces contain a variety of phenols, often lignans. These and the aromatic ester type compounds probably originated as a further step from phenylalanine synthesis giving products with value as bacteriostats or insect repellents. The presence of these simpler aromatic substances in cell walls would then allow the biosynthesis of lignans and lignins by the development of oxidative coupling.

Agathis australis provides an excellent example of the variations possible even for a single biogenetic group of compounds. Here the heartwood, the sapwood, the resin canals of the bark, the resin canals of the primary stem tissue and of the leaves, the photosynthetic tissue, the leaf epidermis and the cones all give different products (Thomas, 1969a). These products can however be rationalized to a considerable extent in terms of protective functions. Thus the leaf epidermis, a layer about three cells thick with a high diterpene content, contains a mixture of *ent*-kaurene (XXIX), *ent*-beyerene (XXX), 8β-hydroxy-isopimarene (XXXI) and the labdatriene (XXXII). This leaf surface resin with its low dialkylbutadiene content and low polarity should spread better on the non-polar leaf cuticle and give a flexible film. It may also contribute to the normal leaf coating.

The main component of the resin canals of the leaf and primary stem tissue is apparently (XXXIII), in accord with a material that must be exuded as an emulsion to spread on the wet end of a broken leaf and give a hard even though less flexible film.

The resin canals produced by the cambium in the phloem, i.e. in the "bark", yield a mixture of labdatrienes, isopimaradienes and abietadienes mostly with carboxyl groups at C-18 or C-19. This gives a polar mixture suitable for wet surfaces but producing a more flexible coat. In the heartwood resin, strength is not required and the resin (up to 10% of the wood) is largely made up of hydroxylated isopimarenes, e.g. (XXXIV) and the phenol agatharesinol. The resin content drops abruptly at the heartwood boundary suggesting that the heartwood resin is analogous to a wound exudate excreted into the adjacent dead space by the cells at the heart-sap boundary.

In *Pinus*, resin canals are formed to the inside of the cambium and in the rays but still produce a labdane–pimarane–abietane resin. Here the canal resin impregnates the heartwood. Since hardness and strength are not required this may be why the labdane content of these resins has become very much reduced. The high abietane content of the resin now however allows it in some cases to solidify by crystallization (Sandermann, 1960).

In the angiosperms, *Trachylobium* (Hugel *et al.* 1966) shows the same sort of difference as *A. australis*. The resin canals of the trunk give a series of labdanes while the seed pod resin, which will be analogous to a leaf resin, gives a mixture of kauranes and trachylobanes, e.g. (XXXV) (Hugel *et al.* 1965).

B. THE TERPENOID CYCLIZATION SITE

The general outlines of the cyclization and coupling reactions involved in the biosynthesis of diterpenes and triterpenes from geranylgeraniol and farnesol are now well known (Clayton, 1965; Cornforth, 1968). It is however worth considering here a few aspects of the origin and relationships of these cyclizations.

Cyclization is usually considered an electrophilic process starting at C-3 (see also Cornforth, 1968; Oehlschlager and Ourisson, 1967; Zander and Caspi, 1969). It is simpler and more interesting to regard it in the present context as a nucleophilic process initiated by the pyrophosphate ion. At physiological pH the pyrophosphate group will be ionized and in kaurene cyclization enzyme activity shows a peak at a pH corresponding to the pyrophosphate pK_a (Upper and West, 1967).

In an aqueous environment the alkyl chain of a geranylgeraniol pyrophosphate molecule will of course coil in a close-packed clump with the polar pyrophosphate group at one side, probably attached to a polar surface by coordination to a metal ion or by hydrogen bonding. In general, the structure of the final molecule will be determined by the packing of the clump, though the environment may alter the relative stability of different packings. In the most stable packing the coils will stack with chair configurations but the alternative boat configuration is also possible.

In a coiled geranylgeraniol pyrophosphate molecule, when the C-17 allylic methyl comes close enough to the pyrophosphate anion, abstraction of a proton would initiate a concerted cyclization giving a labdadienyl pyrophosphate (Fig. 1a). In solution, geranylgeraniol pyrophosphate will be hydrated and is reasonably stable. When the molecule is adsorbed on a surface however, the pyrophosphate group should be able to gain access to the hydrocarbon chain more easily.

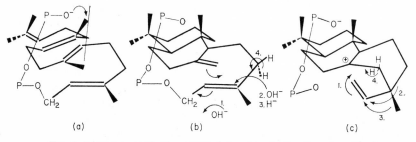

FIG. 1. Diterpene cyclization (C-20 shown here in the β-configuration)

b.		c.	
1	copalic acid	1	*ent*-beyerene (α C-20)
2	manool	1, 3	antikaurene (β C-20)
3	eperuic acid	1, 2	phyllocladene (β C-20)
4	*cis*-communic acid	1, 2	*ent*-kaurene (α C-20)
		4	isopimaradiene (β C-20)
		4, 3	abietadiene
		4, 3	pimaradiene (β C-20)

Elimination of the pyrophosphate group in a β direction from the labdadienyl configuration shown in Fig. 1b will lead to two successive sequences of anti-parallel electron displacements. The first sequence will give all the common labdane types by termination at some point along the sequence or it will lead to an isopimarene ion. The positive charge might well be stabilized by pyrophosphate anion (cf. Cornforth, 1968).

The second sequence of electron displacements (Fig. 1c) will lead to sandara-copimaradiene if ring C changes to a chair form. It will give abietadienes if the C-14 proton is transferred to C-16 either directly or by mediation of pyrophosphate anion. It will give a beyerene if attack of the positive charge on C-16 is followed by proton loss; it will give phyllocladene (or *ent*-kaurene in the *enantio* series) by migration of C-12; and it will give anti-kaurene if the vinyl group is twisted to allow rear attack. It is perhaps somewhat remarkable that all the common diterpene types can be accounted for by this single set of reactions (see also Oehlschlager and Ourisson, 1967; Wenkert, 1955). In compounds of the *enantio* series, Rings B and A will coil below C-9 and the clump will thus have quite different spatial requirements. However, in all the common diterpenes, cyclization will involve the absolute configuration of ring C that is shown in Figs. 1b and 1c.

A number of diterpenes are known in which a "friedo" shift, or transdiaxial migration of a series of methyl groups and protons, has taken place, e.g. in erythroxydiol (XXXVI) (Connolly *et al.* 1966) and kolavenol (XXXVII) (Misra and Dev, 1968). These migrations often are initiated at or near C-14 and also indicate the presence of an acidic proton in the vicinity.

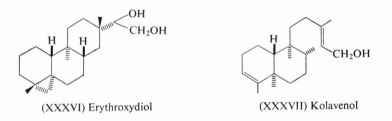

(XXXVI) Erythroxydiol   (XXXVII) Kolavenol

In triterpenoid cyclization, the squalene epoxide precursor will be held on the reaction site only by the aqueous phase above it. The situation is therefore more complicated. The epoxide grouping will require that the C-4 and C-10 axial methyls are still β and in a close-packed array the future ring C will have a chair configuration with the C-8 and C-14 methyls pointing in opposite directions. If the C-8 methyl is α the C-14 methyl will be β and the cyclized product will be a lanostene. If the reverse, the product will have a normal triterpenoid configuration.

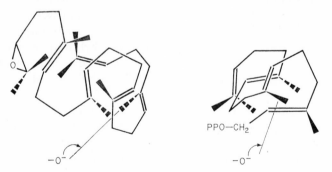

FIG. 2. Cyclization configurations for lanosterol and *ent*-kaurene

In the lanosterol case, ring B will have a boat configuration leaving a space beneath it which for close packing must be filled by the side chain. This will give a rounded clump and form the five-membered ring D with the correct stereochemistry (Fig. 2). Initiation of cyclization by removal of a proton in the direction shown in Fig. 2 will give the antiparallel electron displacements for construction of the five-membered ring. Replacement of the proton after cyclization will then generate the friedo shifts required in steroid and triterpenoid biogenesis. In the pentacyclic triterpenoids the cyclizing configuration will be a long flat clump which for stability might require to lie on a non-polar surface. The large number of ring E variants is not unexpected since it will be much less sensitive than the lanosterol configuration to small changes in packing.

Curiously, the biologically very important molecules kaurene and lanosterol both form a somewhat rounded cyclization clump. If the cyclization mechanism discussed above is actually involved in biosynthesis (there need not necessarily be only one, and in fact several different mechanisms probably occur, cf. Zander and Caspi, 1969), then it seems that it may be possible to explain terpenoid cyclization to a considerable extent in terms of pyrophosphate chemistry and of the closest packing of a polyisoprenoid chain in a primeval oil speck.

These reactions could therefore have arisen at a very primitive stage in evolution. It therefore might be interesting to examine what part this type of isoprenoid synthesis could play in the production of protective layers around very primitive membranes.

C. OXIDATIVE MODIFICATIONS

In addition to giving a harder tougher coat, oxygenation of the constituents of bled resins is probably needed for emulsification of the resin and to aid spreading and adhesion.

The site of oxidation may indicate the point of initiation or termination of cyclization, as in many triterpenes; it may derive from hydrolysis of a pyrophosphate group, e.g. in manool (XIII) or cativic acid (XXI); or it may be introduced by direct insertion after cyclization. In many diterpene resins, direct oxidation usually occurs at one of the C-4 methyl groups by insertion of a hydroxyl group, or possibly an acetoxyl group since the corresponding acetates occur, and proceeds through the aldehyde to the carboxylic acid. The acid may then be further converted to the methyl ester.

While the main products in conifer resins are usually carboxylic acids, some resins, for instance *Trachylobium* trunk resin (Hugel *et al.* 1966), *Pinus silvestris* resin (Westfelt, 1966) and *Agathis australis* canal resin (Thomas, 1966, 1969a) contain compounds at several or all stages between hydrocarbon and methyl ester.

In pimaranes, abietanes, kauranes and beyeranes, oxidation occurs very largely on the α side giving equatorial oxygenation at C-4 in compounds of normal configuration and axial oxygenation in *enantio* compounds. Labdanes and podocarpatrienes however are usually oxidized on the β side though here the specificity may be somewhat less. A number of exceptions to this simple pattern are known but in view of the wide species range and the variety of oxidations that can occur, more might have been expected.

The simplest explanation of the C-4 α-methyl oxidation pattern is that the extra hindrance of ring C requires the tricyclic and tetracyclic diterpenes to be reversed on the active site of the initial oxidizing enzyme. C-4 α-methyl oxidation also occurs in the much older sequences involved in the biosynthesis of gibberellins such as (XXXVIII) and in the removal of the gem-dimethyl group in steroid biosynthesis (Sharpless *et al.* 1968). It is thus a further indication of the general relationship between these biosynthetic routes.

While canal resin diterpenes usually show a simple oxidation pattern, wood resin compounds may show much more extensive oxygenation involving almost any point in the diterpene molecule. For instance *Dacrydium colensoi* gives a series of oxidation products such as (XXXIX) (Carman *et al.* 1966).

(XXXVIII) Gibberellic acid   (XXXIX)

Triterpenes in bled resins usually have a 3β hydroxyl group though 3α also occurs, presumably indicating either some subsequent oxidation step or that the stereospecificity of the terminal epoxidation is not perfect. They frequently have oxygen at C-28 as in oleanolic acid, occasionally at other points; C-28 oxidation may be spatially analogous to C-4 α-methyl oxidation.

V. FOSSIL RESINS

A. OCCURRENCE AND COMPOSITION

Fossil resins occur all over the world, in the soil, in coals and sometimes in rocks. They are found as far back as the Carboniferous era, particularly where conifers were present. In some places they are found in the ground in very considerable quantities and have been mined along the Baltic coast, in the Congo (copal), in Kamchatka, Burma, British Columbia and other places. Quite a small area in New Zealand has yielded over half a million tons of resin.

Major surface deposits such as those in the Congo and New Zealand appear to have accumulated within the past few thousand years. Resin exposed near the surface is, however, gradually destroyed by "weathering" and older resins are more likely to be preserved and to be found in coals. Some brown coals can contain as much as 5% of resin. Coal resins may occur in quite large pieces or may be distributed through the coal.

The occurrence of fossil resins has been admirably reviewed from a botanical viewpoint (Langenheim, 1964, 1969) and this section will deal more with the chemical information available (see also Tschirch and Stock, 1936) and, as examples of fossil resins, with Baltic amber and kauri resins (*Agathis* sp.).

1. Constituents

Fossil resins are often largely polymeric and rather insoluble. "Fossilization" will usually involve vinyl polymerization (in the commoner diterpene resins) and some loss of volatile constituents. The original constituents may also be modified, for instance by hydrogenation, dehydrogenation or decarboxylation. Oxidation, though probably significant in thin films of resin should be unimportant for resin in the ground where it is protected from air and light. In the fossil resin, succinite, the oxygen content is about 10 % but in weathered layers it is much higher (Plonait, 1935).

The extent to which these changes take place will be dependent on the age of the resin and the temperatures and pressures to which it has been subject. The "semi-fossil" resins that are usually regarded as not more than a few thousand years old can contain large amounts of low molecular weight material.

Labdatrienes such as communic acid are converted to polymer and disappear within a year or two (cf. Carman and Cowley, 1967) but agathic acid and pimaradienes can apparently survive for many millions of years (Roxburgh *Agathis* resin). Abietadienes are particularly liable to disproportionation and give rise for instance to fichtelite (XL) (Burghstahler and Marx, 1964), simonellite (XLI) (Ghigi *et al.* 1968) and retene (XLII). Both retene and fichtelite may be found, sometimes together, in association with old pine logs and stumps and in brown coal (see Tschirch and Stock, 1936). The absence of labdatriene polymer and the ease of decarboxylation may however make these high abietane resins much more liable to degradation.

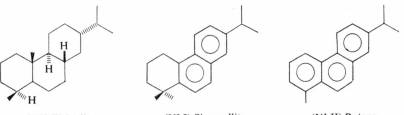

(XL) Fichtelite   (XLI) Simonellite   (XLII) Retene

The angiosperm resins should polymerize less readily. Thus betulin (XLIII) and other triterpenes have been isolated from German and Czechoslovakian coal resins (Ruhemann and Raud, 1932; Jarolím *et al.* 1963) and isoarborinol (XLV) has been obtained from a German Eocene shale (Albrecht and Ourisson, 1969). Both these compounds are rather sensitive and therefore suggest that there are excellent prospects of obtaining useful information from these triterpenoid resins. The betulin in the Czechoslovakian resin is accompanied by dehydration products such as (XLIV) and by partially aromatized material.

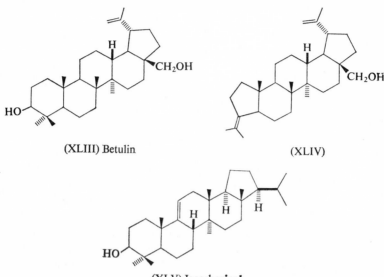

(XLIII) Betulin

(XLIV)

(XLV) Isoarborinol

2. Identification

GLC or preferably GLC-MS identification of the pattern of volatile constituents is in general the best way of characterizing a fossil resin. There can however be difficulties in interpretation until much more is known of the constituents that can survive or that can be formed in a resin.

Very useful additional information is available from infrared spectra which are comparatively insensitive to the degree of polymerization of the resin (Langenheim, 1969). Phenolic resins, some triterpene resins and non-acidic resins such as those from *Dacrydium* can be readily distinguished. Many conifer resins however are mixtures of acids derived from labdanes, pimaranes and abietanes and resins from a variety of sources may show rather similar spectra. Thus, spectra of ancient *Agathis* resins (Thomas, 1969a, 1969b) show a close resemblance to spectra from Baltic ambers and particularly to that of Cretaceous Manitoba resin (Langenheim, 1969). In the case of Chiapas resin (Tertiary, Mexico), infrared spectra provide good evidence that this derives from *Hymenaea courbaril* since other possible sources would not give copalic acid resins (Langenheim, 1969).

Chemical evidence of the nature of the components in a resin polymer can be obtained by selenium dehydrogenation. The amounts of naphthalenes such as agathalene (XLVI) and of pimanthrene (XLVII) and retene (XLII) give an approximate measure of the proportions of labdanes, pimaranes and abietanes in the resin. This has given very useful evidence on the nature of Baltic amber (Gough, 1969, see below) and the formation of agathalene has been used as

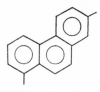

(XLVI) Agathalene   (XLVII) Pimanthrene

evidence (in the local context) that various New Zealand and Australian resins derive from *Agathis* (Brandt, 1939; Brooks and Steven, 1967).

The techniques now available, particularly GLC-MS, should open up a number of interesting geochemical problems. Two of these, the identification of the source of Baltic amber and the development of *Agathis*, are dealt with below. Resins associated with fossil remains assigned to *Araucaria* that occur from the Triassic to the Cretaceous in Europe and North America might offer further evidence on the origin of this material. Doubts have been expressed because of the difficulty of migration from the southern hemisphere (Florin, 1963). Of particular interest of course would be the examination of the triterpenes of the early angiosperms.

B. BALTIC AMBER

Baltic amber, because of its historical associations perhaps the best known fossil resin, occurs in a "blue earth" seam about seven metres thick in a small area on the south-east coast of the Baltic (Bachtofen-Echt, 1949; Czeczott, 1960; Langenheim, 1964, 1969; Tschirch and Stock, 1936). It is apparently Eocene in origin but accumulated in its present site during the Oligocene. Presumably resin deposits from forests around the Baltic area were reworked by river or glacial action and the resin was trapped in an estuary or lake at its present site.

In addition to succinite, the principal type of resin in the Baltic amber deposit, there are several other less common varieties. Similar resins are found over a wide area. Some resin has apparently been redispersed from the Baltic amber deposit but the same resins are to be expected in coals and other Eocene deposits elsewhere.

The plant and insect remains in the resin include both tropical or sub-tropical species (as expected since the European climate was warmer) and also temperate species. This suggests that the resin came from a mountainous area or was supplied by rivers from both north and south. If the resin content in the deposit is uniform, figures commonly quoted (1–2 kg m^{-3} × 7 m × 500 km^2) would indicate that the deposit is of the order of 5–10 million tons. While this has given rise to speculation on disease or catastrophe, it is not unduly remarkable by comparison with the surface deposits of *Agathis australis* resin which have accumulated over, presumably, the past few thousand years. The recovered resin alone taken from the entire present area of *A. australis* amounts to 25 tons km^{-2} and Baltic amber may have accumulated during several million years over an area as much as 10^6 km^2.

The succinite source tree is probably to be sought in southern or central Europe, since the present-day climate is colder. While extinction of what must have been a successful species can be rejected as less probable, at least initially, the species may well be rather modified.

TABLE III

Approximate percentages of labdanes, pimaranes and abietanes
in amber and other resins (Gough, 1969)

	Labdanes	Pimaranes	Abietanes
Baltic amber (succinite)	70	10	20
Yorkshire amber	94	5	1
Kauri resin (*Agathis*)	77	17	6
"Polycommunic acid"	95	5	0·5
Rosin (*Pinus*)	3	16	81

From percentages of labdanes, pimaranes and abietanes estimated from dehydrogenation results (Table III, Gough, 1969) and from infrared spectra, succinite appears to be a resin acid polymer (strong carbonyl absorption) with a large proportion of labdane units. It contains borneol and yields up to 8 per cent succinic acid on distillation or alkali treatment. This is unexpectedly high though other resins can give up to 2% (Tschirch and Stock, 1936). The succinic acid is presumably present as ester and is responsible for the additional carbonyl absorption near 1735 cm^{-1} and other small differences between the infrared spectra of amber and the rather similar *Agathis* resins. Free succinic acid is occasionally also present in succinite (Plonait, 1935).

Although the succinite source is usually termed *Pinus succinifera* Conwentz, the data in Table III exclude any probable modern *Pinus* species since these give high-abietane–low-labdane resins. The source then is probably to be sought in *Picea* or *Abies*, some species of which give large amounts of labdanes. *Picea engleri* and *P. obconica*, mentioned by Tschirch and Stock (1936) in this connection, are of particular interest. Modern *P. abies* gives a high abietane resin (Kahila, 1957) and can be excluded. *Abies balsamifera* which gives 1·7% succinic acid (Tschirch and Stock, 1936) has a high abienol (I) content, like many other *Abies* species (Gray and Mills, 1964) and is also less probable.

Infrared spectra indicate that succinite is reasonably homogeneous (Langenheim and Beck, 1965) but several less common varieties of resin are known from the Baltic amber deposit. One of these, glessite, has been shown to contain crystalline α-amyrin by X-ray diffraction (Frondel, 1967) and is thus an angiosperm resin, probably from the Burseraceae. This is presumably the same resin as or a very similar resin to the Eocene Highgate copalite from southern England which gives the same X-ray pattern. As expected they accord with a tropical Eocene climate. Isoarborinol from German Eocene shale (see above) has been isolated from modern tropical species of the Rutales and could possibly have come from the same or a related species.

C. KAURI RESINS

Kauri or *Agathis* resins afford a particularly good series of recent and fossil resins. *Agathis* is at present found widely spread between New Zealand and Malaya and the Philippines. Fossil resins are found in a number of places in the area and as much as a thousand kilometres to the south in both Australia and New Zealand. It is also possible that *Agathis* resin might be found in other southern or formerly southern areas such as Antarctica and India. It is thus of interest to be able to identify the fossil resins with modern or older *Agathis* species and to distinguish them from other labdane–pimarane–abietane resins. This complex problem is of additional interest because of its relationship to the breakup of the southern continent and the way in which the genus has developed under the influence of changing environments and increasing isolation.

Agathis bled resins (Thomas, 1969a) are, in general, mixtures of the common labdanes, pimaranes and abietanes, usually with a carboxyl at C-18 or C-19. Most contain substantial amounts of labdatrienes and harden rapidly. Resins from the island species, *A. australis* (New Zealand), *A. vitiensis* (Fiji) and *A. lanceolata* (New Caledonia), contain all four of the main diterpene types: i.e. labdatriene, agathic, pimarane and abietane. The Australian resins (Carman and Marty, 1966; Carman and Dennis, 1964; Carman 1964) lack pimaranes (*A. microstachya*) or both pimaranes and agathic type compounds (*A. palmerstonii* and *A. robusta*). *A. philippinensis* which apparently contains labdatrienes (since it hardens) and agathic compounds could resemble either the Australian or the island group. Malayan *A. dammara* is non-hardening and contains a large proportion of agathic compounds but very little labdatriene (Thomas, 1969a).

Two of the New Caledonian species, *A. ovata* and *A. moorei*, according to a preliminary survey, differ markedly from the other *Agathis* species investigated. *A. ovata* does not oxidize either of the C-4 methyl groups and apparently gives copalic acid (*ent*-labdadien-15-oic acid) and similar compounds but little or none of the labdatriene type. This may be a late modification, or more interestingly, might be a primitive type of resin. *A. moorei* gives apparently abietanes and agathic compounds including epiabietic acid and epiagathic acid. If the oxidation follows the C-4α pattern then these two compounds should be enantiomeric also. The occurrence of *enantio* diterpenes in conifer canal resin would be most unusual and requires further confirmation.

A number of fossil and semifossil *Agathis* resins are known. The "semifossil" resins will essentially be simplified resins corresponding to the present local

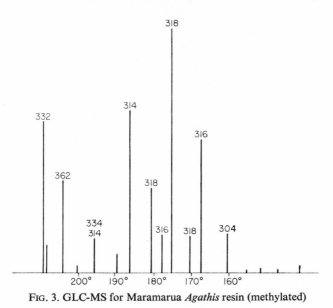

FIG. 3. GLC-MS for Maramarua *Agathis* resin (methylated)

species. The fossil resins however are much older and much more altered. The fossil *Agathis* resins that occur in brown coals in New Zealand and Australia are largely polymeric but usually still contain significant amounts of low molecular weight material. An *Agathis* species is the most probable local source and confirmatory evidence that these are usually high-labdane resins has been obtained from dehydrogenation results and infrared spectra. New Zealand resins extending as far back as the Cretaceous (Greymouth resin, 70 m.y. ?) give mixtures of agathalene and pimanthrene (Brandt, 1939) while fossil resin from southern Australia (Pliocene?) gives retene (Brooks and Steven, 1967), in accord with the differences between modern New Zealand and Australian *Agathis* resins.

Further evidence has been obtained from Roxburgh resin (Oligocene, 25 m.y.) which still has a characteristic *A. australis* odour and contains agathic acid (IV) (Hosking, 1934), sandaracopimaric acid (V) and sandaracopimarol (NMR and GLC, Thomas 1969a). In the older Maramarua resin (upper Eocene, 40 m.y.) degradation is further advanced. The gas chromatogram (Fig. 3) is complex and none of the compounds present in fresh modern resin is now recognizable. Mass spectra indicate that some decarboxylation and dehydrogenation has occurred. Ohai resin (lower Eocene, 60 m.y.) gives a somewhat similar GLC-MS picture (Thomas, 1969b).

REFERENCES

Albrecht, P., and Ourisson, G. (1969). *Science*, **163**, 1192–1193.
Aplin, R. T., Cambie, R. C., and Rutledge, P. S. (1963). *Phytochemistry*, **2**, 205–214.

Bachtofen-Echt, A. (1949). "Der Bernstein und seine Einschlüsse." Springer, Wien.
Baggaley, K. H., Erdtman, H., McLean, N. J., Norin, T., and Eriksson, G. (1967). *Acta Chem. Scand.* **21**, 2247–2253.
Boiteau, P., Pasich, B., and Ratsimamanga, A. R. (1964). "Les Triterpenoides en Physiologie Vegetale et Animale." Gauthier Villars, Paris.
Brandt, C. W. (1939). *N.Z. J. Sci. Technol.* **20**, 306B–310B.
Brooks, J. D., and Steven, J. R. (1967). *Fuel (London)*, **46**, 13–18.
Bruns, K. (1968). *Tetrahedron*, **24**, 3417–3423.
Burgstahler, A. W., and Marx, J. N. (1964). *Tetrahedron Lett.* 3333–3338.
Cambie, R. C., and Weston, R. J. (1968). *J. N.Z. Inst. Chem.* **32**, 105–121.
Carman, R. M. (1964). *Aust. J. Chem.* **17**, 393–394.
Carman, R. M., and Cowley, D. E. (1967). *Aust. J. Chem.* **20**, 193–196.
Carman, R. M., and Dennis, N. (1964). *Aust. J. Chem.* **17**, 390–392.
Carman, R. M., and Marty, R. A. (1966). *Aust. J. Chem.* **19**, 2403–2406.
Carman, R. M., Corbett, R. E., Grant, P. K., McGrath, M. J., and Munro, M. H. (1966). *Tetrahedron Lett.* 3173–3179.
Clayton, R. B. (1965). *Quart. Rev. (London)*, **19**, 168–230.
Connolly, J. D., McCrindle, R., Murray, R. D. H., Renfrew, A. J., Overton, K. H., and Melera, A. (1966). *J. Chem. Soc.* 268–273.
Cornforth, J. W. (1968). *Angew. Chem. Intern. Ed. Eng.* **7**, 903–964.
Czeczott, H. (1960). *Pr. Muz. Ziemi*, **4**, 139–145.
Eglinton, G., and Hamilton, R. J. (1967). *Science*, **156**, 1322–1335.
Engler, A. (1964). "Syllabus der Pflanzenfamilien." Borntraeger, Berlin.
Erdtman, H. (1969). *In* "Perspectives in Phytochemistry" (J. Harborne and T. Swain, eds). Academic, London and New York.
Erdtman, H., and Norin, T. (1966). *Fortschr. Chem. org. NatStoffe*, **24**, 206–287.
Florin, R. (1963). *Acta Horti Bergiani*, **20** (4), 121–312.
Frondel, J. (1967). *Nature*, **215**, 1360–1361.
Ghigi, E., Drusiani, A., Plessi, L., and Cavrini, V. (1968). *Gazz. Chim. Ital.* **98**, 795–799.
Gough, L. H. (1969). Private communication.
Gray, P. S., and Mills, J. S. (1964). *J. Chem. Soc.* 5822–5825.
Henrick, C. A., and Jefferies, P. R. (1965). *Aust. J. Chem.* **18**, 2005–2013.
Hillis, W. E. (1962). *In* "Wood Extractives" (W. E. Hillis, ed.). Academic, London and New York.
Hosking, J. R. (1934). *In* Evans, W. P., *N.Z. J. Sci. Technol.* **15**, 365–385.
Hugel, G., Lods, L., Mellor, J. M., Theobald, O. W., and Ourisson, G. (1965). *Bull. Soc. Chim. France*, 2882–2887.
Hugel, G., Oehlschlager, A. C., and Ourisson, G. (1966). *Tetrahedron Suppl.* **8**, 203–216.
Jarolím, V., Hejno, K., and Sorm, F. (1963). *Colln Czech. chem. Commun. Engl. Edn*, **28**, 2318–2327, 2443–2454.
Kahila, S. K. (1957). *Paperi Puu*, **39**, 7–8.
Kodicek, E. (1954). *Biochem. J.* **57**, XII.
Kubota, T., Matsuura, T., Tsutsui, T., Uyeo, S., Takahashi, M., Irie, H., Numata, A., Fujita, T., Okamoto, T., Natsume, M., Kawazoe, Y., Sudo, K., Ikeda, T., Tomoeda, M., Kanamoto, S., Kosuge, T., and Adachi, K. (1964). *Tetrahedron Lett.* 1243–1256.
Langenheim, J. (1964). *Bot. Mus. Leafl. Harv. Univ.* **20**, 225–287.
Langenheim, J. (1969). *Science*, **163**, 1157–1169.
Langenheim, J., and Beck, C. W. (1965). *Science*, **149**, 52–55.
Margulis, L. (1968). *Science*, **161**, 1020–1022.
Misra, R., and Dev, S. (1968). *Tetrahedron Lett.* 2685–2686.

Oehlschlager, A. C., and Ourisson, G. (1967). *In* "Terpenes in Plants" (J. B. Pridham, ed.), pp. 83–109. Academic, London and New York.

Plonait, C. (1935). *Angew. Chem.* **48**, 184–187, 605–607.

Ponsinet, G., Ourisson, G., and Oehlschlager, A. C. (1968). *In* "Recent Advances in Phytochemistry" (T. J. Mabry, ed.), Vol. 1, pp. 271–302. North-Holland, Amsterdam.

Rowe, J. W., Bower, C. L., and Wagner, E. R. (1969). *Phytochemistry*, **8**, 235–241.

Ruhemann, S., and Raud, H. (1932). *Brennst.-Chem.* **13**, 341–345.

Sandermann, W. (1960). "Naturharze Terpentinöl Tallöl." Springer, Berlin.

Sharpless, K. B., Snyder, T. E., Spencer, T. A., Maheshwari, K. K., Guhn, G., and Clayton, R. B. (1968). *J. Am. Chem. Soc.* **90**, 6874–6875.

Thomas, B. R. (1966). *Acta Chem. Scand.* **20**, 1074–1081.

Thomas, B. R. (1969a). *In* "Organic Geochemistry" (G. Eglinton and M. Murphy, eds). Springer, Heidelberg (In press).

Thomas, B. R. (1969b). Unpublished.

Tschirch, A., and Stock, E. (1936). "Die Harze." Borntraeger, Berlin.

Upper, C. D., and West, C. A. (1967). *J. biol. Chem.* **242**, 3285–3292.

Wenkert, E. (1955). *Chem. Ind.* (*London*), 282.

Westfelt, L. (1966). *Acta Chem. Scand.* **20**, 2829–2840.

Zander, J. M., and Caspi, E. (1969). *Chem. Commun.* 209–210.

CHAPTER 5

Cell Wall Composition and Other Biochemical Markers in Fungal Phylogeny*

S. BARTNICKI-GARCIA

Department of Plant Pathology, University of California, Riverside, California, USA

I. PHYLOGENY WITHOUT FOSSILS

A. PALAEOMYCOLOGY

In the near absence of a fossil record (Martin, 1968), the phylogeny of fungi has remained a fertile subject for speculation.

When did fungi first evolve? Precambrian specimens dated 1·9 b.y. old contain some of the oldest fossil forms suggestive of primitive fungi (Tyler and Barghoorn, 1954; see also Echlin, Chapter 1). In a fossilized fern from the Carboniferous period, Dennis (1969) found hyphae with clamp connections, a reliable diagnostic feature of a basidiomycete. This finding demonstrated that members of what is universally regarded as the most advanced class of fungi were already in existence as early as 300 m.y. ago. Thus the major groups of fungi probably emerged sometime during the broad time span encompassed by the above datings. What their ancestors were and how the various classes

---

* The emphasis on cell walls is not to be construed as a claim of superiority over other markers; it merely reflects the research bias of the author.

evolved are questions whose answers must be conjectured almost entirely on the basis of comparative studies on the structure and behaviour of contemporary fungi.

B. PHYLOGENETIC SCHEMES BASED ON MORPHOLOGY

Since the latter part of the last century, mycologists have been speculating on the phylogenetic relationships among the fungi and between the fungi and their plausible ancestors. Often criticized, these speculative excursions can be amply justified by the need for establishing a natural system of fungal classification based on evolutionary development. Almost all schemes of fungal phylogeny have been based largely if not entirely on comparative morphology of extant species.* Although morphology is admittedly the best available criterion to reconstruct the course of evolution of the fungi in the absence of palaeontological data, the phylogenetic trees proposed in the past must be regarded as purely hypothetical. There is no definitive way to prove whether the presumed morphological homologies are in fact a true indication of phylogenetic relatedness or the result of convergent or parallel evolution. The arrangement of contemporary fungi into evolutionary sequences is similarly suspect since it is impossible to decide unequivocally whether species with simple morphology and life cycles represent archetypes or regressive forms of the more highly evolved groups.

In recent years, however, the prospects of attaining a more satisfactory evolutionary scheme of the fungi have brightened. It has become increasingly clear that auxiliary markers, such as the biochemical characters discussed below, may be employed to begin assessing the validity of phylogenetic schemes based on morphology.†

II. BIOCHEMICAL APPROACH TO FUNGAL PHYLOGENY

Adorning the remarkable unity of biochemical activities displayed by the fungi (and the rest of the living world) there is a host of biochemical characters which identify and separate individual organisms. Some of these distinctive traits, because of their constancy and distribution, may be of considerable phylogenetic significance. Although it would be difficult to erect phylogenetic trees of the fungi based on single biochemical properties (see however Fitch and Margoliash, 1967), these properties can be very useful in evaluating phylogenetic relationships derived from other considerations. Biochemical

---

* For accounts of fungal phylogeny see Bessey (1950); Gäumann (1964); Martin (1968); Savile (1968) and Raper (1968).

† See Klein and Cronquist (1967) for an extensive review of biochemical and physiological characters with possible significance in the phylogeny of thallophytes, including the Fungi; note however that their biochemical data did not include most of the material reviewed here and some of their conclusions with regard to fungal evolution, e.g. position of Zygomycetes and Hemiascomycetes, are in disagreement with those proposed herein.

comparisons serve, so to speak, as diagnostic tests to support or dispute evolutionary relationships proposed on the basis of morphological development.

Strictly speaking the value of these biochemical comparisons is a taxonomic one—they may be used to support the validity of drawing symbolic lines between groups of organisms to denote their relatedness. However, the placing of arrowheads on these lines, to connote evolutionary direction, remains a

TABLE I

Fungal groups

Lower fungi		
Slime moulds	Acrasiales	Amoeboid; pseudoplasmodia
	Myxomycetes	Amoeboid and flagellate; true plasmodia
Phycomycetes	Oomycetes	Coenocytic thallus/mycelium; oospores; biflagellate (whiplash + tinsel) zoospores
	Hyphochytridiomycetes	Coencytic thallus; anteriorly uniflagellate (tinsel) zoospores
	Chytridiomycetes	Coenocytic thallus/mycelium; posteriorly uniflagellate (whiplash) zoospores
	Zygomycetes	Coenocytic mycelium; zygospores; sporangiospores
Higher fungi		
Ascomycetes	Hemiascomycetidae	Yeasts and yeastlike forms; asci naked
	Euascomycetidae	Septate mycelium; asci in ascocarps
Basidiomycetes	Heterobasidiomycetidae	Yeast forms/smuts/rusts/jelly fungi; septate basidia
	Homobasidiomycetidae	Septate mycelium; non-septate basidium

speculative task. And, for this purpose, we shall continue to rely on the intuition of morphologists to point out the general direction of fungal evolution.

The various biochemical properties to be discussed have been found useful in assessing relationships amongst the major groups of fungi at the Class or Sub-class level, though occasionally the biochemical tests may differentiate a taxon of lower rank. The groups of fungi to be discussed are listed in Table I together with some of their morphological idiosyncrasies.

A. TYPES OF BIOCHEMICAL CRITERIA

Three different levels of biochemical criteria will be recognized: chemical, metabolic and enzymic. At the *chemical* level, the relatedness of two groups of organisms, A and B, may be attested by the common presence of certain chemical compounds. Obviously, compounds which appear erratically in a given group of organisms have little or no phylogenetic value. Other substances which occur typically and consistently in the groups under study may have considerable phylogenetic significance. Some of the cell wall components of fungi appear to fall in the latter category. At the *metabolic* level, is the case in which two groups of organisms, A and B, may appear to be closely related by being able to elaborate the same metabolite, yet, by showing that different metabolic routes are used for the synthesis of this metabolite, a fundamental discrepancy in their evolutionary development may be disclosed. Finally, at the *enzymic* level, phylogenetic differences may be disclosed between groups of organisms showing identical metabolic reactions but differing in some of the fine properties of their enzymes. These differences are a direct reflection of alterations in the structure of the genome and may be experimentally recognized by biochemical or genetic manipulations.

One example of each of these three criteria will be discussed in detail. In addition, there have been various other important studies attempting to correlate biochemical properties with fungal phylogeny. For instance, Cantino (1955) proposed evolutionary relationships within the aquatic Phycomycetes based on their nutritional requirements. Storck (1965, 1966) suggested that the base composition of fungal DNA and RNA may have phylogenetic significance. Fitch and Margoliash (1967), proposed a method to erect phylogenetic trees based on mutation distances estimated from cytochrome *c* sequences. Three fungi were included in their evolutionary scheme of living organisms. The phylogenetic significance of the occurrence of α- and γ-linolenic acid in fungi was considered by Shaw (1965).

B. CHEMICAL MARKERS: CELL WALL COMPOSITION

The diversity of cell wall composition exhibited by fungi began to be recognized last century (Van Wisselingh, 1898; Mangin, 1899). The old distinction of fungi as either cellulosic or chitinous was employed by Von Wettstein (1921) as a taxonomic and phylogenetic criterion to support the division of the aquatic Phycomycetes into two groups. But, because of contradictory findings, the use of cell wall composition in the taxonomy and phylogeny of fungi failed to gain acceptance. The confusion stemmed from the use of unreliable staining reactions to characterize cell wall polymers. For instance, Thomas (1943) claimed the presence of chitin in a typically cellulosic fungus such as *Phytophthora* (Bartnicki-Garcia, 1966; Novaes-Ledieu *et al.* 1967; Aronson *et al.* 1967); conversely, the noncellulosic fungus *Mucor rouxii* (Bartnicki-Garcia and

Nickerson, 1962) was believed to have cellulose in its walls (Hopkins, 1929). Questions on the occurrence of chitin and cellulose in fungi were largely dispelled by the X-ray work of Frey (1950). Von Wettstein's correlation of cell wall composition and taxonomy of aquatic Phycomycetes is still tenable, as modified by Fuller and Barshad (1960) who added a third category of water moulds with both cellulose and chitin in their walls. The three cell wall categories correspond to the three different types of zoospore flagellation in the aquatic Phycomycetes (see Tables I and III).

The advent of techniques for preparing pure cell walls has greatly stimulated research on the structure of the fungal cell wall. The survey studies made by Kreger (1954) and by Crook and Johnston (1962) deserve special mention. These studies, plus a number of other investigations, many of a quantitative nature on selected species, have made it possible to appreciate both the chemical complexity of an individual cell wall and the diversity of cell wall composition encountered in the fungal kingdom (for recent reviews, see Aronson, 1965; and Bartnicki-Garcia, 1968). The fungal cell wall is no longer believed to be an inert envelope of chitin or cellulose but a dynamic, highly complex structure composed of proteins, lipids and other substances, in addition to various polysaccharides which are, of course, the principal wall constituents. Except for some polysaccharides, other wall components have not been well characterized and their taxonomic value is unknown. The taxonomic and phylo-

Sugars of Fungal Wall Polysaccharides

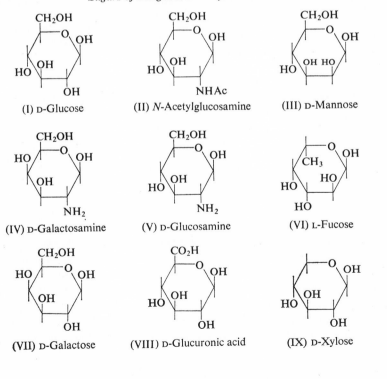

(I) D-Glucose   (II) *N*-Acetylglucosamine   (III) D-Mannose

(IV) D-Galactosamine   (V) D-Glucosamine   (VI) L-Fucose

(VII) D-Galactose   (VIII) D-Glucuronic acid   (IX) D-Xylose

genetic potential of the fungal cell wall presently resides in its carbohydrate variety. Fungal wall polysaccharides are made of hexoses, aminosugars, hexuronic acids, methylpentoses and pentoses. Of these, glucose (I) and N-acetylglucosamine (II) are the most common and often abundant building blocks. D-Mannose (III) is universally found in small amounts in most mycelial fungi and abundantly in yeasts. Other sugars, e.g. D-galactosamine (IV), D-glucosamine (V), L-fucose (VI), D-galactose (VII), D-glucuronic acid (VIII) and D-xylose (IX), are restricted to certain groups of fungi. And, finally certain sugars are usually present so minutely that their role as genuine wall com-

TABLE II

Polysaccharides of fungal walls

	Linkages	Monomers	References
Aminopolysaccharides			
Chitin	β-1,4-	N-acetylglucosamine	(Aronson, 1965)
Chitosan	β-1,4-	D-glucosamine	(Kreger, 1954)
Neutral polymers			
Cellulose	β-1,4-	D-glucose	(Aronson, 1965)
β-Glucan	β-1,3; 1,6-	D-glucose	(Bartnicki-Garcia, 1968)
α-Glucan	α-1,3; 1,4-	D-glucose	(Johnston, 1965)
Glycogen	α-1,4-	D-glucose	(Toama and Raper, 1967)
Mannan	α-1,2; 1,6; 1,3-	D-mannose	(Lee and Ballou, 1965)
Polyuronides			
Pullularia hetero-polysaccharide	α-1,6; 1,5- β-1,3; 1,6-	D-glucose, D-glucuronic acid, D-galactose, D-mannose	(Brown and Lindberg, 1967)
Mucoran		L-fucose, D-mannose, D-galactose, D-glucuronic acid	(Bartnicki-Garcia and Reyes, 1968)

ponents is debatable, e.g. rhamnose, arabinose, ribose. This collection of sugars occurs in a variety of different polymers some of which have been partly characterized (Table II). For taxonomic and phylogenetic purposes, an outstanding feature of these polysaccharides is their distribution—they occur neither universally nor sporadically throughout the entire spectrum of fungi but appear consistently and characteristically in certain groups of fungi. By selecting dual combinations of those polysaccharides which appear to be the main components of fungal cell walls, a classification of fungi into a minimum of eight cell wall categories was recently proposed (Bartnicki-Garcia, 1968). By so doing, a close correlation could be seen between cell wall chemistry and the conventional morphological classification of the fungi (Table III).

The chemical complexity of the wall and the suspected intricacy of the process of cell wall fabrication may be invoked to explain the constancy of cell wall composition observed among related fungi. This constancy probably makes cell wall composition a reliable criterion for phylogenetic studies. Cell wall polysaccharides do not appear to exist free in the cell wall but are probably chemically complexed with proteins (see Nickerson *et al.* 1961) and possibly

TABLE III

Cell wall composition and taxonomy of fungi[a]

Cell wall category	Taxonomic group	Representative genera
I. Cellulose–Glycogen	Acrasiales	*Polysphondylium, Dictyostelium*
II. Cellulose–β-Glucan	Oomycetes	*Phytophthora, Pythium, Saprolegnia*
III. Cellulose–Chitin	Hyphochytridiomycetes	*Rhizidiomyces*
IV. Chitin–Chitosan	Zygomycetes	*Mucor, Phycomyces, Zygorhynchus*
V. Chitin–β-Glucan	Chytridiomycetes	*Allomyces, Blastocladiella*
	Euascomycetes	*Aspergillus, Neurospora, Histoplasma*
	Homobasidiomycetes	*Schizophyllum, Fomes, Polyporus*
VI. Mannan–β-Glucan	Hemiascomycetes	*Saccharomyces, Candida*
VII. Chitin–Mannan	Heterobasidiomycetes	*Sporobolomyces, Rhodotorula*
VIII. Galactosamine–Galactose polymers	Trichomycetes	*Amoebidium*

[a] For specific examples and references see Bartnicki-Garcia (1968). Seemingly, the β-glucan is always 1,3- and 1,6- linked. Deuteromycetes are included in their corresponding groups of perfect fungi.

lipids also. Cell wall formation is probably a very elaborate process involving a constellation of specific enzymes related to the manufacture of wall precursors, their transport across membranes, and their polymerization into a three-dimensional structure with its own characteristic shape. In view of this complexity one would intuitively suspect that the mechanism of cell wall construction could not tolerate any erratic departures in the chemical make-up of the wall, if a cell wall is to be formed with the typical shape of the organism in question. Pertinent examples of this delicate balance are certain morphological mutants of *Neurospora crassa* in which a single-gene mutation causes alterations in carbohydrate metabolism which affect the quantitative composition of the wall and drastically changes the morphology of the fungus (Mahadevan and Tatum, 1965).

The impression must not be left that cell wall composition is entirely rigid; a certain latitude for quantitative and even qualitative changes has been observed in response to environmental alterations or associated with the onto-genetic development of the fungus (Bartnicki-Garcia, 1968).

It may not be superfluous to restate that Table III is a tentative scheme of taxonomic correlations assembled from a very limited and fragmentary collection of data on fungal cell walls. Important families, orders and even some classes of fungi (e.g. Myxomycetes) have not been critically examined. Often the information refers to only one representative of a given taxon. Also, the degree of analytical refinement has been widely variable—from a few qualitative tests to rather thorough attempts at making a quantitatively complete determination of the cell wall composition of a single fungus. Notwithstanding these shortcomings, and the inevitability of later revisions, there is sufficient consistency in the current data to merit much confidence on the proposed correlation between wall chemistry and morphology. Accordingly, the following conclusions with phylogenetic bearing have been made.

1. Diversity of Wall Composition in the Phycomycetes

The four classes of fungi formerly grouped together under the class Phyco-mycetes, namely Oomycetes, Hyphochytridiomycetes, Chytridiomycetes and Zygomycetes, display a great diversity in cell wall composition, each group having its own peculiar chemical combination (Table III). This multiplicity of wall chemistry probably reflects the taxonomic disparity of the Phycomycetes, and lends further justification to the modern tendency to dissolve the class Phycomycetes* and to elevate its former subclasses to full class rank (Alexo-poulos, 1962).

2. Uniformity of Wall Composition in Higher Mycelial Fungi

In contrast to the chemical diversity found in the walls of Phycomycetes, the mycelial members of the higher fungi display considerable similarity in wall composition. Thus all species of Euascomycetes and Homobasidiomycetes examined have chitin and non-cellulosic glucans as principal wall components. Both β- and α-glucans may be present (Johnston, 1965; Bacon et al. 1968). Euascomycetes and Homobasidiomycetes probably differ in minor components as suggested by the tendency to find D-galactose and D-galactosamine in Euascomycetes cell walls (Crook and Johnston, 1962) whereas the Homo-basidiomycetes usually contain xylose and L-fucose (O'Brien and Ralph, 1966). A detailed characterization of cell wall polymers is needed to assess with greater confidence the degree of similarity of cell wall structure between Ascomycetes and Basidiomycetes.

---

* Even if its taxonomic value is questionable, the term phycomycete has been retained as a traditional and convenient way of denoting a coenocytic fungus.

3. The Specialized Wall of Yeasts

Within the higher fungi, the major divergency in cell wall composition occurs in the yeast forms of the Hemiascomycetes and the Heterobasidiomycetes (Kreger, 1954; Crook and Johnston, 1962; Phaff, 1963). In both instances, the change from the mycelial to the yeast habit is accompanied by a substantial increase in the mannan content of the wall (Garzuly-Janke, 1940). In ascomycetous yeasts or their imperfect counterparts (e.g. *Candida*), the increase in mannan is at the expense of chitin (mannan–glucan category), whereas in the basidiomycetous yeasts or their imperfect counterparts (e.g. *Rhodotorula*), the chitin is retained and the glucan is reduced or lost (mannan–chitin category).

4. The Relationship of Chytridiomycetes, Euascomycetes and Homobasidiomycetes

Of all the lower fungi, only the Chytridiomycetes have a cell wall composition which resembles that of the higher mycelial fungi, i.e. Euascomycetes and Homobasidiomycetes. The cell walls of these three groups of fungi have in common two important polymers: chitin and non-cellulosic alkali-insoluble glucan. The common occurrence of other polysaccharides is not clear. Aronson and Machlis (1959) and Skucas (1967) found no hexoses other than glucose in the walls of Chytridiomycetes of the genus *Allomyces* but Fultz (1965) reported the presence of D-galactose and D-mannose in *Allomyces* spp., and also in *Blastocladiella* spp. The presence of D-galactose may be taken as a preliminary indication of a closer relationship between Ascomycetes and Chytridiomycetes. A final conclusion on the degree of cell wall similarity among these three groups of fungi must of course await further characterization of the polysaccharides. At present, the analogies observed agree with the idea that the Chytridiomycetes, Euascomycetes and Homobasidiomycetes belong to the same phylogenetic trunk line.

5. The Distinctive Wall Composition of the Zygomycetes

The Zygomycetes (Mucorales) appear to be a group of fungi with unique chemical features in their cell walls. Their mycelial walls contain chitin but lack, in contrast to all other true fungi, glucose polymers (Bartnicki-Garcia and Nickerson, 1962; Crook and Johnston, 1962); instead, they have chitosan (Kreger, 1954), a polymer not evident in other fungi. Depending on the morphological stage, their cell walls may also contain large amounts of polyuronides of D-glucuronic acid, D-mannose, L-fucose and D-galactose (Bartnicki-Garcia and Reyes, 1968). Another seemingly exclusive feature of the Zygomycetes is that their asexual spore walls have a chemical composition vastly different from that of vegetative walls; the possible significance of this disparity is discussed below.

6. Ontogenetic Recapitulation of Cell Wall Phylogeny?

In most of the few fungi examined asexual spores exhibit a cell wall composition similar to that of the mycelium. This is true of the conidial walls of

Ascomycetes and the sporangial walls of Chytridiomycetes (see examples in Bartnicki-Garcia, 1968) as well as the sporangial walls of the oomycete *Phytophthora palmivora* (Tokunaga and Bartnicki-Garcia, in preparation). In the Zygomycetes, however, a seemingly unique situation exists. Sporangiospore walls of *Mucor rouxii* (Bartnicki-Garcia and Reyes, 1964) or *M. ramannianus* (Jones *et al.* 1968) are of almost entirely different composition compared to their mycelial walls (Bartnicki-Garcia and Nickerson, 1962; Bartnicki-Garcia and Reyes, 1968). Thus in every turn in the life cycle of these fungi the cell wall composition shifts drastically from the glucan–polyglucosamine (chitin?)–melanin type of spore walls to the chitin–chitosan–polyuronide type of vegetative cell walls. Together with this chemical discontinuity there is a structural discontinuity between spore walls and vegetative walls: first, the spore wall is formed at a distance from the vegetative walls; second, when the spore germinates, a new vegetative wall is separately synthesized underneath the spore wall (Hawker and Abbott, 1963; Bartnicki-Garcia *et al.* 1968). As one of the plausible explanations for the drastic cyclic change in cell wall composition (from spore to mycelium and vice versa), it was proposed that the changeover represents an ontogenetic recapitulation of the phylogenetic history of the cell walls of Mucorales (Bartnicki-Garcia, 1968). Accordingly, the sporangiospore wall of a zygomycete may be an ancient structure whose composition betrays the chytridiomycete origin of the Zygomycetes. This supposition is in accord with a partial chemical resemblance between the sporangiospore walls of *M. rouxii* and the sporangial walls of the chytridiomycete *Allomyces neo-moniliformis* (Skucas, 1967). In both cases, the cell wall contains non-cellulosic glucan, melanin and chitin (the identity of chitin in the spore wall of *Mucor* is only tentative). During the ontogenetic development of *Mucor* a sporangiospore gives rise to a hypha which possesses what may be a more advanced type of cell wall composed of chitin–chitosan–polyuronide.

7. Polyphylogeny of the Mycelium

A comparison of the mycelial walls of a typical zygomycete (e.g. *Mucor rouxii*) and an oomycete (e.g. *Phytophthora cinnamomi*) reveals two totally different chemical compositions (Table IV), as different as any two mycelial fungi are known to be. Yet, the gross morphology of the mycelia of these two fungi may be the same depending on growth conditions. This wide disparity in wall composition is a strong indication that these two classes of Phycomycetes evolved independently. Consequently, the development of similar coenocytic mycelia in different types of Phycomycetes may be considered as a case of separate parallel evolution.

8. Relationships with Borderline Fungal Groups

The dubious affinities of certain groups of organisms to the true fungi are reflected in some unusual features in cell wall composition. The differences may be construed as further indication of phylogenetic heterogeneity. Thus, the

Trichomycetes, a group of disputable taxonomic standing, show in *Amoebidium parasiticum* (Trotter and Whisler, 1965), the only representative so far studied, a cell wall composition entirely unlike that of typical fungi (Table III). The Acrasiales or cellular slime moulds contain cellulose as other lower fungi but have neither chitin nor the alkali-insoluble non-cellulosic glucan characteristic of true fungi (Toama and Raper, 1967).

TABLE IV

Composition of mycelial walls of an oomycete vs. a zygomycete

	% *Phytophthora cinnamomi*[a]	% *Mucor rouxii*[b]
D-Glucose	88[c]	0
D-Glucosamine	0·3	44·5[d]
D-Glucuronic acid	0	11·8
L-Fucose	0	3·8
D-Galactose	0	1·6
D-Mannose	0·6	1·6
Phosphate	0·3	23·3
Lipid	2·4	7·8
Protein	3·6	6·3

[a] From Bartnicki-Garcia (1966).
[b] From Barnicki-Garcia and Nickerson (1962); Bartnicki-Garcia and Reyes (1968).
[c] Estimated to be roughly 25% cellulose (wall dry wt) and the remainder a β1,3-β1,6-glucan.
[d] Composed of approximately 9·4% chitin, 32·7% chitosan and 2·4% unidentified.

9. Relationships to Other Microorganisms

Cell wall properties may be helpful in disclosing the affinities of fungi to other organisms. The lack of close kinship between fungi and actinomycetes is manifested by, among other characters, their drastically different cell wall composition. Actinomycetes have bacterial type components not found in fungi: muramic acid, meso-diaminopimelic acid, D-amino acids, etc. (Becker *et al.* 1965; Rogers and Perkins, 1968).

A comprehensive systematic study of the cell walls might help us clarify the proposed phylogenetic relationships between algae and fungi. Presently, our knowledge on cell walls is too fragmentary to favour or oppose this presumed relationship, though the tendency is to find more dissimilarity than homology in cell wall structure. For example, poorly crystalline cellulose is common to certain algal (Kreger, 1962; Roelofsen, 1959) and fungal groups (Table III, categories I, II, III), but two of the most characteristic polymers of fungal walls, chitin and non-cellulosic β1,3-β1,6-glucan, have not been found in algae.*

* A preliminary report on the presence of chitin in red algae (Quillet and Priou, 1963) has not been confirmed by reliable techniques such as X-ray diffractometry or chitinase digestion.

Parker *et al.* (1963) undertook a study of cell wall composition of several species of the siphoneous alga *Vaucheria* and several genera of saprolegniaceous fungi (class Oomycetes) to see if this comparison could be helpful in disclosing relationships between these two groups of organisms which have long been considered phylogenetically related. Both groups were found to have cellulose but there were a number of important differences which would seem to negate any close phylogenetic tie. For instance, the fungal cellulose was poorly crystalline and represented only a minor portion of the cell wall whereas the cellulose of *Vaucheria* was highly crystalline and constituted nearly 90 % of the cell wall mass. Mannose was present in the fungal polysaccharides but not in *Vaucheria*. The major component of saprolegniaceous cell walls was an alkali-insoluble non-cellulosic polysaccharide, which is probably a β1,3-β1,6-linked glucan characteristic of oomycete cell walls (Aronson *et al.* 1967). This glucan was absent from *Vaucheria* walls.

C. METABOLIC MARKERS: LYSINE PATHWAYS

Living organisms synthesize lysine by either one of two different pathways (Fig. 1). Some organisms utilize the α,ϵ-diaminopimelic acid route (DAP)

DAP	AAA
L-Aspartic β-Semialdehyde + Pyruvate	α-Ketoglutarate + Acetyl Coenzyme A
↓	↓
Dihydrodipicolinic acid	Homocitric acid
↓	↓
Δ^1-Piperideine-2,6- dicarboxylic acid	Homoisocitric acid
↓	↓
N-Succinyl-L-α-amino- ϵ-ketopimelic acid	α-Ketoadipic acid
↓	↓
N-Succinyl-L-α,ϵ- diaminopimelic acid	L-α-Aminoadipic acid
↓	↓
	Activated intermediate
L-α,ϵ-Diaminopimelic acid	↓
↓	L-α-Aminoadipic semialdehyde
meso-α,ϵ-Diaminopimelic acid	↓
↓	Saccharopine
	↓
L-Lysine	L-Lysine

FIG. 1. The α,ϵ-diaminopimelic acid (DAP) and α-aminoadipic acid (AAA) pathways of lysine biosynthesis

whereas others employ an entirely different pathway via α-aminoadipic acid (AAA) (Vogel, 1963). The reasons for the existence of two entirely different modes of lysine synthesis in nature are not clear. One can only conclude that the two pathways probably arose as independent answers to the synthesis of an essential metabolite in groups of organisms that were probably phylogenetically unrelated. Vogel's (1964) studies disclosed the potential value of the lysine pathway as a marker for fungal phylogeny. According to the route of lysine synthesis, the fungi may be clearly divided into two camps cutting across the old taxon of Phycomycetes (Table V). The Oomycetes and the Hyphochytridiomycetes are the only fungi with the DAP pathway. Significantly, this is the pathway found in vascular plants, green algae, blue-green algae and also in bacteria. On the other hand, the majority of the fungi namely, the Chytridiomycetes, Zygomycetes and all higher fungi have the AAA pathway. This route

TABLE V

Distribution of lysine pathways among fungi[a]

DAP (Diaminopimelic acid)	
Oomycetes	*Achlya, Sapromyces, Sirolpidium, Pythium, Traustotheca*
Hyphochytridiomycetes	*Hyphochytrium, Rhizidiomyces*

AAA (Aminoadipic acid)	
Chytridiomycetes	*Rhizophlyctis, Phlyctochytrium, Allomyces, Monoblepharella*
Zygomycetes	*Rhizopus, Cunninghamella, Syncephalastrum*
Hemiascomycetes	*Dipodascus, Taphrina, Saccharomyces*
Euascomycetes	*Penicillium, Neurospora, Gibberella, Venturia, Morchella, Sclerotinia*
Heterobasidiomycetes	*Ustilago*
Homobasidiomycetes	*Polyporus, Coprinus, Calvatia*

[a] Selected examples of each taxon (Vogel, 1963).

of lysine biosynthesis seems to be almost exclusively restricted to the fungi and it has only been found elsewhere in euglenids. It seems improbable (Vogel, 1963) that one of the lysine pathways could emerge sporadically in organisms having the other, mainly because of the numerous enzymes involved. This relative immutability may be invoked to explain the constancy of occurrence of only one of the pathways in a group of related organisms and to support the value of lysine biosynthesis as a reliable indicator of the phylogenetic dichotomy of the fungi.

D. ENZYMIC MARKERS: TRYPTOPHAN BIOSYNTHESIS

Tryptophan is synthesized by the same set of enzymic reactions in all organisms studied (Fig. 2) but genetic and biochemical analyses have revealed important differences in the behaviour of these enzymes, differences which depend on the nature of the source organism (DeMoss, 1965). Hütter and DeMoss (1967) made use of this enzymic heterogeneity to explore the phylogenetic relationships amongst fungi. Their simple and elegant technique consisted in determining the sedimentation behaviour of the following enzymes: anthranilate

Intermediates	*Enzymes*
Chorismic acid	
↓	Anthranilate synthetase
Anthranilic acid	
↓	PR-Transferase
N-(5′-Phosphoribosyl) anthranilic acid	
↓	PRA-Isomerase
1-(o-Carboxyphenylamino)-1-deoxyribulose-5-phosphate	
↓	InGP-Synthetase
Indole-3-glycerol phosphate	
↓	Tryptophan synthetase
Tryptophan	

FIG. 2. Tryptophan biosynthesis pathway

synthetase, phosphoribosyl transferase, phosphoribosyl anthranilate isomerase and indole-glycerophosphate synthetase, by sucrose density gradient centrifugation. In this manner, four different sedimentation patterns were recognized in the fungi (Fig. 3). These patterns are indicative of mutual associations between the different enzymes involved in tryptophan biosynthesis. The occurrence of the four different sedimentation patterns amongst fungi is shown

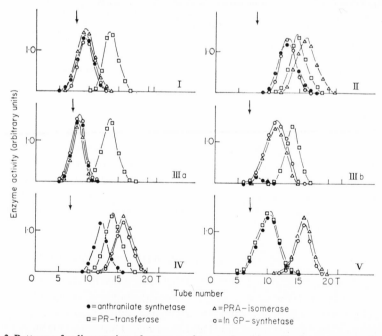

FIG. 3. Patterns of sedimentation of enzymes of tryptophan biosynthesis on a sucrose density gradient. I. *Neurospora crassa*; II. *Saccharomyces cerevisiae*; IIIa. *Mucor hiemalis*, +10⁻² M L-glutamine, +10⁻⁴ M EDTA; IIIb. *Mucor hiemalis*, no additions; IV. *Saprolegnia* species; V. *Escherichia coli*. Reproduced with permission from Hütter and DeMoss (1967)

in Table VI. Although few representatives of each taxon were tested, related organisms consistently exhibited the same pattern.

The distribution of sedimentation patterns of tryptophan biosynthesis enzymes amongst the fungi bears some striking correlations with the distribution of cell wall polymers. Thus the three groups of Phycomycetes examined—Oomycetes, Chytridiomycetes and Zygomycetes—all show different sedimentation patterns in accord with their dissimilar cell wall composition. Most significantly, the Chytridiomycetes, Euascomycetes and Homobasidiomycetes,

TABLE VI

Distribution of sedimentation patterns of tryptophan
biosynthesis enzymes in the fungi[a]

Type I	
Myxomycetes	*Physarum*
Chytridiomycetes	*Allomyces, Rhizophlyctis*
Euascomycetes	*Aspergillus, Neurospora, Byssochlamys, Morchella, Gibberella*
Homobasidiomycetes	*Coprinus, Polyporus*
Type II	
Hemiascomycetes	*Dipodascus, Saccharomyces, Endomyces*
Type III	
Zygomycetes	*Rhizopus, Mucor, Phycomyces*
Heterobasidiomycetes	*Sporobolomyces, Rhodotorula, Ustilago, Tremella*
Type IV	
Oomycetes	*Pythium, Saprolegnia*

[a] From Hütter and DeMoss (1967). The four types of sedimentation patterns are illustrated in Fig. 3.

which have been shown to have similar cell wall components (chitin–glucan category) also display the same pattern of sedimentation of tryptophan-biosynthesis enzymes. The higher fungi with specialized cell walls, i.e. the Hemiascomycetes and the Heterobasidiomycetes, manifest different sedimentation patterns from the rest of the higher fungi. The only major inconsistency is the finding that both Zygomycetes and Heterobasidiomycetes share the same sedimentation pattern of tryptophan enzymes even though their cell wall compositions are different. The significance of their homology in sedimentation pattern is not clear. It may be merely a case of convergent biochemical evolution or may lend some credence to old proposals, e.g. Brefeld's (cf. Bessey, 1950), on the possible ties between Zygomycetes and Basidiomycetes.

III. Conclusions

A. BIOCHEMICAL SUPPORT FOR EVOLUTIONARY RELATIONSHIPS

The overall phylogenetic value of the biochemical properties discussed in the preceding sections is probably best appreciated by testing, as Hütter and DeMoss (1967) did, the relationships proposed in some well known scheme of fungal phylogeny derived largely from morphological considerations, e.g. Gäumann's (1964), shown in abbreviated form in Fig. 4. A revised scheme consistent with available biochemical data is shown in Fig. 5.

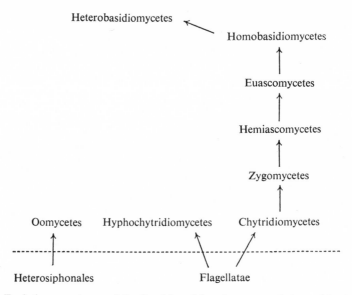

FIG. 4. Evolutionary scheme of the fungi based largely on morphology. Adapted from Gäumann (1964)

1. Most Fungi Belong to a Monophyletic Series

The biochemical data discussed above, in particular lysine biosynthesis, support in most details the partition of fungi indicated in Gäumann's scheme (Fig. 4). There are two unequal phylogenetic series: the major one comprises all fungi having the AAA pathway of lysine synthesis, and is primarily an assembly of chitinous fungi—Chytridiomycetes, Zygomycetes and all higher fungi. The minor series consists solely of fungi with the DAP lysine pathway. This includes the cellulosic water moulds—Oomycetes—and the cellulosic and chitinous water moulds—Hyphochytridiomycetes. Because of their common pathway of lysine synthesis, the Hyphochytridiomycetes should be considered, contrary to Fig. 4, more closely allied to the Oomycetes than to the Chytridiomycetes (Fig. 5). These biochemical considerations also throw some conciliatory light onto the old argument of whether fungi are monophyletic or

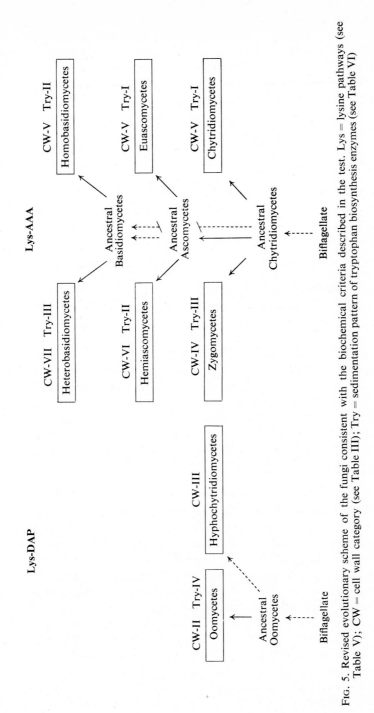

Fig. 5. Revised evolutionary scheme of the fungi consistent with the biochemical criteria described in the test. Lys = lysine pathways (see Table III); Try = sedimentation pattern of tryptophan biosynthesis enzymes (see Table VI); CW = cell wall category (see Table V); CW = cell wall category (see Table V).

polyphyletic. As stated above, fungi appear to be at least diphyletic (Klein and Cronquist, 1967), and may even be polyphyletic if borderline groups, for which little or no biochemical information is available, are included, e.g. Acrasiales, Trichomycetes, Myxomycetes, Labyrinthulales, etc. At the same time, the consensus of biochemical data indicate that the large majority of the species classed as fungi probably belong to a single phylogenetic series. And, perhaps, strictly speaking these are the only organisms which should be considered as true fungi (Klein and Cronquist, 1967; see further discussion in Whittaker, 1969). The argument of whether fungi are monophyletic or polyphyletic may resolve itself either way depending on which organisms are regarded as fungi.

2. The Phycomycetous Origin of the Higher Fungi

There have been two principal schools of thought on the origin of the higher fungi (Bessey, 1950). One school maintains that they evolved from the Phycomycetes; the other one, which has been more severely criticized (e.g. Savile, 1968) assumed that the higher fungi descended from various groups of red algae (floridean ancestry).

Chytridiomycetes, Euascomycetes and Homobasidiomycetes share three seemingly unrelated sets of biochemical properties: similar cell wall composition, same pattern of organization of tryptophan biosynthesis enzymes and same lysine pathway. All this collective evidence suggests these three groups of fungi belong to the same phylogenetic line and hence provides strong support for the widespread belief that higher fungi evolved from Phycomycetes (Atkinson, 1915; Gäumann, 1964; Savile, 1968) as is implicit in Fig. 4. Furthermore, the biochemical data single out the Chytridiomycetes as the direct ancestors of the higher fungi (see next sextion). By supporting the hypothesis of a chytridiomycetous origin of the higher fungi, the alternative and highly controversial hypothesis on the floridean ancestry of higher fungi (Bessey, 1950) becomes further discredited.

3. The Connection between Phycomycetes and Higher Fungi

In various schemes of fungal phylogeny (Atkinson, 1915; Gäumann, 1964; Klein and Cronquist, 1967) the evolution of higher mycelial fungi from water moulds is depicted via Zygomycete- and Hemiascomycete-like forms. This evolutionary sequence (Fig. 4), however, fails to gain biochemical support. All members of the Zygomycetes and Hemiascomycetes that have so far been examined display their own sedimentation pattern of tryptophan biosynthesis enzymes plus distinctive cell wall traits, unlike those in the Chytridiomycetes, Euascomycetes and Basidiomycetes. These three groups of organisms appear to constitute a natural series of fungi. It seems improbable that the postulated progression from Chytridiomycetes to Euascomycetes and Basidiomycetes took place in roundabout fashion, via other organisms such as Mucorales, Endomycetales or Saccharomycetales, which would have implied a drastic and temporary change in biochemical identity. Most likely, ancestral Chytridiomycetes gave rise directly to primitive Ascomycetes from which the present-day Euascomycetes were derived (Fig. 5). The available biochemical information does not discriminate whether Basidiomycetes arose from Ascomycetes, as is universally believed (Bessey, 1950; Gäumann, 1964), or whether the derivation might have been directly from some ancestral Chytridiomycetes.

4. *The Position of Zygomycetes, Hemiascomycetes and Heterobasidiomycetes*

Because of their idiosyncrasies in tryptophan enzyme organization and in cell wall structure, the Zygomycetes, the Hemiascomycetes and the Heterobasidiomycetes, may be considered as highly specialized groups which evolved as off-shoots from the main evolutionary trunk (Chytridiomycetes–Euascomycetes–Homobasidiomycetes) as depicted in Fig. 5. Conceivably, during their morphological evolution concomitant biochemical changes took place in the organization of tryptophan biosynthesis enzymes, and in cell wall composition, but they retained some cell wall components (e.g. chitin) and the same pathway of lysine synthesis characteristic of their presumed ancestors.

In older schemes (e.g. Bessey, 1950) the Zygomycetes have been considered phylogenetically related to some Oomycetes. This relationship could be supported, biochemically, by the fact that Oomycetes and Zygomycetes, have γ-linolenic acid (XI) in their lipids, whereas all of the higher fungi have α-linolenic acid (X) (Shaw, 1965). However, in view of the enormous differences

$$CH_3-(CH_2-CH=CH)_3-(CH_2)_7-CO_2H$$
(X) α-Linolenic Acid

$$CH_3-(CH_2)_3-(CH_2-CH=CH)_3-(CH_2)_4-CO_2H$$
(XI) γ-Linolenic Acid

in cell wall structure (Table IV) and drastic differences in the organization of tryptophan biosynthesis enzymes, as well as dissimilar lysine pathways, it seems improbable that Oomycetes and Zygomycetes are phylogenetically related. The common occurrence of γ-linolenic acid in these fungi may be considered as a case of convergence in biochemical evolution. As suggested earlier (Section II.B.6) considerations of cell wall ontogeny of Zygomycetes support the view that they are descendants from the Chytridiomycetes, as is more commonly believed (Gäumann, 1964).

B. FUNGAL ANCESTORS

What group or groups of organisms gave rise to the fungi? This is a formidable question for which no satisfactory answer is available. It is likely that the properties we now ascribe to a fungus: eucaryotic, nonphotosynthetic, cell-walled, mycelial, etc., arose gradually from procaryotic ancestors but the evolutionary path(s) is far from clear. Various groups of algae (Bessey, 1950; Klein and Cronquist, 1967) and protozoa (Martin, 1968) have been favoured as progenitors of the fungi.

The position of the fungi in the living world is shown in Fig. 6. A major gap in the evolutionary history leading eventually to the fungi is the lost link between procaryotes and eucaryotes (Stanier and van Niel, 1962). There is no evidence suggesting that the gap was bridged directly from heterotrophic procaryotes (bacteria, actinomycetes) to eucaryotic fungi; if such connection ever existed the transition forms have disappeared.

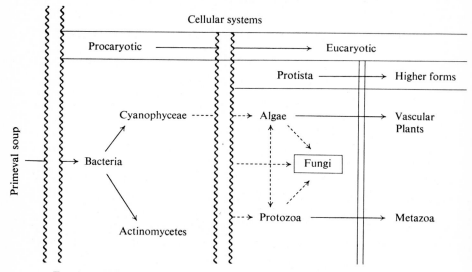

FIG. 6. Plausible phylogenetic relationships of the fungi with other living organisms

The zoospore is commonly regarded as the most rudimentary cellular form of aquatic Phycomycetes from which cell-walled structures later developed; hence the widespread tendency to treat zoospores as primordial cells of the true fungi, and to search for clues to fungal ancestors among algae or protozoa with similar flagellated cells. Furthermore, it now appears that all aquatic Phycomycetes, with either biflagellate *or* uniflagellate zoospores, arose from biflagellated ancestors. This is a straightforward proposal for the biflagellated Oomycetes whose origin has been sought in algal forms exhibiting both tinsel and whiplash flagella similar to the oomycete zoospore, for instance in xanthophycean algae (Bessey, 1950; Klein and Cronquist, 1967). The biflagellate origin of both groups of uniflagellate Phycomycetes—Chytridiomycetes and Hyphochytridiomycetes—is less obvious but it seems well supported by the finding of two kinetosomes in their zoospores. One is associated with the single flagellum while the other is considered to be a non-functional vestigial kinetosome betraying the biflagellated ancestry of these fungi (Koch, 1956; Olson and Fuller, 1968). Because of their different pathways of lysine biosynthesis, it seems doubtful that Chytridiomycetes and Hyphochytridiomycetes descended from related biflagellated ancestors (contrary to Fig. 4). Presumably, the Hyphochytridiomycetes evolved from biflagellated organisms with the capacity to synthesize lysine by the DAP pathway, possibly from the Oomycetes by loss of the whiplash flagellum (Fig. 5), whereas the origin of the Chytridiomycetes probably lies outside the fungi in a different kind of biflagellated organism having the AAA pathway of lysine synthesis (cf. Klein and Cronquist, 1967). So little is actually known about the comparative biochemistry of flagellated algae, protozoa and fungi, that it would be unwise to propose, at this time, any extant organisms as contemporary representatives of the ancestors from which the presumed two phylogenetic series of fungi evolved.

C. PROBLEMS OF INTERPRETATION

The difficulty in correlating the distribution of linolenic acids with other biochemical properties exhibited by the various groups of Phycomycetes (Section II.A.4) clearly demonstrates that the use of a single biochemical criterion could lead to conflicting interpretations of evolutionary relationships. Therefore, in order to render a verdict of phylogenetic relationships, it seems essential to appraise the total picture of morphological and biochemical characters. Barring an unforeseen breakthrough in palaeomycology, the use of morphology in conjunction with biochemical data represents the only hope for reconstructing key events in the evolutionary history of the Fungi.

ACKNOWLEDGEMENTS

I thank P. H. Tsao for his suggestions and valuable criticism of this manuscript. The experimental work from the author's laboratory reported herein was supported in part by generous research grants (AI-05540 and AI-06205) from the National Institutes of Health, U.S. Public Health Service.

REFERENCES

Alexopoulos, C. J. (1962). "Introductory Mycology," 2nd edition. Wiley, New York.
Aronson, J. M. (1965). *In* "The Fungi" (G. C. Ainsworth and A. S. Sussman, eds.), Vol. 1, pp. 49–76. Academic, New York and London.
Aronson, J. M., Cooper, B. A., and Fuller, M. S. (1967). *Science*, **155**, 332–335.
Aronson, J. M., and Machlis, L. (1959). *Am. J. Bot.* **46**, 292–300.
Atkinson, G. F. (1915). *Ann. Mo. bot. Gdn*, **20**, 315–376.
Bacon, J. S. D., Jones, D., Farmer, V. C., and Webley, D. M. (1968). *Biochim. Biophys. Acta*, **158**, 313–315.
Bartnicki-Garcia, S. (1966). *J. Gen. Microbiol.* **42**, 57–69.
Bartnicki-Garcia, S. (1968). *Ann. Rev. Microbiol.* **22**, 87–108.
Bartnicki-Garcia, S., Nelson, N., and Cota-Robles, E. (1968). *Arch. Mikrobiol.* **63**, 242–255.
Bartnicki-Garcia, S., and Nickerson, W. J. (1962). *Biochim. Biophys. Acta*, **58**, 102–119.
Bartnicki-Garcia, S., and Reyes, E. (1964). *Arch. Biochem. Biophys.* **108**, 125–133.
Bartnicki-Garcia, S., and Reyes, E. (1968). *Biochim. Biophys. Acta*, **170**, 54–62.

Becker, B., Lechevalier, M. P., and Lechevalier, H. A. (1965). *Appl. Microbiol.* **13**, 236–243.

Bessey, E. A. (1950). "Morphology and Taxonomy of Fungi". Hafner, New York.

Brown, R. G., and Lindberg, B. (1967). *Acta Chem. Scand.* **21**, 2383–2389.

Cantino, E. C. (1955). *Quart. Rev. Biol.* **30**, 138–149.

Crook, E. M., and Johnston, I. R. (1962). *Biochem. J.* **83**, 325–331.

DeMoss, J. A. (1965). *Biochem. Biophys. Res. Commun.* **18**, 850–857.

Dennis, R. L. (1969). *Science*, **163**, 670–671.

Fitch, W. M., and Margoliash, E. (1967). *Science*, **155**, 279–284.

Frey, R. (1950). *Ber. schweiz. bot. Ges.* **60**, 199–230.

Fuller, M. S., and Barshad, I. (1960). *Am. J. Bot.* **47**, 105–109.

Fultz, S. A. (1965). *Dissertation Abstr.* p. 2433.

Garzuly-Janke, R. (1940). *Zentbl. Bakt. Parisitenk. Abt. II*, **102**, 361–365.

Gäumann, E. A. (1964). "Die Pilze," 2nd edition. Birkhäuser, Basel.

Hawker, L. E., and Abbott, P. M. (1963). *J. Gen. Microbiol.* **32**, 295–298.

Hopkins, E. W. (1929). *Trans. Wisconsin Acad. Sci.* **24**, 187–196.

Hütter, R., and DeMoss, J. A. (1967). *J. Bacteriol.* **94**, 1896–1907.

Johnston, I. R. (1965). *Biochem. J.* **96**, 659–664.

Jones, D., Bacon, J. S. D., Farmer, V. C., and Webley, D. M. (1968). *Antonie van Leeuwenhoek*, **34**, 173–182.

Klein, R. M., and Cronquist, A. (1967). *Quart. Rev. Biol.* **42**, 105–296.

Koch, W. J. (1956). *Am. J. Bot.* **43**, 811–819.

Kreger, D. R. (1954). *Biochim. Biophys. Acta*, **13**, 1–9.

Kreger, D. R. (1962). *In* "Physiology and Biochemistry of Algae" (R. A. Lewin, ed.), pp. 315–335. Academic, New York and London.

Lee, Y-C., and Ballou, C. E. (1965). *Biochemistry*, **4**, 257–264.

Mahadevan, P. R., and Tatum, E. L. (1965). *J. Bacteriol.* **90**, 1073–1081.

Mangin, L. (1899). *J. Bot., Paris*, **13**, 276–286.

Martin, G. W. (1968). *In* "The Fungi" (G. C. Ainsworth and A. S. Sussman, eds.), Vol. 3, pp. 635–648. Academic, New York and London.

Nickerson, W. J., Falcone, G., and Kessler, G. (1961). *In* "Macromolecular Complexes" (M. V. Edds, ed.), pp. 205–228. Ronald Press, New York.

Novaes-Ledieu, M., Jimenez-Martinez, A., and Villanueva, J. R. (1967). *J. Gen. Microbiol.* **47**, 237–245.

O'Brien, R. W., and Ralph, B. J. (1966). *Ann. Bot.* **30**, 831–843.

Olson, L. W., and Fuller, M. S. (1968). *Arch. Mikrobiol.* **62**, 237–250.

Parker, B. C., Preston, R. D., and Fogg, G. E. (1963). *Proc. R. Soc. Ser. B*, **158**, 435–445.

Phaff, H. J. (1963). *Ann. Rev. Microbiol.* **17**, 15–30.

Quillet, M., and Priou, M. L. (1963). *C.r. hebd. Séanc. Acad. Sci., Paris*, **256**, 2903–2905.

Raper, J. R. (1968). *In* "The Fungi" (G. C. Ainsworth and A. S. Sussman, eds.), Vol. 3, pp. 677–693. Academic, New York and London.

Roelofsen, P. A. (1959). "The Plant Cell Wall". Borntraeger, Berlin.

Rogers, H. J., and Perkins, H. R. (1968). "Cell Walls and Membranes". Spon Ltd., London.

Savile, D. B. O. (1968). *In* "The Fungi" (G. C. Ainsworth and A. S. Sussman, eds.), Vol. 3, pp. 649–675. Academic, New York and London.

Shaw, R. (1965). *Biochim. Biophys. Acta*, **98**, 230–237.

Skucas, G. P. (1967). *Am. J. Bot.* **54**, 1152–1158.

Stanier, R. Y., and van Niel, C. B. (1962). *Arch. Mikrobiol.* **42**, 17–35.

Storck, R. L. (1965). *J. Bacteriol.* **90**, 1260–1264.

Storck, R. L. (1966). *J. Bacteriol.* **91**, 227–230.

Thomas, R. C. (1943). *Ohio J. Sci.* **43**, 135–138.
Toama, M. A., and Raper, K. B. (1967). *J. Bacteriol.* **94**, 1016–1020.
Trotter, M. J., and Whisler, H. C. (1965). *Can. J. Bot.* **43**, 869–876.
Tyler, S. A., and Barghoorn, E. S. (1954). *Science,* **119**, 606–608.
Vogel, H. J. (1963). *In* "Evolving Genes and Proteins" (V. Bryson and H. J. Vogel, eds.), pp. 25–40. Academic, New York and London.
Vogel, H. J. (1964). *Am. Nat.* **98**, 435–446.
Von Wettstein, F. (1921). *Sber. Akad. Wiss. Wien. Abt.* 1, **130**, 3–20.
Whittaker, R. H. (1969). *Science,* **163**, 150–160.
Van Wisselingh, C. (1898). *Jb. wiss. Bot.* **31**, 619–687.

CHAPTER 6

Comparative Lipid Biochemistry of Photosynthetic Organisms

B. W. NICHOLS

Unilever Research Laboratory,
Colworth House, Sharnbrook, Bedford, England

I. INTRODUCTION

Although the basic function of all chloroplasts is similar, their chemical composition frequently varies significantly and in a systematic manner which has frequently been related to the positions of the organisms in the pattern of plant evolution (Goodwin, 1966).

The photosynthetic apparatus of plants and bacteria contain a high proportion (frequently 30–40%) of their weight as lipophilic substances, and this review summarizes the salient differences which exist between the composition and metabolism of the lipids and fatty acids in different classes of photosynthetic organisms. It will refer only briefly to carotenoids and other terpenoid compounds which have already been the subject of recent reviews (Weissmann, 1966; Goodwin, 1966).

II. HIGHER PLANTS

Some of the most marked variations in the lipid composition of leaves from different plant classes occur within their cuticular waxes, and the application of cuticular lipid analyses to the taxonomic classification of plants is referred to in a recent review by Mazliak (1968).

Cuticular waxes apart, the lipid and fatty acid compositions of whole leaves or their chloroplasts show a remarkably consistent pattern over a wide variety of plant families, despite the considerable differences which may occur in the fatty acid composition of the seeds from the same group of plants (Nichols and James, 1968).

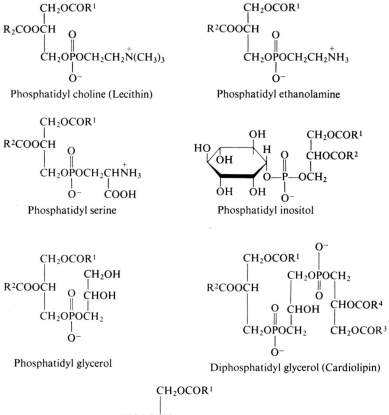

Phosphatidyl choline (Lecithin)

Phosphatidyl ethanolamine

Phosphatidyl serine

Phosphatidyl inositol

Phosphatidyl glycerol

Diphosphatidyl glycerol (Cardiolipin)

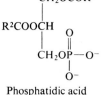

Phosphatidic acid

FIG. 1. Major phospholipids of plant photosynthetic tissues

The structure of the major acyl lipids and fatty acids of leaves are given in Figs. 1 and 2 and Table I. Benson and Maruo (1958) showed that chloroplasts possess a much simpler acyl lipid composition than the intact cells of leaves (Table II) and contain four major acyl lipids, namely mono- and di-galactosyl

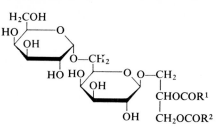

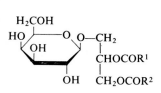

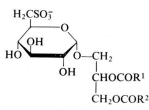

(I) Monogalactosyl diglyceride [β-D-galactosyl-(1-1′) − 2′, 3′-diacyl-D-glycerol]

(II) Digalactosyl diglyceride [α-D-galactosyl-(1-6)-β-D-galactosyl-(1-1′)-2′, 3′-diacyl-D-glycerol]

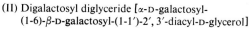

(III) Sulphoquinovosyl diglyceride [6-Sulpho-α-D-quinovosyl-(1-1′)-2′, 3′-diacyl-D-glycerol]

FIG. 2. Major glycolipids of plant photosynthetic tissue

diglyceride, sulphoquinovosyl diglyceride (the plant sulpholipid) and phosphatidyl glycerol. Minor proportions of phosphatidyl choline have also been observed in some chloroplast preparations (Allen *et al.* 1966).

The fatty acid composition of broad bean leaves and their chloroplasts are given in Table III and are representative of those found in the photosynthetic apparatus of many plants. Despite the similarity between the lipid composition of the leaves of most higher plants, two departures from this general pattern have been established. In 1961, Shenstone and Vickery reported that the cyclo-

TABLE I

Major fatty acids of photosynthetic tissue

Name	Abbreviated structure[a]
Palmitic	16:0
Palmitoleic	16:1 (9)
Oleic	18:1 (9)
Linoleic	18:2 (9,12)
α-Linolenic	18:3 (9,12,15)
γ-Linolenic	18:3 (6,9,12)
Arachidonic	20:4 (5,8,11,14)

[a] The figure before the colon denotes the number of carbon atoms, and that after the colon the number of double bonds. The positions of the double bonds are denoted by the figures in parentheses.

propenoid acids which characterize the seed fats from members of the Malva-
ceae and Sterculiaceae are also present in extracts from their leaves. Although
the location of these acids within the leaf cell is uncertain, the chloroplast does
not appear to be the primary site of synthesis (Yano *et al.* 1969). More recently,
Jamieson and Reid (1968, 1969) have shown that the 6,9,12-octadecatrienoic
acid (γ-linolenic acid) and the octadecatetraenoic acid which are characteristic

TABLE II

The major acyl lipids of leaves

	Total leaf	Chloroplast
Glycolipids		
Monogalactosyl diglyceride		Present
Digalactosyl diglyceride		Present
Sulphoquinovosyl diglyceride		Present
Cerebroside		—
Sterol glycoside		—
Phospholipids		
Phosphatidyl glycerol		Present
Phosphatidyl choline		?
Phosphatidyl ethanolamine		—
Phosphatidyl inositol		—

of seed fats from members of the Boraginaceae also occur in the leaves of these
plants. It is perhaps surprising that while the γ-linolenic acid of such leaves
does not occur in the chloroplast lipids to a significant extent, the galactosyl
diglyceride fractions from the same tissue contain a large proportion of the
octadecatetraenoic acid.

TABLE III

The fatty acid composition of broad bean tissues (Crombie, 1958)

	Fatty acid %						
	16:0	16:1	18:0	18:1	18:2	18:3	22:0
Etiolated leaf	16·7	—	4·7	—	33·5	39·4	4·6
Green leaf	11·7	6·9	3·2	3·4	14·3	56·4	4·0
Chloroplasts	7·4	9·2	1·2	5·2	2·6	72·0	1·2

III. FERNS AND MOSSES

Although the constituent acyl lipids of the photosynthetic tissues of ferns
and mosses are qualitatively similar to those of higher plants, they differ
dramatically in fatty acid composition. Schlenk and Gellerman (1965) have

shown that they contain substantial quantities of 5,8,11,14-eicosatetraenoic acid (arachidonic acid) and other polyenoic acids of the C_{20} and C_{22} series, while Nichols (1965a) and Wolf *et al.* (1966) demonstrated that these fatty acids are largely concentrated in the chloroplasts.

IV. ALGAE

Qualitatively, the major difference between the lipid composition of algal and leaf tissue is that algae synthesize neither cerebrosides nor sterol mono-glycosides. Other differences are mainly of a quantitative nature.

A. GREEN (FRESHWATER) ALGAE

Apart from the absence of cerebrosides and sterol glycosides, lipid extracts from whole cells of green algae are qualitatively very similar in acyl lipid and fatty acid composition to those from the leaves of higher plants. In quantitative terms the lipids of green algae contain much lower proportions of 9,12,15-octadecatrienoic acid (α-linolenic acid) than leaves, but the concentration of polyenoic acids in algae may vary considerably according to the conditions employed for their culture. Thus green algae grown photoautotrophically produce enhanced quantities of those acids which are synthesized by the chloroplast, and photoautotrophic cells of *Chlorella vulgaris* and *Euglena gracilis* contain much higher proportions of α-linolenic acid than similar cells grown heterotrophically (Nichols, 1965; Hulanicka *et al.* 1964) (Table IV).

TABLE IV

Fatty acid composition of cells of *Chlorella vulgaris*
grown under different conditions (Nichols, 1965b)

| | Fatty acid % | | | | | | | |
	16:0	16:1	16:2	16:3	18:0	18:1	18:2	18:3
Photoautotrophic	26	8	7	2	2	2	34	20
Heterotrophic	26	11	4	—	4	18	36	1

B. OTHER (HIGHER) ALGAE

Several classes of algae, including most of the marine algae and diatoms, synthesize C_{20} and C_{22} polyenoic acids, including arachidonic acid (Klenk *et al.* 1963; Kates and Volcani, 1966). These acids predominate in the galac-tosyl diglyceride fractions of such algae (Table V) suggesting that, as in ferns and mosses, arachidonic acid is a product of the photosynthetic apparatus (Radunz, 1968; Nichols and Appleby, 1969).

In this last respect marine algae differ from some other algal forms which

TABLE V

Fatty acid compositions of corresponding lipid fractions from three algae (Nichols and Appleby, 1969)

	Fatty acid (%)									α- 18:3	γ- 18:3							
	14:0	16:0	16:1	16:2	16:3	16:4	18:0	18:1	18:2			18:4	20:2	20:3	20:4	20:5	22:5	22:6
Euglena gracilis (heterotrophic)																		
Monogalactosyl diglyceride	1	8	3	17	10	14	—	1	14	27	—	—	—	—	—	—	—	—
Digalactosyl diglyceride	—	11	4	20	15	t[a]	4	4	23	20	—	—	—	—	—	—	—	—
Phosphatidyl choline	—	10	1	—	—	—	1	4	1	1	—	—	10	3	25	32	7	3
Ochromonas danica (heterotrophic)																		
Monogalactosyl diglyceride	5	2	3	—	—	—	t[a]	7	31	9	25	—	—	1	1	—	—	—
Digalactosyl diglyceride	18	11	5	—	—	—	t[a]	4	27	15	10	8	—	1	1	—	—	—
Phosphatidyl choline	8	9	t[a]	—	—	—	2	15	25	2	16	4	—	5	13	—	4	—
Porphyridium cruentum																		
Monogalactosyl diglyceride	1	24	1	—	—	—	1	2	4	—	—	—	—	—	26	40	—	—
Digalactosyl diglyceride	1	38	1	—	—	—	2	3	5	—	—	—	—	—	18	34	—	—
Phosphatidyl choline	1	22	1	—	—	—	1	3	4	1	—	—	—	6	59	10	—	—

[a] t = trace.

also synthesize arachidonic acid. Thus both *Euglena gracilis* and *Ochromonas danica* synthesize this acid, but analysis of their chloroplast lipids shows that these organelles contain little or no C_{20} polyenoic acids (Table V). Therefore, although total fatty acid analyses may imply possible relationships between different algal classes, more detailed study often shows the resemblance to te a superficial one. In fact, the arachidonic acid of *E. gracilis* and *O. danica* is a product of the mammalian form of metabolism which is particularly evident during the heterotrophic growth of these cells; photoautotrophic cultures of these algae accumulate considerably smaller proportions of C_{20} polyenoic acids (Hulanicka *et al.* 1964).

Another factor to be considered when using arachidonic acid as a taxonomic or phylogenetic marker is that it can be synthesized by at least two pathways, one involving γ-linolenic acid and the other involving 11,14-eicosadienoic acid (Fig. 3). In higher plants, marine algae and *O. danica* the γ-linolenate pathway

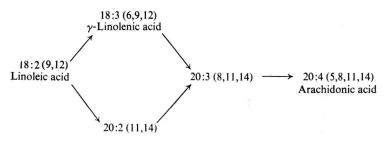

FIG. 3. Alternate pathways in the biosynthesis of arachidonic acid

appears to be the preferred one (Nichols and Appleby, 1969) whereas that involving the C_{20} dienoic acid is the major route in *Euglena* (Hulanicka *et al.* 1964).

The chloroplasts of *O. danica* differ from those of most other algae, apart from the blue-green alga *Spirulina platensis*, in accumulating appreciable quantities of γ-linolenic acid, and *O. danica* and *O. malhamensis* also differ from all other algae so far investigated in containing appreciable quantities of two novel lipids, the structures of which are still unknown. One of them (lipid A, Fig. 4) contains nitrogen, but no phosphorus, and appears to be a diglyceride derivative of a nitrogenous alcohol. It is not a major component of the chloroplast lipids of these algae, but appears to play an important role in fatty acid metabolism during heterotrophic growth (Nichols *et al.*, 1969).

C. BLUE-GREEN ALGAE

Among photosynthetic organisms which perform the Hill reaction, the blue-green algae possess a uniquely simple acyl lipid composition. They do not synthesize phosphatidyl choline, phosphatidyl ethanolamine or phosphatidyl inositol, which are common to all other algae, and their major acyl lipids are those four classes ubiquitous to the photosynthetic apparatus of leaves and algae, namely mono- and di-galactosyl diglyceride, sulphoquinovosyl diglyceride and phosphatidyl glycerol (Nichols *et al.* 1965) (Fig. 5).

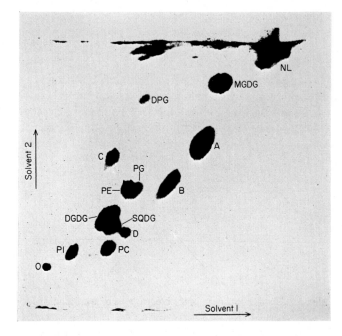

FIG. 4. 2-Dimensional thin layer chromatography of a lipid extract from *Ochromonas danica*. Stationary phase : silica gel containing 10% gypsum. Mobile phases: Solvent 1, chloroform–methanol–7N ammonium hydroxide (65:25:4, v/v); Solvent 2, chloroform–methanol–acetic acid–water (170:25:25:4 v/v). Visualization: 25% sulphuric acid at 220° (5 min). Identification of components: Unidentified lipids (A, B, C and D); neutral lipid and pigments (NL); monogalactosyl diglyceride (MGDG); digalactosyl diglyceride (DGDG); sulphoquinovosyl diglyceride (SQDG); cardiolipin (DPG); phosphatidyl glycerol (PG); phosphatidyl ethanolamine (PE); phosphatidyl choline (PC); phosphatidyl inositol (PI)

Some members of the Cyanophyceae contain lipids which are as yet uncharacterized but which have not been observed in other algae (Nichols *et al.* 1965; Allen *et al.* 1966). In particular, Nichols and Wood (1968a) noted that algae which have fixed nitrogen during growth synthesize a novel class of glycolipid which not only appears to be absent from all other photosynthetic organisms but is also absent from N_2-fixing algae when they are cultured in media containing ammonium salts (Fig. 5). Walsby and Nichols (1969) later demonstrated that the glycolipid is specific to the heterocyst structures peculiar to N_2-fixing algae. The structure of none of these glycolipids is known in detail, but in general they appear to comprise a series of molecules in which mono-saccharide units are linked to long-chain polyhydroxy alcohols. In *Anabaena cylindrica* the chief sugar moiety is glucose with smaller quantities of galactose, and the alcohol is thought to be a C_{26} trihydric alcohol, containing a 2,3-dihydric alcohol group, i.e. $CH_3CHOH(CH_2)_{21}CHOHCHOHCH_3$.

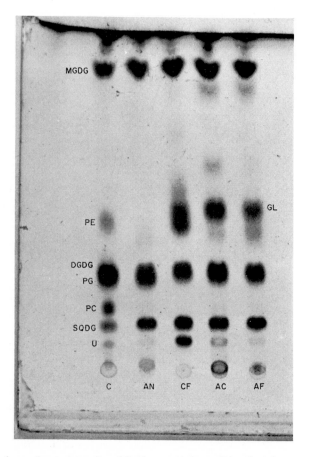

FIG. 5. Thin layer chromatography of lipid extracts from *Chlorella vulgaris* (C), *Anacystis nidulans* (AN) *Chlorogloea fritschii* (CF) *Anabaena cylindrica* (AC) and *Anabaena flos-aquae* (AF). Stationary phase: silica gel containing 10% gypsum. Mobile phase: Chloroform–methanol–acetic acid–water (85:15:10:3·7 v/v). Visualization 25% sulphuric acid at 220° (5 min). Identification of components: monogalactosyl diglyceride (MGDG); alcohol gly-cosides (GL); phosphatidyl ethanolamine (PE); digalactosyl diglyceride and phosphatidyl glycerol (not separated) (DGDG, PG); phosphatidyl choline (PC); sulphoquinovosyl di-glyceride (SQDG); an unidentified lipid (U).

Long chain alcohols such as these will clearly be of some interest to scientists concerned with the biological origin of hydrocarbons found in geological specimens. The blue-green algae are often considered likely contributors to the hydrocarbon content of such materials, but the hydrocarbon chain-lengths of many of the blue-green algae known today are not always compatible with those of many geologically important alkanes and alkenes (Gelpi *et al.* 1968; Han *et al.* 1968; Winters *et al.* 1969). Alcohols like that in *A. cylindrica* could clearly make a significant contribution to the hydrocarbon content of ancient oils and sediments, especially as they may occur in algae at more than ten times the concentration of alkanes and alkenes in the same cell.

TABLE VI

Percentage fatty acid composition of some blue-green algae

Algal species	Fatty acid %							
	16:0	16:1	16:2	18:0	18:1	18:2	α-18:3	γ-18:3
Spirulina platensis	43·4	9·7	t	2·9	5·0	12·4	t	21·4
Myxosarcina chroococcoides	38·2	8·6	1·2	4·0	6·8	9·2	33·3	—
Chlorogloea fritschii	42·3	4·9	t	5·4	14·3	17·2	15·8	—
Anabaena cylindrica	46·0	6·4	5·6	3·6	6·0	24·0	11·2	—
Anabaena flos-aquae	39·5	5·5	4·3	1·0	5·2	36·5	10·7	—
Mastigocladus laminosus	38·5	42·5	—	t	16·8	2·1	—	—
Anacystis nidulans[a]	47·0	38·8	—	1·4	10·0	—	—	—

[a] Holton *et al.* 1964.

Within the blue-green algae there occur much greater qualitative variations in fatty acid composition than are observed within the other algal classes, and because nearly all the acyl lipids of the blue-greens originate from the photosynthetic apparatus, these differences clearly represent variations in the synthesizing capacities of the lamellae (Table VI). The one generalization that might be made concerning the fatty acid composition of members of this class of alga is that none has been found which synthesizes arachidonic acid. Consequently, the capacity of certain chloroplasts to make this acid must have developed during the evolutionary steps which resulted in the genesis of the red algae.

Reference to Table VI shows that some blue-green algae are unique among cells which perform the Hill reaction in containing no polyenoic acids (Holton *et al.* 1964; Parker *et al.* 1967). Others contain linoleic acid but no trienoic acids, whereas a third group synthesizes α-linolenic and has fatty acid compositions not unlike those of green freshwater algae. One blue-green alga, *Spirulina platensis*, synthesizes γ-linolenic acid rather than α-linolenic acid (Nichols and Wood, 1968b), and because γ-linolenic acid is the precursor of arachidonic acid in some of the higher algae, this class of blue-green could represent an important step in plant evolution.

The photosynthetic apparatus of blue-green algae also differ from those of all other plants in not synthesizing *trans*-3-hexadecenoic acid (Nichols *et al.* 1965; Nichols and Wood, 1968b).

V. Photosynthetic Bacteria

Photosynthetic bacteria differ considerably from algae in both acyl lipid and in fatty acid composition. Only one lipid of plant chloroplasts, phosphatidyl glycerol, occurs in all photosynthetic bacteria studied to date (Nichols and James, 1965) although monogalactosyl diglyceride appears to be a minor constituent to many of them (Constantopoulos and Bloch, 1967). Digalactosyl diglycerides have not been detected in any bacterium and the plant sulpholipid in only two (Nichols and James, 1965). Photosynthetic bacteria do not synthesize polyenoic acids (Wood *et al.* 1965).

VI. The Evolutionary Pattern of Algae, Based on Their Lipid Metabolism

The information summarized above and in Table VII can be employed to draw up a pattern of plant evolution which fits in fairly well with views based on more conventional forms of assessment. Nevertheless, certain modifications in the general pattern are suggested by fatty acid analyses in particular and these will be referred to here, even though they may be unacceptable on other grounds.

If one proposes that some forms of bacteria were the precursors of the most primitive algae, i.e. the blue-greens, then clearly this evolutionary step was accompanied by marked changes in lipid metabolism. For example, the major glycolipids of the blue-green algae are found infrequently in bacteria, and even then in only minor proportions. Also, the bacteria do not synthesize the polyenoic acids which are found in many blue-greens. On the other hand the plastids of some blue-green algae, unlike those of other plant forms, synthesize only saturated and monoenoic acids and it seems most likely that it was blue-green algal species such as these which were originally derived from bacteria.

It is possible to arrange certain species (or groups of species) of blue-green algae in order of their capacity to synthesize fatty acids of increasing degrees of unsaturation, and it is not unreasonable to suppose that such an arrangement is also consistent with their order of evolutionary development (Fig. 6). Blue-greens which synthesize α-linolenic acid and γ-linolenic acid represent the most highly developed algae in this class (in terms of fatty acid synthesizing capacity) and can be regarded as the logical ancestors of the remaining classes of algae and higher plants, which may be divided into two major groups

TABLE VII

A comparison of the lipid composition of photosynthetic tissues

	Monogalactosyl diglyceride	Digalactosyl diglyceride	Sulphoquinovosyl diglyceride	Phosphatidyl glycerol	Phosphatidyl choline	Cerebroside	Sterol glycoside	Acids 3-16:1	9,12,15-18:3	6,9,12-18:3	5,8,11,14-20:4	5,8,11,14-20:4 in chloroplast	Sterols
Photosynthetic bacteria	P	a	S	P	S	a	a	a	a	a	a	a	?
Blue-green algae (1)	P	P	P	P	a	a	a	a	a	a	a	a	S
Blue-green algae (2)	P	P	P	P	a	a	a	a	P	a	a	a	S
Blue-green algae (3)	P	P	P	P	a	a	a	a	a	P	P	a	S
Red algae	P	P	P	P	P	a	a	P	P	P	P	P	P
Other Marine algae	P	P	P	P	P	a	a	P	P	a	P	a	P
Euglenaceae	P	P	P	P	P	a	a	P	P	P	P	a	P
Chrysophyceae	P	P	P	P	P	a	a	P	P	a	a	a	P
Green (freshwater) algae	P	P	P	P	P	a	a	P	P	P	P	P	P
Ferns and Mosses	P	P	P	P	P	a	a	P	P	a	P	P	
Leaves													
Most higher plants	P	P	P	P	P	P	P	P	P	a	a	a	P
Boraginaceae	P	P	P	P	P	P	P	P	P	P	a	a	P
Chloroplasts	P	P	P	P	?	a	a	P	P	a	a	a	P

P = present; a = absent; S = present in some members.

depending on whether their chloroplasts synthesize γ-linolenic acid (and possibly arachidonic acid) or α-linolenic acid.

Thus the red algae (and other marine algae) and the Chrysophyceae could be regarded as having developed from those species of blue-green algae represented by *Spirulina platensis*, while the other plant classes, in which the most highly unsaturated chloroplast fatty acid is α-linolenic acid, may be considered as derived from the group of which *Anabaena variabilis* is a member.

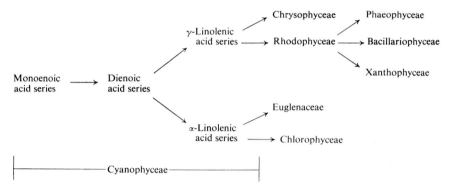

FIG. 6. Phylogenetic classification of algae based on their lipid metabolism

It must be stressed that the scheme represented in Fig. 6 is more a plant classification based on the lipid composition and metabolism of chloroplasts than a definition of evolutionary development. In particular, it assumes that the capacity to synthesize certain acids such as the linolenic acids and arachidonic acid is not lost once it has been acquired, and this in itself is possibly a gross simplification of the evolutionary processes involved. For example, Goodwin (1966) and Dougherty and Allen (1960) have suggested a pattern of evolution in which the Chrysophyceae are derived from the red algae. This is only acceptable in terms of lipid metabolism if the capacity to synthesize arachidonic acid is lost by the chloroplast during this evolutionary step.

It is nevertheless hoped that the data presented here may be of some value to those interested in plant taxonomy and the evolution of plant forms, at least by supplying extra support for already proposed relationships and in suggesting other areas worthy of more intensive investigation.

REFERENCES

Allen, C. F., Hirayama, O., and Good, P. (1966). *In* "Biochemistry of Chloroplasts" (T. W. Goodwin, ed.), Vol. 1, p. 165. Academic, New York and London.
Benson, A. A., and Maruo, B. (1958). *Biochim. biophys. Acta,* **27**, 189.
Constantopoulos, G., and Bloch, K. (1967). *J. Bacteriol.* **93**, 1788.
Crombie, W. M. (1958). *J. exp. Bot.* **9**, 254.

Dougherty, E. C., and Allen, M. B. (1960). *In* "Comparative Biochemistry of Photo-reactive Pigments" (M. B. Allen, ed.), p. 102. Academic, New York and London.
Gelpi, E., Oro, J., Schneider, H. J., and Bennett, E. O. (1968). *Science*, **161**, 700.
Goodwin, T. W. (1966). *In* "Comparative Phytochemistry" (T. Swain, ed.), p. 121. Academic, London and New York.
Han, J., McCarthy, E. D., Calvin, M., and Benn, M. H. (1968). *J. Chem. Soc.* 2785.
Holton, R. W., Blecker, H. H., and Onore, M. (1964). *Phytochemistry*, **3**, 595.
Hulanicka, D., Erwin, J., and Bloch, K. (1964). *J. biol. Chem.* **239**, 2778.
Jamieson, G. R., and Reid, E. H. (1968). *J. Sci. Food Ag.* **19**, 628.
Jamieson, G. R., and Reid, E. H. (1969). *Phytochemistry*, **8**, 1489.
Kates, M., and Volcani, B. E. (1966). *Biochim. biophys. Acta*, **116**, 264.
Klenk, E., Knipprath, W., Eberhagen, D., and Koof, H. D. (1963). *Hoppe-Seyler's Z. physiol. Chem.* **334**, 44.
Mazliak, P. (1968). *In* "Progress in Phytochemistry" (L. Reinhold and Y. Liwschitz, eds.), Vol. 1, p. 49. Interscience, New York.
Nichols, B. W. (1965a). *Phytochemistry*, **4**, 769.
Nichols, B. W. (1965b). *Biochim. biophys. Acta*, **106**, 274.
Nichols, B. W., and Appleby, R. S. (1969). *Phytochemistry*, **8**, 1907.
Nichols, B. W., Bryce, T., and Robinson, M. (1969). Unpublished work.
Nichols, B. W., Harris, R. V., and James, A. T. (1965). *Biochim. biophys. Acta*, **20**, 256.
Nichols, B. W., and James, A. T. (1965). *Biochem. J.* **94**, 22P.
Nichols, B. W., and James, A. T. (1968). *In* "Progress in Phytochemistry" (L. Reinhold and Y. Liwschitz, eds.), Vol. I, p. 1. Interscience, New York.
Nichols, B. W., and Wood, B. J. B. (1968a). *Nature*, **217**, 767.
Nichols, B. W., and Wood, B. J. B. (1968b). *Lipids*, **3**, 46.
Parker, P. L., Van Baalen, C., and Maurer, L. (1967). *Science*, **155**, 707.
Radunz, A. (1968). *Hoppe-Seyler's Z. physiol. Chem.* **349**, 1091.
Schlenk, H., and Gellerman, J. L. (1965). *J. Am. Oil Chem. Soc.* **42**, 504.
Shenstone, F. S., and Vickery, J. R. (1961). *Nature*, **190**, 168.
Walsby, A., and Nichols, B. W. (1969). *Nature*, **221**, 673.
Weissmann, G. (1966). *In* "Comparative Phytochemistry" (T. Swain, ed.), p.97. Academic, London and New York.
Wolf, F. T., Coniglio, J. G., and Bridges, R. B. (1966). *In* "Biochemistry of Chloroplasts" (T. W. Goodwin, ed.), Vol. I, p. 187. Academic, New York and London.
Winters, K., Parker, P. L., and Van Baalen, C. (1969). *Science*, **163**, 467.
Wood, B. J. B., Nichols, B. W., and James, A. T. (1965). *Biochim. biophys. Acta*, **106**, 261.
Yano, I., Morris, L. J., Nichols, B. W., and James, A. T. (1969). Unpublished work.

CHAPTER 7

The Relationship between Bacteria, Blue-green Algae and Chloroplasts

N. G. CARR AND I. W. CRAIG

*Department of Biochemistry, University of Liverpool,
Liverpool, England*

I. INTRODUCTION

The existence of features common to the blue-green algae and to bacteria was observed in the nineteenth century, but the presence of a typical "green plant," O_2-evolving photosynthesis and the possession of the appropriate photopigments by the Cyanophyceae led to their being grouped as primitive examples of the higher algae and green plants. With increasing knowledge of fine structure the disparity between cell organization in higher plants and blue-green algae, and the similarity of the latter to bacteria, became more apparent. Certain colourless, hence non-photosynthetic, species of Cyanophyceae, such as *Beggiatoa*, are especially difficult to separate taxonomically from the bacteria. In a series of reviews van Niel and Stanier (Stanier and van Niel, 1941,

1962; Stanier, 1964) have emphasized the relationship between bacteria and blue-green algae and have suggested that these collectively comprise one of the two basic types of cell organization. In their early analysis Stanier and van Niel (1941) proposed three, major, negative characteristics which were held in common by blue-green algae and bacteria and which separated them from all other cells: the absence of nuclei, the absence of plastids and the absence of sexual reproduction. The latter was soon shown not to be so in the case of bacteria by Lederberg and Tatum (1946) and the last two decades have seen the accumulation of an impressive description of sexual reproduction albeit in a somewhat limited range of bacterial species (Hayes, 1968).

Only recently has the unequivocal demonstration of recombination by mutant species of a blue-green alga, *Anacystis nidulans*, been described and the process of gene-material transfer is not yet known (Bazin, 1968). The two other characteristic features shared by bacteria and the blue-green algae remain valid and have been documented and amplified by modern techniques of cell analysis. The basic morphological feature of all bacteria and blue-green algae is their lack of membrane-bound, sub-cellular organelles that are specialized for a particular physiological function. This is the primary criterion on which Stanier and van Niel (1962), extending earlier terminology and concepts, proposed that all living cells may be divided into procaryotic and eucaryotic types. Bacteria and blue-green algae comprise the Procaryotes; all higher algae, fungi, protozoa, plants and animals are Eucaryotes. Thus all the latter possess well-defined nuclei, and different species have mitochondria, chloroplasts, Golgi apparatus, flagella basal cells as examples of intra-cellular, morphologically specialized organelles. The metabolic activities of many eucaryotic sub-cellular organelles have a functional, but not morphological, equivalent in procaryotic organisms. Thus, for example, essentially the same primary processes of photosynthesis are carried on the cytoplasmic lamellae which are located around the periphery of the blue-green alga, *A. nidulans*, as occur in chloroplasts of higher algae and green plants. The crucial difference is that in the blue-green alga the site of photo-energy conversion is in intimate contact with the rest of the cytoplasm, or at least is not delineated from it by a lipoprotein membrane such as the chloroplast outer membrane. In procaryotic organisms there is no nucleus as such; the genetic material may be detected as an amorphous, low electron-dense area within the cytoplasm and without any form of nuclear membrane. A more detailed examination of procaryotic fine structure is presented in the next section.

The division of all cells into procaryotic or eucaryotic types is absolute; there are no intermediate forms known and the assignment of any particular group of organism can be made unequivocally. This point is discussed by Stanier and van Niel (1962) who also examined the special case of viruses, which cannot be considered to be cellular entities. The separation of all cells into one of two possible types presents a major step in any account of existing species in phylogenic and evolutionary terms. As Stanier (1964) has said, the separation of procaryotic and eucaryotic microorganisms "is without doubt the largest single evolutionary discontinuity." Such a discontinuity demands that attempts to arrive at a synthesis that would relate satisfactorily these two types of cellular structure are necessary and that all possible mechanisms, some of which may hitherto be thought naive, should be considered. We intend in this article to collect recent information about the structure, and metabolism, of bacteria and blue-green algae and compare it with aspects of our knowledge of green plant chloroplasts. This may allow a reassessment of an early view on chloroplast origins (Mereschkowsky, 1905) and of the validity of possible experimental approaches.

II. Fine Structural Similarities Between Bacteria and Blue-green Algae

A comparison of sections of both bacteria and blue-green algae in the electron microscope amply confirms the general similarities in structural organization suggested by earlier morphological studies. Some species of photosynthetic bacteria, such as *Rhodomicrobium vannielii*, in which the photosynthetic apparatus is located as peripheral lamellae (Fig. 1a,b) bear a striking resemblance to certain blue-green algae. As in *Anacystis nidulans* (Fig. 2a), the photosynthetic membrane systems are clearly separate, with cytoplasm containing ribosome-like material interspersed between them. In both organisms a greater amount of photomembrane structures, containing more photo-pigments, are found after growth in low (1100 lux) illumination when compared with growth in high (11,000 lux) light. This phenomenon has been reported in many species of both types of organism with respect to photo-pigments and accessory electron transport components and it has been demonstrated that adjustment of pigment level to different light intensities by *Rhodopseudomonas spheroides* is obligatorily linked to photosynthetic membrane formation (Sistrom, 1962). Of course, not all photosynthetic bacteria and blue-green algae possess photosynthetic structures that bear so close a resemblance; the bacteria in particular present variation in organization, but all possess their pigments located onto a membrane that is itself not differentiated into a unit membrane-bound organelle (see Cohen-Bazire and Sistrom, 1966; Lang, 1968).

The photosynthetic unit membrane may arise in blue-green algae from an invagination of the cytoplasmic membrane (Pankratz and Bowen, 1963) as in bacteria, or alternatively it has been suggested that lamellae are formed *de novo* in the central part of the cell and migrate to the periphery (Echlin and Morris, 1965). Recent electron micrographs of Allen (1968a) which show continuity of photosynthetic lamellae and cytoplasmic membrane support the former suggestion.

The low electron-dense nucleoplasm is in all cases seen as an ill-defined area in the cytoplasm quite distinct from the nucleus of eucaryotic cells. Earlier workers had shown this with Feulgen staining in both bacteria (Stille, 1937)

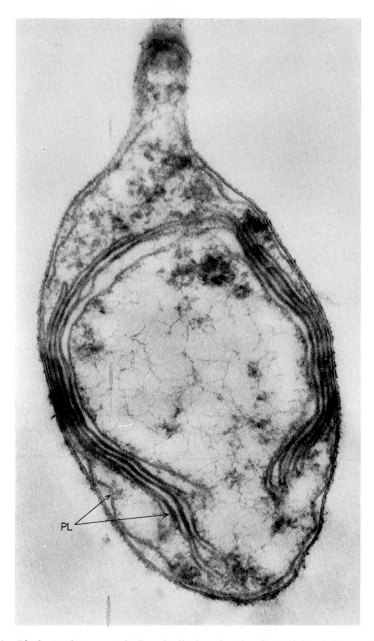

FIG. 1a. *Rhodomicrobium vannielii*. Longitudinal section showing peripheral photosynthetic lamellae (PL) ×89,000. Reproduced by kind permission from E. S. Boatman and H. C. Douglas, Fine Structure of the Photosynthetic Bacterium *Rhodomicrobium vannielii*, *J. Biophys. Biochem. Cytol.* **11**, 469–483 (1961)

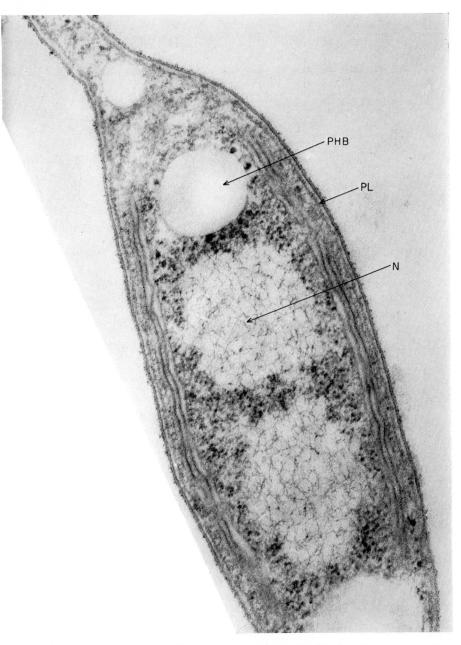

FIG. 1b. *Rhodomicrobium vannielii.* Section illustrating poly-β-hydroxybutyrate granule (PHB), nuclear area (N) and photosynthetic lamellae (PL) approx. ×102,000. Reproduced by kind permission from W. C. Trentini and M. P. Starr, Growth and Ultrastructure of *Rhodomicrobium vannielii* as a Function of Light Intensity, *J. Bact.* **93**, 1669–1704 (1967)

and blue-green algae (Fogg 1951; Cassel and Hutchinson, 1954). The nucleo-plasm is usually central in blue-green algae and the DNA is thought to be present as fibrillar structures 2–5 nm wide (Ris and Singh, 1961; Pankratz and

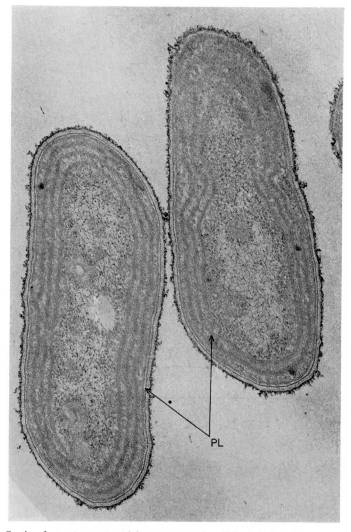

FIG. 2a. Section from *Anacystis nidulans* grown at low light (100 ft cm) showing extensive peripheral photosynthetic lamellae (PL) development ×60,000. Reproduced by kind permission from M. M. Allen, Photosynthetic Membrane System in *Anacystis nidulans, J. Bact.* **96**, 836–841 (1968)

Bowen, 1963). The association of Feulgen-staining nucleoplasmic areas of *Anabaena* sp. with localization of thymidine (methyl-T) incorporation has been shown (Leak, 1965). An extensive review of the nucleo-areas of both

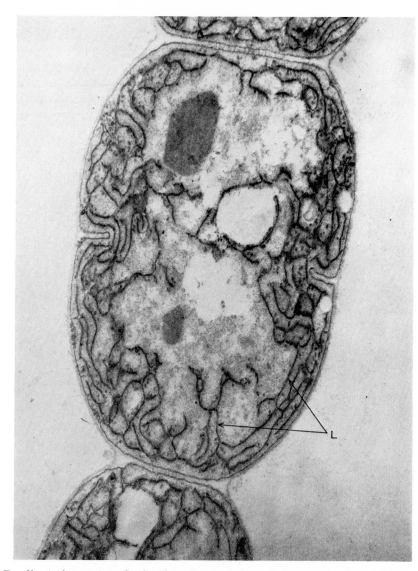

FIG. 2b. *Anabaenopsis* sp. Section through a vegetative cell, the extensive anastomizing network of lamellae is shown (L). Approx. ×48,000. Reproduced by kind permission from A. Peat and B. W. Whitton, Vegetative Cell Structure in *Anabaenopsis* sp., *Arch. Mikrobiol.* **63**, 170–176 (1968)

bacteria and blue-green algae has been presented by Fuhs (1969). Some comparative chemical properties of procaryotic DNA will be discussed in Section III.

For some years a feature of interest in blue-green algal fine structure has been the presence, in some species, of gas vacuoles which may be clearly seen in the light microscope as areas of low refractive index (Fig. 3a,b). They have been characterized by electron microscopy as regular, cylindrical bodies arranged in three-dimensional clusters (Bowen and Jensen, 1965; Smith and Peat, 1967). They vary in size and may, in some cases, be associated with the photosynthetic lamellae, and up to 39 % of the total cell volume of *Oscillatoria agardhii* is occupied in these vacuoles (Smith and Peat, 1967). Recently Walsby (1969) has shown that these are self-erecting structures which are permeable to

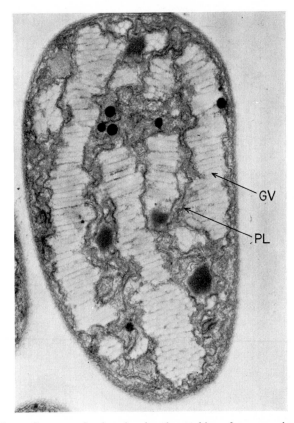

FIG. 3a. *Anabaena flos-aquae*. Section showing the stacking of gas vacuoles (GV) and the photosynthetic lamellae (PL). Approx. ×20,000. Reproduced by kind permission of R. V. Smith, A. Peat and B. A. Whitton, Dept. of Botany, University of Durham

gases and that they have an important role in maintaining the buoyancy of the blue-green alga. The presence of gas vacuoles in some bacteria is an important point of comparison. The photosynthetic bacterium *Pelodictyon clathratiforme* has large electron transparent vessels bounded by a non-unit membrane (Pfennig and Cohen-Bazire, 1967) (Fig. 3b); gas vacuoles similar to those of blue-green algae have also been reported in a species of halophilic bacteria (Larsen *et al*. 1967).

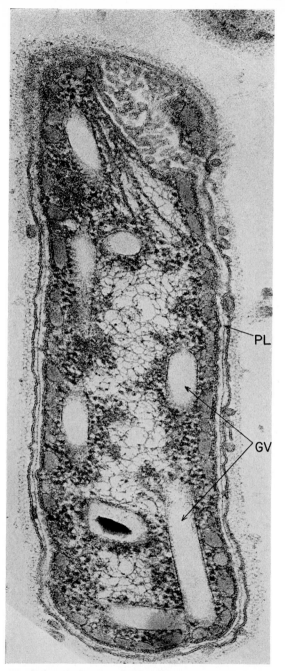

Fig. 3b. *Pelodictyon clathratiforme.* Longitudinal section showing gas-vacuoles (GV) and photosynthetic lamellae (PL), ×96,000. Reproduced by kind permission of N. Pfennig and G. Cohen-Bazire, Some Properties of the Green Bacterium *Pelodictyon clathratiforme, Arch. Mikrobiol.* **59**, 226–236 (1967)

Many species of blue-green algae secrete around their cell wall a mucilaginous sheath comparable to the capsule produced by some bacteria. A fibrillar nature has been detected in sheathes from both types of organism; although the dimensions of component fibres vary markedly (Echlin and Morris, 1965) there is, at least, morphological similarities between sheathes and capsules with polysaccharide components comprising a large percentage of the structural material. The procaryotic cell wall is the shape defining structure and has been extensively examined in bacteria and the blue-green algae (see Salton, 1964). In the latter it would appear to consist of two, or possibly three, layers and is structurally, as well as functionally, similar to that found in some gram-positive bacteria (Glauert, 1962). In gram-negative bacteria species, in addition to having a chemically less complex structure, the cell wall is usually thinner; Ogura (1963) has shown a two-layer wall structure in *Escherichia coli* and a similar arrangement has been found in some photosynthetic species (Cohen-Bazire, 1963). An electron microscopic analysis of the cell wall *Anacystis nidulans*, which has four layers outer to the cytoplasmic membrane, has been described by Allen (1968b) who concludes the structure and mode of division to be comparable to the bacterial model, and unlike any cell division observed in plants. The presence of fibres in the blue-green algae cell wall, not found in bacteria but similar to the fibres of red algae, have been observed by Drawert and Metzner (1956).

III. CHEMICAL AND METABOLIC FEATURES SHARED BY PROCARYOTIC MICROORGANISMS

A. CELL WALL STRUCTURE

The analysis of the cell wall has provided perhaps the most detailed comparison between bacteria and blue-green algae at the chemical level. The nature of the bacterial cell wall is described in reviews by Salton (1964) and that of blue-green algae by Holm-Hansen (1968); Echlin and Morris (1965) have pointed to the basic chemical similarities. The presence of diaminopimelic acid in many species of bacteria, its occurrence in three samples of Cyanophyceae and its absence from protozoa, fungi and higher algae was reported by Work and Dewey (1953) and proved an early indication of the similarity between bacteria and blue-green algae cell walls. Subsequent work (Hoare and Work, 1957) has shown trace amounts of diaminopimelic acid in *Chlorella ellipsoidea*, but this presumably represents its role as an intermediate in the biosynthesis of lysine. Vogel (1959) has shown that the diaminopimelic pathway of lysine synthesis is operative in bacteria, green algae, vascular plants and certain fungi, whilst other fungi use an α-aminoadipic acid route (Fig. 4). Holm-Hansen *et al.* (1965) have confirmed the presence of diaminopimelic acid in *C. pyrenoidosa* and *Cyanidium caldarium* at levels of 10^{-1} to 10^{-3}, of that found in procaryotic Flexibacteria, *Beggiatoa* and Cyanophyceae species.

A mucopeptide containing muramic acid, glucosamine and several amino acids was found in the cell wall of *Phormidium uncinatum* by Frank *et al.* (1962), who showed it to be similar in composition to two gram-negative bacterial species (Table I). The presence in cell walls of *Anacystis nidulans* and *Chloro-*

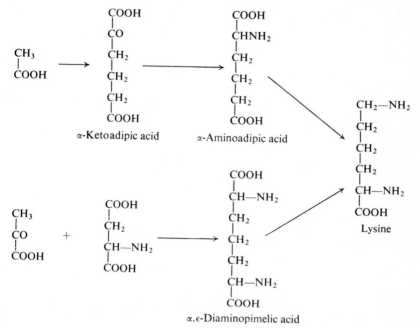

FIG. 4. The two routes of lysine biosynthesis

gloea fritschii of typical bacterial cell wall constituents such as *N*-acetylglucos-amine, *N*-acetylmuramic acid and ten amino acids including diaminopimelic acid was noted by Drews and Meyer (1964). Absence of neutral sugars and uronic acids confirmed the bacterial nature of the blue-green algal wall. The

TABLE I

Amino sugars and major amino acids present in a blue-green algal and two gram-negative bacterial cell walls[a]

	Organism		
Component	*Phormidium uncinatum*	*Escherichia coli*	*Spirrillum* sp.
Muramic acid	0·63	1	1
Glucosamine	1·00	1	1
Diaminopimelic acid	1·00	1	1·2
Glutamic acid	0·94	1·1	1·2
Alanine	2·0	1·8	2·6

[a] Figures indicate the molar ratio of each component (After Frank *et al.* 1962).

5

effect of lysozyme on blue-green algal cell walls, leading to the production of protoplasts under the appropriate conditions, further documents their similarity to bacterial walls (Fuhs, 1958; Crespi *et al.* 1962). The similar specificity of penicillin and chloramphenicol toxicity to bacteria and blue-green algae, relative to eucaryotic organisms, has been discussed by Echlin and Morris (1965).

B. POLY-β-HYDROXYBUTYRIC ACID

The presence of a large storage body consisting of poly-β-hydroxybutyrate is clearly seen in the electron micrographic picture of *Rhodomicrobium vannielii* (Fig. 1b). This material is a product widespread in bacterial species especially after growth of the organism on acetate (Stanier *et al.* 1959). When the blue-green alga *C. fritschii* is grown in the presence of acetate, significant amounts of poly-β-hydroxybutyrate are found (Carr, 1966). The presence of this storage material has not been reported in any eucaryotic cell.

C. STEROLS

A hitherto major point of similarity between bacteria and blue-green algae, namely the absence of sterols in these organisms, has recently required reassessment. Notwithstanding the failure to detect sterols in Cyanophyceae reported by earlier workers, Reitz and Hamilton (1968) have detected cholesterol and β-sitosterol in *A. nidulans* and *Fremyella diplosiphon* and de Souza and Nes (1968) have found 24-ethyl-Δ^7-cholestenol as the major sterol component of *Phormidium luridum*. A sterol component in *Anabaena variabilis* has been shown to be fractionated, by centrifugation, with the photosynthetic lamellae, along with chlorophyll *a* and carotenoids (N. G. Carr and J. W. Pennock, unpublished). The "harsh" methods of extraction employed suggested to Reitz and Hamilton (1968) that the sterols were bound tightly onto a cell membrane or wall. In a similar manner, recent workers have detected sterols in several species of bacteria. Schubert *et al.* (1967) reported that cholesterol could be isolated from *Streptomyces olivaceus* and Schubert *et al.* (1968) found several sterols present in this organism and in *Azotobacter chroococcum* and *Escherichia coli*. The presence of squalene has been recorded in a *Staphylococcus* species (Suzue *et al.* 1968) and in *Halobacterium cutirubrum* (Tornabeae *et al.* 1969). Clearly the presence and role of sterols in procaryotic organisms must be examined anew; it is perhaps significant that cholesterol has been detected in chloroplasts and is a major sterol of Rhodophyceae and that β-sitosterol is widely distributed in higher plants (Goad, 1967).

D. RIBOSOMES

All cells contain particulate ribonucleoprotein bodies of molecular weight in the range $2 \cdot 7$–$4 \cdot 5 \times 10^6$, containing 40–65% RNA by weight and very little lipid. Ribosomes are characterized by their sedimentational properties in the analytical ultracentrifuge and considerable evidence exists to show that two forms exist of 70 s and 80 s dimensions. The former are encountered in extracts of bacteria whilst the latter appear in preparation from animals, plants and yeast (Petermann, 1964). Taylor and Storck (1964) have investigated the sedimentational behaviour of about 25 examples of procaryotic (bacterial) and eucaryotic (yeast) cells; they found that the ribosomal s values fell into two distinct classes which have mean values of s_{20}^0 W 68·4 and 81·3 respectively. Taylor and Storck (1964) found that the blue-green algae *Anacystis nidulans* and *Anabaena cylindrica* both possessed ribosomes of the 70 s type, and suggested that ribosomal category could form a basic means of identification of procaryotic and eucaryotic cells.

A. variabilis has a 70 s type ribosome and the behaviour of its ribosomal sub-units examined by sucrose density gradient and analytical ultracentifugation was also similar to that of bacteria (Craig and Carr, 1968). When Mg^{++} concentration is reduced to $10^{-4}M$, ribosomes split into 40 s and 60 s (eucaryotic) and 30 s and 50 s (procaryotic) units. For typical procaryotic behaviour of *A. variabilis* ribosomes see Fig. 5 in which a major peak at 70 s is apparent with minor sub-unit peaks at 50·5 s and 32 s. The delineation of the 32 s sub-unit is obscured by being incompletely resolved from the broad slow moving peak representing the accessory photopigment phycocyanin. Analysis of *A. variabilis* ribosomal preparation at 10^{-4} M Mg^{++} led to complete breakdown into 50 s and 30 s fractions. There was a marked tendency for the 70 s particle to lose a protein portion, becoming 60 s, and the 50 s sub-unit to undergo the same process becoming 39 s. Cell disintegration by ultrasonic treatment, rather than by the French Pressure Cell, and preparation in tris rather than phosphate buffer accelerated this loss of protein moiety (Craig and Carr, 1968). This instability was not observed in ribosomes from *Escherichia coli* and leads to ribonucleoprotein particle sizes similar to those found in some chloroplast preparations. The clear difference between 70 s and 80 s ribosomes is reflected in their constituent ribosome–RNA species; this will be discussed further in Section V in relation to the ribosomal–RNA from chloroplasts.

IV. THE ENDOSYMBIOTIC VIEW OF CHLOROPLAST ORIGIN

The concept that all living cells are of two basic types with fundamentally different intracellular morphological arrangements offers a new perspective of living organisms. When organisms are considered at the cellular level the difference between disparate tissues are less apparent and the division into unicellular and multicellular forms becomes, at the intracellular level, of less importance than the separation of unicellular forms into procaryotic and eucaryotic types. The structural criteria, which form the basis for the separation into two cell types, are crucial in that no intermediate form of cell has yet been found; for example, there is no known cell which lacks a nucleus but

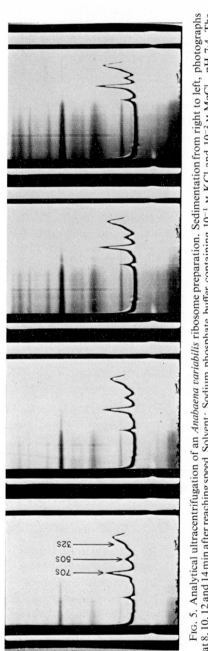

FIG. 5. Analytical ultracentrifugation of an *Anabaena variabilis* ribosome preparation. Sedimentation from right to left, photographs at 8, 10, 12 and 14 min after reaching speed. Solvent: Sodium phosphate buffer containing 10^{-1} M KCl and 10^{-2} M MgCl$_2$, pH 7·4. The photographs were taken using Schlieren optics, concentration 2–4 mg/ml. Peak a = 70·2 s, b = 50·5, c = 32. (After Craig and Carr, 1968)

possesses mitochondria. This discontinuity in cellular arrangement invites suggestions as to the mechanism and sequence of development that produced it. A satisfactory account can be made if the endosymbiotic origin of cell plastids is adopted. Mereschkowsky (1905) suggested that chloroplasts of plant cells were derived from blue-green algae which had entered into, and become established in, the cytoplasm of another cell. Since this early suggestion, other workers, over the years, have presented arguments and examples in its favour, although generally the endosymbiotic view of cell origin has not been accepted. Many of the earlier suggestions have been reviewed by Lederberg (1952) and their relation to cytoplasmic inheritance discussed. The range of endosymbiotic organisms is wide and a certain difficulty is experienced in differentiating them from the distinct cytoplasmic organelles which are to some extent autonomous. Thus, several blue-green algae which are endosymbiotic within cells of other algae have been considered to be intracellular organelles (Orenski, 1966). Perhaps the best studied example is that of *Glaucocystis nostochinearum* in which a blue-green alga carries out the function of a chloroplast in an otherwise colourless alga (Echlin, 1967). The exact role of the host in this relationship is not clear, although obviously the blue-green alga is protected from much environmental change. The electron microscope studies of Echlin (1967) and of Hall and Claus (1963) with *Cyanophora paradoxa* show a marked reduction in the cell wall of the endosymbiotant and are in agreement with the failure to detect diaminopimelic acid in *Glaucocystis* (Holm-Hansen *et al.* 1965). The pigments of plastids from *C. paradoxa* and *G. nostochinearum* include chlorophyll *a*, β-carotene and *C*- and allophycocyanin, and as such, are characteristic of free living blue-green algae (Chapman, 1966).

Small particles within the cells of certain *Paramecium* species may be seen in the electron microscope and these have been associated with the "killing" property of those species when conjugated with other strains. The probable bacterial nature of these bodies was suggested when Van Wagtendonk *et al.* (1963) grew the λ particle of *Paramecium* stock 299 in a nutrient medium, outside the host. The detection of the characteristic bacterial wall constituent, diaminopimelic acid, in nm particles from *P. aurelia* is further evidence of their bacterial nature (Stevenson, 1967). The presence of intracellular symbiotic organisms which are consistently passed to daughter cells on division of the host, offers a possible model for the origin of true organelles, which have lost entirely the capacity for host-free existence.

Recently, Sagan (1967) has restated the hypothesis that the three organelles of eucaryotic cells, the mitochondria, chloroplasts and the basal bodies of flagella, are derived from once free living procaryotic organisms and has presented a theory of the origin of eucaryotic cells based on this supposition. Sagan (1967) paid particular attention to the role of the (9 + 2) basal bodies of flagella and has attempted to relate endosymbiotic development to both fossil records and geochemical change (Fig. 6). The possible origin of mitochondria has been discussed by Roodyn and Wilkie (1968) and the recent demonstration of circular DNA in mitochondria from yeast is noteworthy (Van Bruggan *et al.* 1968). Electron microscopic examination of DNA from chloroplasts of *Chlamydomonas moewussi*, as well as those from blue-green algae, showed areas of low density with 25 A fibrils. This and other similarities led Ris and Plaut (1962) to suggest that the endosymbiotic origin of complex cell systems must be seriously re-examined. Some of the features of chloroplast structure and autonomy that are consistent with this view are examined below.

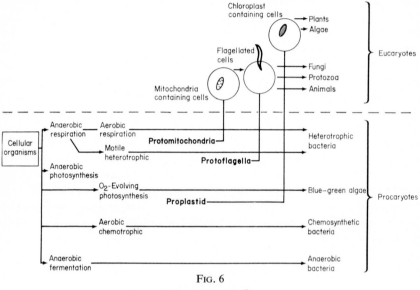

FIG. 6

(After Sagan, 1967.)

V. ASPECTS OF CHLOROPLAST BIOCHEMISTRY THAT SHOW SIMILARITIES TO THE BLUE-GREEN ALGAE

A. LIPID COMPOSITION

The similarity of structures that carry out O_2-evolving photosynthesis (Menke, 1966) extends to their chemical composition, in that the major groups of compounds are present in all such organisms; in addition, there exists an interesting variety of species differences which may reflect phylogenic origin (see Nichols, Chapter 6). Blue-green algae, in addition to possessing pigments characteristic of higher plants such as chlorophyll-*a* and β-carotene, and as such differing from other procaryotic organisms, have an acyl lipid composition normally associated with leaf chloroplasts, thus containing galactosyl diglycerides, sulphoquinovosyl diglyceride and phosphatidyl glycerol (Nichols *et al.* 1965; Allen *et al.* 1966; Nichols and James, 1968). Similarly, the naphthoquinone and benzoquinone composition of blue-green algae closely resembles those of chloroplasts, and is quite distinct from other photosynthetic procaryotic organisms (Carr *et al.* 1967).

B. NUCLEIC ACIDS AND PROTEIN SYNTHESIS

The indication of some degree of chloroplast autonomy arose from the presence of plastid clones which divide without mitosis and maintain their identity when the cell divides. The discovery of DNA, RNA and ribosomes in chloroplasts provides the means for the assembly of proteins separate from that directed by the nucleus, and earlier workers had observed the incorporation of ^{14}C-amino acids into chloroplast protein (Sissakian and Flippovitch, 1957). The incorporation of labelled amino acids into protein of chloroplasts which had been rigorously separated from cytoplasmic material was shown by Eisenstadt and Brawerman (1964), who also succeeded in effecting *in vitro* protein synthesis with ribosomes prepared from chloroplasts. The autoradiographic demonstration of *in vivo* RNA synthesis at the site of chloroplast-DNA indicated that transcription as well as translation occurred in the chloroplast organelle (Gibbs, 1967).

C. DNA

The presence of DNA in chloroplasts is now well established and properties of organelle DNA, including that of mitochondria and centrioles, have been reviewed (Granick and Gibor, 1967); the base ratios of chloroplast nucleic acids are summarized by Brawerman (1966) and their role in protein synthesis by Eisenstadt (1967).

Because of the variation in G + C ratio of chloroplast-DNA, from 25% in the *Euglena* chloroplasts to 60% from those obtained from spinach, no useful comparison can be made with blue-green algal DNA at this level. What is clear is the distinct difference in base ratios of cytoplasmic and chloroplast-DNA from several different algae and green plants (Brawerman, 1966). The total amount of DNA present in the blue-green algae *Anacystis nidulans* and *Anabaena variabilis* have been determined and is of the same order as that present in chloroplasts of three plant species, whilst chloroplasts of *Acetabularia* contain significantly less (Table II). Very little is known about the physical form of chloroplast-DNA; Gibbs (1967) observed a ring-shaped area in *Ochromonas* chloroplasts after ^{3}H-thymidine incorporation, but no circular units of DNA in *Acetabularia* chloroplasts were reported by Werz and Kellner (1968).

D. RIBOSOMES AND RNA

Abundant evidence exists that there are two classes of ribosomes in plant cells. The major ribonuclear component is a particle of 80 s size comparable to the ribosomes of other eucaryotic cells; in addition there are present in chloroplasts 70 s ribosomes which are the size of all procaryotic ribosomes (Taylor and Storck, 1964). Lyttleton (1960, 1962) showed clearly the presence of 80 s and 70 s types in clover and spinach and the derivation of the 70 s species from chloroplasts. An approximately 60 s ribosome was found in *Euglena gracilis* chloroplasts (Eisenstadt and Brawerman, 1964) and a 62 s species in pea chloroplast (Sissakian *et al.* 1965). It is presumed that the smaller units are breakdown products of a 70 s type. Sagar and Hamilton (1967) have, in fact, observed that chloroplasts from *Chlamydomonas* contained 70 s ribosomes that were very unstable and broke down to yield a 50 s unit. Blue-green algae contain ribosomes of the 70 s type (Taylor and Storck, 1964; Craig and Carr, 1968). The latter authors have shown that there is an instability in *Anabaena variabilis* ribosomes, not present in *E. coli* ribosomes, tending to the production of a 60 s unit. Chloroplast ribosomes also differ from those of the cytoplasm

Table II

DNA content of blue-green algae and isolated chloroplasts

	DNA/cell or organelle
Anabaena variabilis	$3\cdot6 \times 10^{-14}$ g[a]
Anacystis nidulans	$3\cdot0 \times 10^{-14}$ g[a]
Chloroplast from broad bean	$1\cdot5 \times 10^{-14}$ g[b]
Chloroplast from tobacco	$0\cdot80 \times 10^{-14}$ g[b]
Chloroplast from *Euglena*	$1\cdot1 \times 10^{-14}$ g[b]
Chloroplast from *Acetabularia*	$0\cdot01 - 0\cdot10 \times 10^{-14}$ g[b]

[a] From Craig, *et al.* (1969).
[b] After Granick and Gibor (1967).

by the ease with which they dissociate in low Mg^{++} concentration (Boardman, *et al.* 1965), and in this they conform to procaryotic ribosome pattern. A further degree of similarity between chloroplast and procaryotic ribosomes are the presence of cleavage furrows, detected by electron microscopy, which are not found in eucaryotic cytoplasmic ribosomes (Bayley, 1964; Miller *et al.* 1966; Bruskov and Odintsova, 1968).

The analysis of component species of RNA has been greatly advanced by the refinement of electrophoretic techniques, and earlier reports of failure to detect differences between cytoplasmic and chloroplast RNA's have not been confirmed (see Loening, 1968). Loening and Ingle (1967), using several plant species, showed that four ribosomal-RNA species could be found which were described as 16 s, 18 s, 23 s and 25 s. The 16 s and 23 s species were markedly enriched in preparations of chloroplasts and these were considered to be of chloroplast origin. Both *E. coli* and a species of the blue-green alga, *Oscillatoria*, contain 16 s and 23 s RNA and these samples were electrophoretically identical with chloroplast bean RNA (Loening and Ingle, 1967). The similarity of RNA produced by chloroplasts of bacteria-free, enucleated *Acetabularia mediterranea* to *E. coli* ribosomal-RNA has been shown by sucrose density gradient analysis (Schweiger *et al.* 1967). Chloroplast ribosomal-RNA from

Chlorella protothecoides has been shown to be of bacterial type by MAK column chromatography (Oshio and Hase, 1968).

The analytical data suggesting that chloroplast ribosomal-RNA is of a different type to cytoplasmic ribosomal-RNA are consistent with the evidence that chloroplast-DNA directs the synthesis of chloroplast-RNA. It was shown that *Euglena gracilis* chloroplast-DNA annealed with high efficiency to chloroplast ribosomal-RNA and that this process of hybridization was specific for ribosomal-RNA from chloroplasts (Scott and Smillie, 1967).

E. EFFECT OF INHIBITORS

Chloramphenicol is a well known inhibitor of protein synthesis and is effective in preventing chloroplast protein synthesis (Spencer, 1965) as well as on the development of the chloroplast itself (Smillie *et al.* 1963). The synthesis of enzymes of the Calvin cycle and of photosynthetic electron transport components *in vivo* are inhibited by chloramphenicol at concentrations that permitted cell division (Smillie *et al.* 1967). Similarly Aaronson *et al.* (1967) showed that chloramphenicol, and ethionine, inhibited the regreening process of bleached *E. gracilis* while cell division proceeded.

[14]C-Labelled chloramphenicol has been shown by Anderson and Smillie (1966) to bind to the ribosomes of chloroplasts and ribosomes from the blue-green alga, *Oscillatoria*, but not to eucaryotic cytoplasmic ribosomes. Of the four stereoisomers of chloramphenicol, the D-*threo* is the naturally occurring antibiotic and the inhibition of bacterial protein synthesis is specific for this isomer (Rendi and Ochoa, 1962); Ellis (1969) has shown that chloroplast ribosomes exhibit the same specificity for D-*threo* isomer. Since other aspects of plant metabolism are inhibited by several isomers of chloramphenicol (Ellis, 1969) the specificity for the D-*threo* form underlines the similarity of bacterial and chloroplast ribosomes. Some evidence exists which shows that cycloheximide is preferentially active against cytoplasmic, but not chloroplast, protein synthesis (Smillie *et al.* 1967; Ellis, 1969).

F. ENZYMES

As yet there is little information about the comparative enzymology of chloroplasts and procaryotic organisms. It is, of course, true that the reductive fixation of carbon dioxide by ribulose-1,6-diphosphate carboxylase is restricted to chloroplasts of plants and to the procaryotic chemosynthetic and photosynthetic bacteria and blue-green algae. Recent work with glyceraldehyde-3-phosphate dehydrogenase has indicated distinct similarities between the chloroplast enzyme and that from *Anabaena variabilis* in that both NAD and NADP limited activities are carried on one protein (Ziegler *et al.* 1968; Hood and Carr, 1969). The exclusive localization of NADP dependent L-glutamate dehydrogenase in chloroplasts of *Vicia faba* (Leech and Kirk, 1968) is perhaps noteworthy in that *Anabaena variabilis* contains only the NADP, not the NAD, linked enzyme (Pearce *et al.* 1969).

VI. Experimental Approaches

The relationship, if any, between chloroplasts and blue-green algae is open to examination by various approaches and techniques. For example, further information about the operative function of DNA and the sequential arrangement of genetic information within both organelles and blue-green algae would be most valuable. The application of two biochemical procedures will be briefly examined, firstly an immunochemical examination of a group of related proteins and secondly the hybridization of single stranded nucleic acids from different species.

An important start in the study of phylogenic relationships of blue-green algae by immunological studies has been made by Berns (1967). Using purified and crude preparations of biliproteins from Cyanophyta, Rhodophyta and Cryptophyta, this author examined their antigenic and immunological inter-relationships. The *C*-phycocyanin from all algal sources is immunologically and antigenically closely related and gives some evidence of being related to allophycocyanin whilst the phycoerythrins were apparently antigenically unrelated to the other biliproteins. However, phycoerythrins from all algal sources were quite closely inter-related with each other. The fact that biliprotein species from widely differing cell types, spanning the discontinuity between procaryotic and eucaryotic organisms, are immunologically closely related was noted by Berns (1967) and is in contrast to the situation found with other protein antigens. The examination of antigenic relationships between other proteins of blue-green algae and eucaryotic algal chloroplasts will be a great interest. The discovery of phytochrome in higher plants and its identification as a biliprotein extends this range of protein pigments right through the O_2-evolving photosynthetic species; the tentative establishment of some antigenic reaction by phytochrome to antisera to several phycocyanins is of particular relevance to the subject of this review (Berns, 1967).

The technique of nucleic acid hybridization, in which separated single strands of DNA are permitted to recombine with similar DNA from another organism, offers to date perhaps the most precise means of measuring the relatedness of two organisms. In principle the procedure measures the genetic homology between the two organisms; clearly the extent to which two different species share common base-sequences in their DNA may be taken as the most meaningful measure of relatedness (McCarthy and Bolton, 1962). When the DNA from each species is labelled with a different radioisotope the hybridization process may be conveniently followed. Some of the possible sources of error in this procedure, which include random adsorption of the second DNA to the first, and the renaturation of the challenge DNA, may be obviated by DNA:RNA hybridization (Hall and Spiegelman, 1961; Nygaard and Hall,

1963; Gillespie and Spiegelman, 1965). This technique in which messenger-RNA, ribosomal-RNA or transfer-RNA may be annealed to complementary DNA has several advantages, the absence of renaturation of the challenge RNA being one of the most important.

A difficulty in the application of nucleic hybridization procedures is the non-uniformity of chloroplast-DNA from different species; likewise the wide range of blue-green algal DNA base ratios makes the choice of experimental material difficult (Edelman *et al.* 1967). Preliminary results which indicate that DNA from *Nostoc muscorum* shows a certain degree of hybridization with DNA from *Euglena gracilis* chloroplasts have been reported (Craig *et al.* 1969). However, considerable specific binding of *E. gracilis* chloroplast-DNA to cytoplasmic-DNA has been reported (Richards, 1967). Tewari and Wildman (1968) showed that whilst chloroplast-DNA hybridized with chloroplast ribosomal-RNA and not with cytoplasmic ribosomal-RNA, the cytoplasmic-DNA hybridized with both cytoplasmic ribosomal-RNA and, to a lesser degree, with chloroplast ribosomal-RNA. This indicates a role for the nucleus in chloroplast development as indeed does the presence of nuclear mutations affecting the biosynthesis of chloroplast material (Kirk, 1966).

VII. Synopsis

In the last decade increasing evidence has indicated the functional autonomy of eucaryotic plastid organelles, but although this may be demonstrable *in vivo* by cytoplasmic inheritance the extracellular survival of such organelles has not been achieved. What has been possible is the clarification of the position of intracellular endosymbionts from true cell organelles; the degree of inter-dependence between these and the host-cell, and the manner in which endosymbionts have diverged from free-living forms, presents a challenging prospect for study at the biochemical level. The possession by eucaryotic organelles of the means for genetic informational transfer, its transcription and translation has been shown, in the case of chloroplasts, in a convincing manner. Gibor (1967) has discussed the inheritance of eucaryotic organelles and their possible evolutionary origin.

In the last year or so two articles have stated essentially opposing views on the origin of lower plants. Klein and Cronquist (1967) have described in great detail the phylogenic relations of numerous metabolic and structural attributes of plants, which together with morphological data, cause these authors to suggest a linear development of lower plants by a process of evolutionary change from photosynthetic procaryotes. This is also the basis of the generally accepted view of plant phylogeny. The opposite approach is to accept an endosymbiotic origin of eucaryotic organelles on the lines discussed in Section IV. Sagan (1967) has presented considerable evidence to support such an explanation and has produced a critique (Margulis, 1968) of Klein and Cronquist's (1967) view. The essential basis of the Sagan approach is the weight attached to the procaryotic–eucaryotic discontinuity. A third solution is offered by Haldar *et al.* (1966) who suggest that the bacterial mesosome, a specialized region of the cytoplasmic lamellae, could serve as a model for the evolutionary origin of cytoplasmic organelles. This is discussed by Gibor (1967) in the light of evidence of synthesis, under certain conditions, of only a portion of the DNA of a cell. Such an explanation, however, would not explain the failure to detect any intermediate form between procaryotic and eucaryotic cell types; indeed, such intermediates would be expected by the Haldar *et al.* (1966) explanation of organelle development.

An important criticism of an endosymbiotic origin of eucaryotic organelles is the variety of base-ratios their DNA contains; this would suggest that chloroplasts, for example, of different species had arisen from different symbionts and had evolved to an identical pattern (Gibor, 1967). It may be that not all chloroplast-DNA is functional, and any non-functional material could acquire mutagenic alteration without affecting the viability of the organelle. Kirk (1966) has calculated that values for DNA in chloroplasts would be sufficient to code for 160 at lowest, to approximately 16,000 at highest, different proteins. In the higher portions of this range clearly some non-functional DNA sequences could be present.

In this short article we have presented some of the biochemical evidence that supports the basic similarity of bacteria and blue-green algae and that is in accord with the endosymbiotic origin of chloroplasts from some procaryotic cell not unlike a blue-green alga. For further details, and alternative views, readers should consult authorities already cited. Klein and Cronquist (1967) dismiss the theory of the endosymbiotic origin of eucaryotic organelles as lacking "chemical, structural or phylogenetic basis for this belief." Whilst disagreeing with this statement and suggesting that the theory is supported by some evidence and is open to experimental procedures, we also take note of Lederberg's (1952) caution against "mistaking probabilities for certainties."

REFERENCES

Aaronson, S., Ellenbogen, B. B., Yellon, L. K., and Hutner, S. H. (1967). *Biochem. Biophys. Res. Commun.* **27**, 535.

Allen, C. F., Hirayama, O., and Good, P. (1966). *In* "Biochemistry of Chloroplasts" (T. W. Goodwin, ed.), Vol. 1, p. 193. Academic, London and New York.

Allen, M. M. (1968a). *J. Bact.* **96**, 836.

Allen, M. M. (1968b). *J. Bact.* **96**, 842.

Anderson, L. A., and Smillie, R. M. (1966). *Biochem. Biophys. Res. Commun.* **23**, 535.

Bayley, S. T. (1964). *J. Molec. Biol.* **8**, 231.

Bazin, M. J. (1968). *Nature*, **218**, 282.

Berns, D. S. (1967). *Pl. Physiol.* **42**, 1569.

Boardman, N. K., Francki, R. I. B., and Wildman, S. G. (1965). *Biochemistry*, **4**, 872.

Bowen, C. C., and Jensen, T. (1965). *Science*, **147**, 1460.

Brawerman, G. (1966). *In* "Biochemistry of Chloroplasts" (T. W. Goodwin, ed.), Vol. 1, p. 302. Academic, London and New York.

Bruskov, V. I., and Odintsova, M. S. (1968). *J. Molec. Biol.* **32**, 471.

Carr, N. G. (1966). *Biochim. biophys. Acta*, **120**, 308.

Carr, N. G., Exell, G., Flynn, V., Hallaway, M., and Talukdar, S. (1967). *Arch. Biochem. Biophys.* **120**, 503.

Cassel, W. A., and Hutchinson, W. G. (1954). *Exp. Cell Res.* **6**, 134.

Chapman, D. J. (1966). *Arch. Mikrobiol.* **55**, 17.

Cohen-Bazire, G. (1963). *In* "Bacterial Photosynthesis" (H. Gest, A. San Pietro and L. P. Vernon, eds.), p. 89. Antioch, Yellow Springs.

Cohen-Bazire, G., and Sistrom, R. W. (1966). *In* "The Chlorophylls" (L. P. Vernon and G. R. Seely, eds.), p. 313. Academic, New York and London.

Craig, I. W., and Carr, N. G. (1968). *Arch. Mikrobiol.* **62**, 167.

Craig, I. W., Leach, C. K., and Carr, N. G. (1969). *Arch. Mikrobiol.* **65**, 218.

Crespi, H. J., Mandeville, S. E., and Katz, J. J. (1962). *Biochem. Biophys. Res. Commun.* **9**, 569.

Drawert, H., and Metzner, I. (1956), *Ber. Bot. Ges.* **69**, 291.

Drews, G., and Meyer, H. (1964). *Arch. Mikrobiol.* **48**, 259.

Echlin, P. (1967). *Br. Phycol. Bull.* **3**, 225.

Echlin, P., and Morris, I. (1965). *Biol. Rev.* **40**, 143.

Edelman, M., Swinton, D., Schiff, J. A. Ehstein, M. T., and Zeldin, B. (1967). *Bact. Rev.* **31**, 315.

Ellis, R. J. (1969). *Science*, **163**, 477.

Eisenstadt, J. N. (1967). *In* "Biochemistry of Chloroplasts" (T. W. Goodwin, ed.), Vol. 2, p. 341. Academic, London and New York.

Eisenstadt, J. N., and Brawerman, G. (1964). *J. Molec. Biol.* **10**, 392.

Fogg, G. E. (1951). *Ann. Bot.* **15**, 23.

Frank, H., Lefort, M., and Martin, H. H. (1962). *Biochem. Biophys. Res. Commun.* **7**, 322,

Fuhs, G. W. (1958). *Arch. Mikrobiol.* **29**, 51.

Fuhs, G. W. (1969). "The Nuclear Structures of Procaryotic Organisms (Bacteria and Cyanophyceae)." Springer Verlag, Vienna.

Gibbs, S. P. (1967). *Biochem. Biophys. Res. Commun.* **28**, 653.

Gibor, A. (1967). *In* "Formation and Fate of Cell Organelles" (K. B. Warren, ed.), p. 305. Academic, New York and London.

Gillespie, D., and Spiegelman, S. (1965). *J. Molec. Biol.* **12**, 829.

Glauert, A. M. (1962). *Br. Med. Bull.* **18**, 245.

Goad, L. J. (1967). *In* "Terpenoids in Plants" (J. B. Pridham, ed.), p. 159. Academic, London and New York.

Granick, S., and Gibor, A. (1967). *Prog. Nucleic Acid Res. Mol. Biol.* **6**, 143.

Haldar, D., Freeman, K., and Work, T. S. (1966). *Nature*, **211**, 9.

Hall, B. D., and Spiegelman, S. (1961). *Proc. natn. Acad. Sci.* **47**, 137.

Hall, W. T., and Claus, G. (1963). *J. Cell. Biol.* **19**, 551.

Hayes, W. (1968). "The Genetics of Bacteria and their Viruses," 2nd edition. Blackwell, Oxford.

Hoare, D. S., and Work, E. (1957). *Biochem. J.* **65**, 441.

Holm-Hansen, O. (1968). *A. Rev. Microbiol.* **22**, 47.

Holm-Hansen, O., Prasad, R., and Lewin, R. A. (1965). *Phycologia*, **5**, 1.

Hood, W., and Carr, N. G. (1969). *Planta*, **86**, 250.

Kirk, J. T. O. (1966). *In* "Biochemistry of Chloroplasts" (T. W. Goodwin, ed.), Vol. 1, p. 319. Academic, London and New York.

Klein, R. M., and Cronquist, A. (1967). *Quart. Rev. Biol.* **42**, 105.

Lang, N. (1968). *A. Rev. Microbiol.* **22**, 15.

Larsen, H., Omang, S., and Steensland, H. (1967). *Arch. Mikrobiol.* **59**, 197.

Leak, L. V. (1965). *J. Ultrastruc. Res.* **12**, 135.

Lederberg, J. (1952). *Physiol. Rev.* **32**, 403.

Lederberg, J., and Tatum, E. L. (1946). *Cold Spring Harb. Symp. quant. Biol.* **11**, 113.

Leech, R. M., and Kirk, P. R. (1968). *Biochem. Biophys. Res. Commun.* **32**, 685.

Loening, U. E. (1968). *A. Rev. Pl. Physiol.* **19**, 37.

Loening, U. E., and Ingle, J. (1967). *Nature*, **215**, 363.

Lyttleton, J. W. (1960). *Biochem. J.* **74**, 82.

Lyttleton, J. W. (1962). *Expl. Cell Res.* **26**, 312.

Margulis, L. (1968). *Science*, **161**, 1020.

McCarthy, B. J., and Bolton, E. T. (1962). *Carnegie Institute Washington Yearbook*, **61**, 244.

Menke, W. (1966). *In* "Biochemistry of Chloroplasts" (T. W. Goodwin, ed.), Vol. 1, p. 3. Academic, London and New York.

Mereschkowsky, C. (1905). *Biol. Zbl.* **25**, 593.

Miller, A., Karlsson, U., and Boardman, N. K. (1966). *J. Molec. Biol.* **17**, 487.

Nichols, B. W., Harris, R. V., and James, A. T. (1965). *Biochem. Biophys. Res. Commun.* **20**, 256.

Nichols, B. W., and James, A. T. (1968). *In* "Progress in Phytochemistry" (L. Reinhold and Y. Liwschitz, eds.), Vol. 1, p. 1. Interscience, London.

Nygaard, A. P., and Hall, B. D. (1963). *Biochem. Biophys. Res. Commun.* **12**, 98.

Ogura, M. (1963). *J. Ultrastruct. Res.* **8**, 251.

Orenski, S. W. (1966). *In* "Symbiosis" (S. M. Henry, ed.), Vol. 1, p. 1. Academic, New York and London.

Oshio, Y., and Hase, E. (1968). *Pl. Cell Physiol.* **9**, 69.

Pankratz, H. S., and Bowen, C. C. (1963). *Am. J. Bot.* **50**, 387.

Pearce, J., Leach, C. K., and Carr, N. G. (1969). *J. gen. Microbiol.* **55**, 371.

Pfennig, N., and Cohen-Bazire, G. (1967). *Arch. Mikrobiol.* **59**, 226.

Petermann, M. L. (1964). "The Physical and Chemical Properties of Ribosomes." Elsevier, Amsterdam.

Reitz, R. C., and Hamilton, J. G. (1968). *Comp. Biochem. Physiol.* **25**, 401.

Rendi, R., and Ochoa, S. J. (1962). *J. Biol. Chem.* **237**, 3711.

Richards, O. C. (1967). *Proc. natn. Acad. Sci.* **57**, 156.

Ris, H., and Plaut, W. (1962). *J. Cell Biol.* **13**, 383.

Ris, H., and Singh, R. N. (1961). *J. Biophys. Biochem. Cytol.* **9**, 63.

Roodyn, D. G., and Wilkie, D. (1968). "The Biogenesis of Mitochondria." Methuen, London.

Sagan, L. (1967). *J. Theoret. Biol.* **14**, 225.

Sagar, R., and Hamilton, M. G. (1967). *Science*, **157**, 709.

Salton, M. R. J. (1964). "Microbial Cell Walls." Wiley, New York.

Schubert, K., Rose, G., and Hörhold, C. (1967). *Biochim. biophys. Acta*, **137**, 168.

Schubert, K., Rose, G., Wachtel, H., Hörhold, C., and Ikekawa, N. (1968). *Eur. J. Biochem.* **5**, 246.

Schweiger, H. G., Dillard, W. L., Gibor, A., and Berger, S. (1967). *Protoplasma*, **64**, 1.

Scott, N. S., and Smillie, R. M. (1967). *Biochem. Biophys. Res. Commun.* **28**, 598.

Sissakian, N., and Flippovitch, I. (1957). *Biokhimiya*, **22**, 375.

Sissakian, N. M., Flippovitch, I. I., Svetailo, E. N., and Aliyev, K. A. (1965). *Biochim. biophys. Acta*, **95**, 474.

Sistrom, R. W. (1962). *J. Gen. Microbiol.* **28**, 599.

Smillie, R. M., Evans, W. R., and Lyman, H. (1963). *Brookhaven Symp. Biol.* **16**, 89.

Smillie, R. M., Graham, D., Dwyer, M. R., Grieve, A., and Tobin, N. F. (1967). *Biochem. Biophys. Res. Commun.* **28**, 604.

Smith, R. V., and Peat, A. (1967). *Arch. Mikrobiol.* **57**, 111.

Spencer, D. (1965). *Arch. Biochem. Biophys.* **111**, 381.

Stanier, R. Y. (1964). *In* "The Bacteria" (I. Gunsalus and R. Y. Stanier, eds.), Vol. 5, p. 445. Academic, New York and London.

Stanier, R. Y., Doudoroff, M., Kunisawa, R., and Contopoulou, R. (1959). *Proc. natn. Acad. Sci.* **45**, 1246.

Stanier, R. Y., and van Niel, C. B. (1941). *Bact. Rev.* **42**, 437.

Stanier, R. Y., and van Niel, C. B. (1962). *Arch. Mikrobiol.* **42**, 17.

Stevenson, I. (1967). *Nature*, **215**, 434.

Stille, B. (1937). *Arch. Mikrobiol.* **8**, 124.

de Souza, N. J., and Nes, W. R. (1968). *Science*, **162**, 363.

Suzue, G., Tsukada, K., Nakai, C., and Tanaka, S. (1968). *Arch. Biochem. Biophys.* **123**, 644.

Taylor, M. M., and Storck, R. (1964). *Proc. natn. Acad. Sci.* **52**, 958.

Tewari, K. K., and Wildman, S. G. (1968). *Proc. natn. Acad. Sci.* **59**, 569.

Torbabene, T. G., Kates, M., Gelpi, E., and Oro, J. (1969). *J. Lipid Res.* **10**, 294.

Van Bruggan, E. F. J., Runner, C. M., Borst, P., Ruttenberg, C. J. C. M., Kroon, A. M., and Schuurmans Stekhoven, F. M. A. H. (1968). *Biochim. biophys. Acta*, **161**, 402.

Van Wagtendonk, W. J., Clark, J. A. D., and Godoy, G. A. (1963). *Proc. natn. Acad. Sci.* **50**, 835.

Vogel, H. J. (1959). *Proc. natn. Acad. Sci.* **45**, 1717.

Walsby, A. E. (1969). *Proc. R. Soc. Ser. B.* **173**, 325.

Werz, G., and Kellner, G. (1968). *J. Ultrastruct. Res.* **24**, 109.

Work, E., and Dewey, D. L. (1953). *J. gen. Microbiol.* **9**, 394.

Ziegler, H., Ziegler, I., and Schmidt-Clausen, H. J. (1968). *Planta*, **81**, 181.

CHAPTER 8

Proteins and Plant Phylogeny

R. L. WATTS

Department of Biochemistry,
Guy's Hospital Medical School, London, England

I. INTRODUCTION

To study proteins within an evolutionary context is doubly beneficial. It is of use to the phylogeneticist because proteins offer a unique combination of evolutionary stability and variability in their chemical structure. It is equally of use to the protein biochemist, since a knowledge of permissible natural variation makes it possible to define which parts of the molecule are essential for function and may give a pointer to the chemical means by which function is carried out. During the rapid increase in interest in the evolution of proteins, over the last few years, there have been some fierce clashes of opinion between organismal evolutionists and protein biochemists concerning both usefulness and basic principles, largely because neither entirely understood the speciality of the other. Zuckerkandl and Pauling (1965) have described some of these arguments and have made efforts to bridge the chasm. More is needed, however, than that two groups, poles apart in training and experimental methods, should respect (at least in part!) each other's conclusions. The evolution of organisms has to be seen in terms of the evolution of the proteins in which their individua-

6

lity is based, and the evolution of proteins has to be seen not only as an exercise in chemical versatility but in relation to the evolution of the organisms dependent on them. The evolution of proteins has its own "rules," and some account of these, so far as we at present understand them, will form the basis of this paper. It is a young discipline within biology. Its history has been analogous to what we have inferred for many plant species; it arose as an uneasy hybrid between two distinct disciplines; it has achieved stability by a "polyploidization" (far more than a doubling of information!) and is now in the phase of mutual accommodation of the parental genomes.

It is probably true to claim that chemical investigations have been used considerably more in the study of plant phylogeny than that of animal groups. Nevertheless, this has not been the case for plant as opposed to animal proteins. In their atlas of amino acid sequences, Dayhoff and Eck (1968) illustrate the distribution of published sequences among groups of living organisms. It is plain that the flowering plants (three known sequences) have received negligible attention as compared to higher animals (166 sequences) including 146 sequences from mammals alone. The bryophytes and algae, in common with most invertebrate groups, have not been studied at all. Microorganisms and insects contributed 19 sequences between them.

The reasons for this are not lack of chemical awareness among plant scientists nor even, as is sometimes suggested, that mammals possess greater intrinsic interest. It is technically very much more difficult to prepare from plant materials the purified proteins necessary for sequence determination, largely because they represent a very small percentage of the total material, which is itself immensely more difficult to obtain in large quantities than those popular objects of biochemical investigation, the organs of meat animals or of the laboratory rat. Moreover, although plants display a generally greater chemical versatility than animals, it is in animals that proteins have come to play a greater variety of physiological roles, notably, and conveniently, in the blood. The haemoglobins of higher animals which are readily obtained and purified and are stable, as many proteins are not, have provided a wealth of amino acid sequence data. From this, we have gained much knowledge about how proteins may vary both between and within species and have been able to make a number of generalizations upon the genetic mechanisms involved in protein evolution. Investigation of human haemoglobin variants, using material available as a result of large scale medical screening, has yielded some of the most valuable information.

In describing the principles of an approach to protein evolution, it will be necessary to draw heavily on evidence from the animal kingdom. It will then be possible to discuss some of the much sparser data upon plant proteins and to consider the particular problems which arise in the study of a group of organisms with a much greater diversity and complexity of genetic systems than the animals and which require a different approach to the concept of species barriers.

II. The Genotype and the Phenotype

A. A Definition of Life

To place the matter in perspective, it may be helpful to begin with a definition of life suitable for use when the process of organic evolution is of prime concern. This will also serve as a basis for the contention of this paper that proteins should be a prime concern of evolutionists. I should like to define life as *a part of the natural universe which has both a genotype and a phenotype, and the means to convert the one into the other.*

Before going on to discuss the implications of this definition, I will offer supplementary definitions of the two words used in it which are somewhat specialized genetic terms: *genotype* can be taken to denote the total genetic potential of an organism, any particular character in which may or may not be expressed in one generation, but must be transmissible to the next; *phenotype* to denote the totality of characters "expressed" in an organism, where "expressed" means capable of observation by the investigator. Definition of any particular phenotype may be complicated by the presence of characters dependent upon the occurrence of specific external conditions.

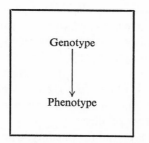

Fig. 1. Definition of a living organism

Being a definition of life, Fig. 1 is not, of course, anything like as simple as it looks. Its four individual components each call for comment.

1. The box containing the definition denotes that the organism is functionally set apart from its environment, including other organisms. It does not necessarily represent any physical barrier, such as a membrane, surrounding the organism but is included to stress the connotation of individuality in the concepts of genotype and phenotype.

2. The genotype must have a physical basis. It should be possible to equate this with the genetic material.

3. The phenotype must have a physical basis. It may be possible to equate this with all the substance of the organism that is not the genetic material.

4. The presence of the arrow excludes entities such as viruses from the definition. They have both genotype and phenotype but have to make use of the processes of actual living organisms to produce the one from the other. This

serves to underline the point that the genotype and the phenotype have no meaning without an arrow, that life is really defined by their relationship, the process the arrow stands for.

Most biochemists interested in evolution appear to hold the view that not only life but the origin of life are very difficult to define. It seems important to contest this position, since evolution by natural selection can only proceed given the existence of the relationship in Fig. 1. Hence, the origin of life can be equated with the origin of the relationship and of the process by which it was achieved. The process, as it is understood in present-day organisms of many kinds, is discussed in section II.B. What is not known is how much of the present-day process can be equated with that which first brought the genotype–phenotype relationship into being.

As a generalization, the definition of life holds true for every organism along a phylogenetic pathway. The individuality of a particular organism resides in both its genotype and its phenotype. Interrelationships between organisms can also be described in these terms. When the concepts of the genotype and the phenotype were adopted into genetics, their connotation was apparently simple, for the recessive alleles (*a*) of major genes had phenotypic expression only in absence of the corresponding dominant alleles (*A*), the latter being always expressed. That is, *a*//*a* was distinguishable from the indistinguishable *A*//*A* and *A*//*a*. It is now clear that there is no such thing as a categorically recessive allele. This follows from the results of genetic experiments using biological assessment alone, as is discussed by Ford (1965). Genetics, understood at the molecular level, is in a position to consider the mechanisms whereby dominance effects are produced (Watts, 1967) and it is equally plain from such theoretical considerations that the useful concepts of dominance and recessivity do not represent absolute properties of alleles.

Description of the genotype and the phenotype must be extended back to its macromolecular basis.

B. THE GENOTYPE AND THE PHENOTYPE IN MOLECULAR TERMS

Figure 2 attempts diagrammatically to show the components and processes involved. The total potential of the genotype only exists because the process exists for its expression in the phenotype. The flow of genetic information outwards from the DNA is shown in Fig. 2 taking place in three steps. In the first, by means of a DNA-dependent RNA polymerase, the information encoded in the deoxyribonucleotide sequence of the DNA is transferred into a complementary ribonucleotide sequence in messenger RNA (*m*RNA). The second transfer of information occurs between two complementary ribonucleotide sequences, between the codons of *m*RNA and the anticodons of transfer RNA (*t*RNA). The first two transfers of information are assumed to work because when nucleotides are lined up in the correct steric arrangement particular pairs are complementary. The mechanism of the third transfer of information, from a ribonucleotide sequence to an amino acid sequence is understood in less detail. The process occurs "off-stage" with respect to Fig. 2, when *t*RNA is recognized by the appropriate amino acid activating enzyme and thus becomes charged with the correct amino acid. The chemical basis of this specificity, unlike that of complementary nucleotide pairs, has yet to be elucidated.

A knowledge of these processes makes it theoretically possible to describe the genotype of an organism completely, given the means to analyse all the DNA of an organism and the means to interpret the analysis. As a practical

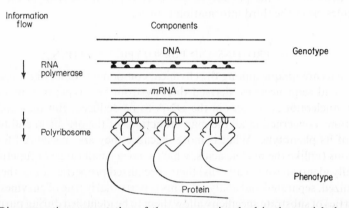

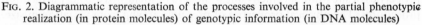

FIG. 2. Diagrammatic representation of the processes involved in the partial phenotypic realization (in protein molecules) of genotypic information (in DNA molecules)

possibility, it is remote, not only because of the enormous technical difficulties involved in separation and analysis of large polymers built of only four different units but because of the overwhelming quantity of information involved. Our present almost complete inability to isolate specific portions of DNA rules out the possibility of studying a small homologous section from different organisms.

There is also a theoretical reason why a complete catalogue of DNA sequences should not be the present primary goal of the evolutionist. This is that the process of natural selection has no direct dealings with genotypes but operates upon the phenotypes derived from them. However clear our understanding of the control mechanisms of the cell, which must ultimately also be encoded in DNA sequences, we shall never be able completely to define a phenotype from this internal information alone, since external conditions may influence what phenotype arises from a particular genotype. DNA sequence information would probably be the most unequivocal means of establishing lineages of organisms, but it would never be able to tell us just what the organisms were like.

The process of discrimination which results in the distinct phenotype of an organism begins with the first information transfer in Fig. 2. Not all DNA sequences are transcribed into *m*RNA. At particular times, particular sequences are transcribed, in accord with the control mechanisms in operation in the organism, actuated by the internal environment of each cell, more or less influenced by the external environment. So, through a multitude of developmental sequences, the total phenotype is built up. Knowledge of the types of mRNA molecules produced might tell us about the first stage of the discriminatory process. It seems probable, in higher organisms at least, that further discriminatory processes occur before the accomplishment of the second and third information transfers. The populations of protein molecules which form the basis of cellular function are not, then, a direct reflection of the populations of mRNA molecules in which their sequences are secondarily encoded. In this sense, definition of the phenotype can only begin at the level of the protein molecules, after the third information transfer.

C. PROTEINS AND PHENOTYPIC VARIATION

Proteins are unique among the chemical components of an organism. Their amino acid sequences are derived directly from genotypic information and permit nucleotide sequences in the DNA to be deduced. But they are, at the same time, concerned in all the many functions of the organism and form the basis of its phenotype. What is more, because they are specialized for many functions (unlike the nucleic acids, which serve a small range of functions in a basically identical way in all cells) their specialized properties allow them to be recognized, separated and analysed. This is particularly true of enzymes, whose properties of substrate specificity allow them to be identified during purification procedures. The physical properties of proteins, arising from variation in size, shape and charge of the molecule, are also immensely useful in analysis.

Studies of proteins can be used to compare distantly related organisms, for many proteins carry out functions so fundamental that they are found in all forms of life. Such proteins, for example, the electron carrier cytochrome c, not only confirm that all forms of life are related to each other but yield information permitting the quantitation of relationships among the major groups. Universal proteins of this kind may undergo only non-radical evolutionary change, because, at all stages, organisms depend for life upon their function. The general properties of such proteins are usually closely similar regardless of their source and comparisons have to be made between amino acid sequences.

In a different category, there are enzymes of specialized function, developed only within certain groups of organisms. These are of particular use in establishing relationships at the level of smaller groups such as genera. They may be as much essential to life as the proteins of fundamental metabolism so far as the organisms now possessing them are concerned, but may be evolutionarily related to proteins of other function in other groups. Some parts of metabolism may be non-essential even in the organisms in which they take place (this may occur more frequently in plants than in animals). The types of evolutionary change which such enzymes can undergo may be more varied (Anfinsen, 1959). With enzymes of specialized function, whether essential or not, it is often possible to make comparisons among organisms without recourse to the determinations of amino acid sequences. Many techniques are available for the characterization of enzymes by measuring the effects of changed conditions on activity. Changes in substrate or inhibitor concentration, of temperature, pH or ionic strength, or of chemical modification of the substrate or the protein itself may all be used to reveal relationships between enzymes from different sources. Such techniques have been successful when examining variation among the genera of a phylum (Watts and Watts, 1968a) and can be used right down to the level of intraspecific variation. It is often unnecessary to purify the enzymes completely before making the comparative experiments, which not only represents a great saving in time but allows comparisons to be made with material too scarce to permit extraction of enough protein for purification.

III. EVOLUTIONARY VARIATION IN PROTEIN STRUCTURE AND FUNCTION

A. GENERAL CONSIDERATIONS

Basic information which is left unstated in this section can be sought in a number of general textbooks of biochemistry of which Karlson (1965) and Mahler and Cordes (1968) are recommended. A very useful account of the chemical structure of proteins and its determination at various levels will be found in Edsall and Wyman (1958).

1. Basic Chemistry and Conventional Terminology

Proteins are polymers of L-amino acids linked by peptide bonds. They thus have a backbone of amino-N, α-C and carboxyl-C atoms and a series of side-chains which may be any of 20 chemical groupings (Fig. 3a, b). The primary structure of a protein is defined by determining the order of the R groups, which is specific and genetically controlled. This information, called the amino acid sequence, is represented conventionally by writing three-letter abbreviations for the amino acid residues from left to right, starting with the amino-terminal residue.

2. Levels of Complexity

(a) Theoretical complexity

Given the concept of the general chemical nature of proteins as polymers containing a number of reactive side-chains, even in absence of data as to their precise chemistry, it was possible to make certain predictions bearing on their natural specificity. Thus Linderstrøm-Lang (1937) proposed that the chemistry of proteins should be considered at three levels, on all of which their capacity for specific biological activity would depend. The primary, secondary and tertiary structure of proteins are enumerated in almost every biochemical textbook, with minor differences in their definition. The primary structure is sometimes taken to include all covalently bonded structures and, thus, disulphide bridges between cysteine residues in different parts of the chain. The presence of such bridges could be determined chemically, but this information alone did not make it possible to assign a tertiary structure. The definition of primary structure assumed in section II.A.1 above has excluded the presence or not of disulphide bridges.

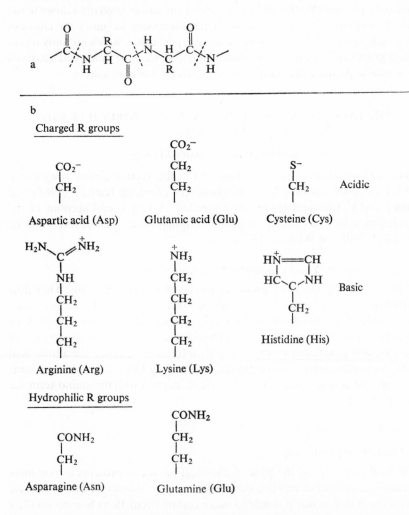

FIG. 3. (a) A section of the polypeptide chain of a protein. Sidechains are shown as —R. Dotted lines delimit the amino acid units (residues)

(b) The 19 —R groups and the prolyl residue. The cyclic amino acid, proline, cannot be represented by means of an —R group for insertion in Fig. 3a. Therefore, a portion of the peptide chain is shown. The α-N atom of a prolyl residue cannot form a hydrogen bond to adjacent ⟩C=O

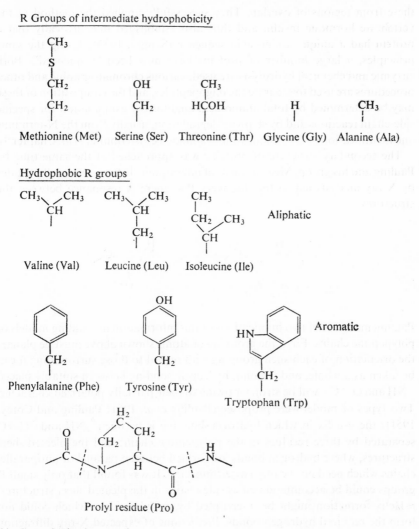

R Groups of intermediate hydrophobicity

Methionine (Met)   Serine (Ser)   Threonine (Thr)   Glycine (Gly)   Alanine (Ala)

Hydrophobic R groups

Valine (Val)   Leucine (Leu)   Isoleucine (Ile)   Aliphatic

Phenylalanine (Phe)   Tyrosine (Tyr)   Tryptophan (Trp)   Aromatic

Prolyl residue (Pro)

(b) Complexity in practice

It was not possible to approach structure determination for proteins until the molecules to be studied could be obtained in pure form, and, for X-ray analysis, crystallized. The attainment of this objective was dependent not only upon technical advances but upon overcoming the prevailing opinion that, like polysaccharides, proteins did not have a specific molecular size and structure.

In the post-war years, both ideas and techniques did advance and the early fifties saw the development of experimental methods for the study of protein chemistry at all its levels. Sanger and his group established the possibility of determining amino acid sequences by chemical characterization of short oligo-peptides produced by controlled hydrolysis and inference of the sequence of these from regions of overlap. They successfully applied the method to the vertebrate hormone insulin and thus also established unequivocally that a protein had a unique amino acid sequence (Sanger, 1955). Using the same principles, a large number of proteins have now been "sequenced". Both enzymic and chemical hydrolysis are used, various chromatographic and other procedures are used to separate the oligopeptides, and the composition of these may be determined by total amino acid analysis, by examination for specific side-chain reactions and by stepwise degradation, usually from the N-terminus and identification of the residues one at a time (see also Boulter *et al.* Chapter 9).

The secondary structure of proteins was approached at the same time by Pauling and his group. Measurement of interatomic distances in small peptides by X-ray analysis led to the discovery that there is resonance between the structures

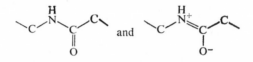

Pauling made use of two implications of this information in building models of polypeptide chains. First, the grouping of atoms shown above must be planar; the orientation of each such grouping with regard to those surrounding it can be taken as a whole, and, second, hydrogen bonding between suitably placed $>$NH and O$=$C$<$ will be stronger owing to their partially polarized character. Two types of model were proposed (Pauling *et al.* 1951; Pauling and Corey, 1951): the α-helix in which hydrogen-bonding is between $>$NH and $>$C$=$O separated by three residues in the polypeptide chain; and the pleated sheet structures, where hydrogen bonds are formed between parallel or antiparallel chains which need not be otherwise connected. It was found that only small R groups could be accommodated as side-chains in the pleated sheet structures. α-Helix formation might be interrupted by prolyl residues which could not form the required hydrogen bonds. Predictions of expected X-ray diffraction patterns based on these models were subsequently confirmed for the fibrous proteins α-keratin, β-keratin and silk fibroin respectively.

Secondary structure, which was thus investigated in fibrous proteins, can be defined as the relationship maintained between adjacent sections of poly-peptide chain by means of hydrogen bonding. Tertiary structure is mainly of interest in the case of globular proteins, and can be defined as the relationships in space of all parts of the molecule, with or without secondary structure, to each other. These are maintained by a variety of interatomic interactions: covalent bonding, ionic bonding, hydrogen bonding, van der Waals' forces.

X-ray diffraction analysis of crystalline globular proteins presented great problems since the entire three-dimensional structure contributed to the patterns obtained. When it proved possible to prepare isomorphous derivatives of haemoglobin and, later, myoglobin containing heavy metal atoms, the phases of the diffracted X-rays could be determined by comparison of these and Fourier transforms could be fitted to the diffraction patterns. The three-dimensional structure of whale myoglobin was first determined at a resolution of 6 Å, by calculation of a Fourier synthesis using the several hundred reflections in the diffraction patterns corresponding to spacings not less than this. After this, electron density maps showing the structure of the molecule to a resolution of 2 Å were obtained by involving in the calculations, performed by a computer, approximately 10,000 reflections (Kendrew *et al.* 1960). At the same time horse haemoglobin, less easily analysed because of its tetrameric nature, yielded a three-dimensional structure at 5·5 Å resolution (Perutz *et al.* 1960). Neither of these analyses was greatly aided by primary structure data, since these were still being determined. The previous work on secondary structure became of use when a large part of the myoglobin polypeptide was found to be in the form of α-helix. At 2 Å resolution, many of the amino acid side-chains could be identified. Sequence data were subsequently found to be in good agreement with the structure.

The same methods of analysis have since been applied to a number of proteins whose primary structures were usually already known at least in part. Lysozyme, the first enzyme whose complete structure was known (Blake *et al.* 1967) has been followed by chymotrypsin, papain, ribonuclease, carboxypeptidase and the electron carrier, cytochrome *c*. Since the last-named will be of some importance in the next section and in Chapter 9, the amino acid sequences for this protein in wheat and man are shown in Fig. 4a, illustrating the large number of differences which may be found between homologous proteins. The three-dimensional picture of this protein was obtained using material from horse heart, so far only at 4 Å resolution (Dickerson *et al.* 1967). Fig. 4b is a drawing of the molecule based on their electron density contour model. The haem group, seen edgeways, is sited in a cleft mainly surrounded by inward-directed non-polar side-chains which form the core of the molecule. The rest of the polypeptide is wound in complex fashion round this, forming a discontinuous shell not more than one amino acid thick. Major "apertures" in the shell give on to the side of the prosthetic group.

3. Present Day View of Protein Structure

Although it has been useful to consider the structure of proteins at different levels, it is clear that an integration of all types of information is necessary in order to form anything resembling a full picture. At this point in time, some generalizations can be made.

(i) It is not possible to predict three-dimensional structures from sequence data alone although features of secondary structure, such as the absence of α-helix from the cytochrome *c* molecule may be predicted (Smith, 1968). For this reason attempts to do so by computer simulation of folding processes fall

```
                                                                          10
Wheat  Acetyl-Ala-Ser-Phe-Ser-Glu-Ala-Pro-Pro-Gly-Asn-Pro-Asp-Ala-Gly-Ala-Lys-Ile-Phe-
Man                            Acetyl -      Asp-Val-Glu-Lys       Lys
                                                                          30
       Lys-Thr-Lys-Cys-Ala-Gln-Cys-His-Thr-Val-Asp-Ala-Gly-Ala-Gly-His-Lys-Gln-Gly-Pro-
       Ile -Met        Ser                          Glu-Lys       Gly-Lys          Thr
                                                                          50
       Asn-Leu-His-Gly-Leu-Phe-Gly-Arg-Gln-Ser-Gly-Thr-Thr-Ala-Gly-Tyr-Ser-Tyr-Ser-Ala-
                                    Lys-Thr      Gln-Asn-Pro                Thr
                                                                          70
       Ala-Asn-Lys-Asn-Lys-Ala-Val-Glu-Trp-Glu-Glu-Asn-Thr-Leu-Tyr-Asp-Tyr-Leu-Leu-Asn-
                        Gly- Ile- Ile      Gly       Asp       Met-Glu          Glu
                                                                          90
       Pro-Lys-Lys-Tyr-Ile-Pro-Gly-Thr-Lys-Met-Val-Phe-Pro-Gly-Leu-Lys-Lys-Pro-Gln-Asp-
                                    Ile       Val       Ile       Lys-Glu-Glu

       Arg-Ala-Asp-Leu-Ile-Ala-Tyr-Leu-Lys-Lys-Ala-Thr-Ser - SerCOOH
                                                Asn-GluCOOH
```

FIG. 4. (a) Amino acid sequence of wheat germ, and human heart cytochrome *c*. The wheat
germ sequence is shown entire; residues which differ in man are indicated beneath

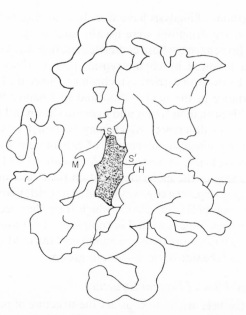

(b) Diagrammatic representation of the three-dimensional structure of horse heart cytochrome *c*, based on the model of Dickerson *et al.* (1967). The complete course of the polypeptide chain cannot be clearly traced at this resolution (4 Å). The haem prosthetic group,
shown mottled, is seen edgeways. The positions of the two thioether bridges from Cys-14 (S)
and Cys-17 (S′) could be identified. The propionyl sidechains of the haem stretch down into
the bottom of the central cleft, possibly in association with basic amino acid sidechains. The
fifth and sixth coordination ligands of the haem iron stretch in from either side of the cleft.
One (H) may be identified with His-18; the other (M), sited on a stretch of polypeptide near
the C-terminus is possibly Met-80

far short of actual determination of tertiary structure. Proteins with a large number of differences in sequence, such as whale myoglobin and horse haemoglobin, may show closely similar tertiary structures.

(ii) Folding of the various sections of polypeptide to form the structure is complex. "Straight" sections may consist of α-helix, or occasionally of pleated sheet structures (between adjacent chains); secondary structure is usually absent in the "bend" regions.

(iii) All proteins so far examined have in common a very closely packed structure. There are few or no water molecules inside. Hence hydrophobic side-chains are mostly arranged so as to be sited inside the molecule; charged side-chains are found on the outside, in the hydrophilic groove or pocket of enzymes or forming bridges within the internal structure.

(iv) Because of the close packing, the nature of almost all side-chains pointing inwards is critical; size and hydrophobicity are both important. In addition, many interactions or lack of them between adjacent side-chains may be important to function.

(v) Conformational changes connected with function may change the gross shape of the molecule very little. However, changes in chemical reactivity of critical regions may involve movement of electrons from chain to adjacent chain to or from quite distant parts of the molecule. An understanding of both the detailed chemistry and the three-dimensional arrangement is essential to the understanding of such processes.

4. How Proteins May Vary

(a) Restrictions

Since proteins of similar function from widely different species such as whale myoglobin and horse haemoglobin have been found to have closely similar three-dimensional structures, the three-dimensional structures of homologous proteins from different species would not be expected to differ. Structure at this level is function-specific rather than organism-specific. Amino acid sequences of related proteins may vary quite widely, but functional restrictions of this variation must also occur. In fact, a hierarchy of variability will be expected, starting with no permitted variation for residues which are uniquely required for function. These include those involved in catalysis, e.g. the essential serine residue of proteolytic enzymes, or those binding prosthetic groups, such as the two cysteines forming thioether bridges with the haem sidechains in cytochrome c. A small group of residues, all expected to be situated on the outside of the molecule, may have no restriction of their variation. Between these extremes will come residues of intermediate variability, mostly with quite severe restriction; these will include those less critically involved in function, including the maintenance of the functionally necessary three-dimensional structure. These intermediate residues, which are more fully discussed by Watts and Watts (1968a), include those of particular interest when homologous proteins are to be compared between organisms.

It is important to note that particular residues should not be considered in isolation, but always within the framework of the total structure. Compensation between mutants can occur within a single polypeptide chain. For instance, occurrence of a second mutation in the gene for the tryptophan synthetase A protein of *Escherichia coli* can restore activity lost as a result of a first mutation; in one case (Yanofsky, 1963), this was dependent upon the amino acid substitutions shown in Fig. 5.

The frequent occurrence of compensating changes of this kind involving two or more residues can be predicted as a consequence of the interdependence of different parts of the structure (Watts and Watts, 1968a). Knowledge of specific interaction between residues in different chains may point the reason and make it possible to scan sequence data for such effects (this seems to never have been attempted in the various computer analyses of sequences). For example, the X-ray picture of ribonuclease has drawn attention to pairs of

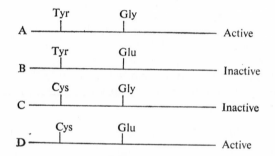

FIG. 5. Tryptophan synthetase variants in *E. coli*. The tryptophan synthetase A protein normally has Tyr and Gly in the positions shown (A). The partial revertant (D) was found to have both these residues changed. Either change alone (B or C) resulted in an inactive protein

compensating differences between the rat and bovine enzymes (Allewell, 1968; Wyckoff, 1969). Many of these could be accounted for in that they result in the same net charge on the molecule. However, two pairs of changes were of particular interest in the context of the model. The first pair, residues 38 and 39 are Met, Thr in the rat. The bovine residue 38, aspartic acid, is positioned so that it might attract Lys-41 away from the active centre, but is in this case screened by the positive charge of an arginine residue at position 39. Of the second pair, residue 79 is methionine in the ox and leucine in the rat, which thus has space for a methyl group in its structure at this point. This is supplied by the isoleucine at residue 57, where the ox has valine.

(b) Process of variation in evolution

It is presumed that the evolution of proteins proceeds by the same process of mutation–recombination–natural selection that has become familiar from studies of organismal evolution. Because changes in protein structure are directly tied to mutational events, the evolutionary process can be examined in as much detail as knowledge of the protein involved permits. One inference of much interest has been the very common occurrence of duplicated genes and parts of genes. An organism may possess a number of proteins whose similarity suggests that they are encoded in structurally homologous genes. On this basis Ingram (1961) proposed a scheme for the evolution of vertebrate haemoglobins by successive gene duplications. Different types of homology between genes and their products have been discussed by Watts and Watts (1968b).

Bridges (1936) proposed a mechanism which is best termed homologous unequal crossing-over (Fig. 6) to account for the ability of a sex-linked *Drosophila* mutant Bar Eye to give rise to Double Bar and normal offspring with a frequency well above the mutation rate. Once a primary gene duplication (Watts and Watts, 1968b) has occurred, the gene can be repeatedly increased in

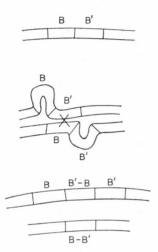

FIG. 6. Change in number of duplicate genes as a result of homologous unequal crossing-over. Where BB′ represents Bar Eye, it is assumed that there has been deletion of parts of each gene at their junction, giving rise to the mutant phenotype which is maintained in BB′–BB′ but lost in B–B′

numbers in this way, and the process applies as well to genes completely duplicated as to partial duplications like Bar Eye. This may be the manner in which the "pseudo-allelic clusters" of the geneticist were built up.

The importance of gene duplication in the evolution of proteins (and thus of organisms) can be appreciated once it is realized that accumulation of compensating mutations of the kind illustrated in Fig. 5 is only likely to occur in the artificial conditions of the culture flask. In the haploid bacterium, the two mutants, both having anomalous tryptophan synthetases, can only survive in absence of the need to make tryptophan. In higher organisms, with their much longer generation times, accumulation of compensating changes by successive mutations might occur in heterozygotes, but so slowly as to be of no evolutionary usefulness. A diploid organism carrying two such alleles, the prerequisite for a recombination restoring activity, would not survive. It is therefore not possible to evolve compensating changes by carrying both deleterious changes in the heterozygote. This situation, it can be seen, is different from that obtaining when recombination is between different genes, the situation usually considered by evolutionists. Duplicate genes bring two solutions to this problem. The first is an all-or-none process where one gene maintains function while the duplicates undergo random variation by mutation–recombination until a functional condition, possibly with a modified activity, is reached. The second involves variation in both or all the duplicates (in at least one, of a non-lethal kind) followed by recombination involving unequal crossing-over. This latter process increases the effective mutation rate for the locus by the number of times the gene is duplicated, besides making use of the shielding effect of duplicate genes.

It has been inferred (Baglioni, 1962) that the human haemoglobin variants, the Lepore haemoglobins, arose by homologous unequal crossing-over between the adjacent loci for β and δ haemoglobin. The two Lepore haemoglobins which have been analysed start with δ-chain sequence at the N-terminus and switch to the β-chain sequence at two different points along the molecule. There is evidence that the process has also occurred in human haptoglobin variation, this time based on a partial gene duplication (Smithies, 1964). It is not known how nearly identical genes must be to enable crossing-over to occur between them, nor the minimum length of homology required. If this is small, and adjacent parts of a gene have some degree of similarity in nucleotide sequence, deletions or insertions of one or a few amino acids might occur as a result of homologous unequal crossing-over (Lehmann and Carrell, 1969). The general stability of genetic behaviour makes it likely that such events would be extremely rare.

B. ANALYSIS OF AMINO ACID SEQUENCE DATA

1. Sequence Homology

Proteins which serve the same function in different organisms may be considered homologous but until their structures can be compared, it is not known unequivocally that the function is performed in the same way in each organism. So far, in all cases where amino acid sequences of proteins of similar function have been compared, very great structural homology has been revealed.

The first step in examination of amino acid sequences for homology is the lining-up process, which may be done by trial and error by the investigator or by computer. The aim is to align the sequences so that the maximum number of residues are identical between the different organisms. Where there are very few differences in the sequences, the lining-up is easy. Where many differences are present, it is often found that a good fit is only obtained, even for proteins with the same total number of residues, when gaps are left at certain places. This acknowledges the occurrence of deletions and insertions in the genes as well as mutations from one amino acid to another. Where the total number of residues differs between proteins, the missing residues can frequently be accounted for at one end of the chain.

Many alignments of a variety of proteins made by computer are given by Dayhoff and Eck (1968). It has been possible in this way to show homologies not only between haemoglobins or cytochrome *c*, but also between related enzymes such as trypsinogen and chymotrypsinogen. Using partial sequence data for the regions surrounding the essential cysteine or serine residues (which can be labelled isotopically by reaction with inhibitor) Watts (1968) has produced alignments for a number of phosphotransferases, dehydrogenases and esterases. These were made by working outwards from the essential cysteine and from a cysteine a few residues away from the essential serine in the esterases.

2. Attempts to Quantitate Homology

When the alignments have been made, how is homology assessed? At its simplest, it can be represented by the percentage of identical residues. The non-identical residues may then be examined in a number of different ways, the simplest of which is perhaps the construction of an implied *m*RNA nucleotide sequence. (The amino acid code, Table I, is usually handled in *m*RNA codons

TABLE I

The amino acid code

UUU	Phe	CUU	Leu	AUU	Ile	GUU	Val
UUC	Phe	CUC	Leu	AUC	Ile	GUC	Val
UUA	Leu	CUA	Leu	AUA	Ile	GUA	Val
UUG	Leu	CUG	Leu	AUG	Met	GUG	Val
UCU	Ser	CCU	Pro	ACU	Thr	GCU	Ala
UCC	Ser	CCC	Pro	ACC	Thr	GCC	Ala
UCA	Ser	CCA	Pro	ACA	Thr	GCA	Ala
UCG	Ser	CCG	Pro	ACG	Thr	GCG	Ala
UAU	Tyr	CAU	His	AAU	Asn	GAU	Asp
UAC	Tyr	CAC	His	AAC	Asn	GAC	Asp
UAA	Chain Termn.	CAA	Gln	AAA	Lys	GAA	Glu
UAG	Chain Termn.	CAG	Gln	AAG	Lys	GAG	Glu
UGU	Cys	CGU	Arg	AGU	Ser	GGU	Gly
UGC	Cys	CGC	Arg	AGC	Ser	GGC	Gly
UGA	Not translated	CGA	Arg	AGA	Arg	GGA	Gly
UGG	Trp	CGG	Arg	AGG	Arg	GGG	Gly

rather than DNA anti-codons). Owing to the degeneracy of the code, there is no certainty in the assignment of codons; for simplicity these can be chosen so as to maximize homology between the sequences. For example, the human β and δ haemoglobins, with 10 residues different in 146, have 93·85% homology on an amino acid basis and a maximum of 97·45% on an *m*RNA basis.

Alternatively the non-identical residues may be analysed into those which are "one-step mutations" (whose codons, at best, differ in only one nucleotide), two-step and three-step mutations. Closely related proteins are generally found to include a high percentage of one-step mutations among their non-identical residues. A numerical value inversely related to the homology between pairs of proteins can be obtained by totalling the number of nucleotide changes required to convert one into the other.

Some standard is also required for the other end of the scale which begins with complete identity of sequence. Epstein (1967) has attempted to provide this using another method of quantitation based on the chemical nature of the amino acid side chains. Each amino acid is assigned a value on a polarity scale reaching from zero for hydrophobic side chains to 1·0 for the charged hydrophilics, with some weighting for size. At each position in an alignment of two sequences, the difference is taken; these are totalled and divided by the number of residues, yielding a mean coefficient of difference. This procedure may be applied to the entire sequence (Δp) or to the residues which are not identical (Δm). A value near 0·5 is expected from the comparison of random sequences, and has been obtained by Epstein for pairs of proteins which are not considered to be homologous, while pairs of proteins considered to be in some degree homologous gave values from 0·04–0·31 for all residues and 0·29–0·40 for the non-identical positions only.

The most closely related proteins considered were human and horse haemoglobin α-chains, which have 17 non-identical residues. If the method is applied to sequences from more closely related animals, using cytochrome c, for instance, the value of the difference coefficient (Δm) being calculated per residue, does not fall but tends to scatter as the number of differences falls. In its present form, that is using the particular assignment of values for pairs of amino acids, the method would seem unsuitable for the quantitation of homology although it may be useful in testing whether homology exists or not or whether an alignment is optimal. Epstein (1968) considered that some of the values might be too high and pointed out the difficulty of working with values which take no account of the location (and therefore function) of changed residues within the structure.

3. Unquantitated Aspects of Homology

Attempts to quantitate sequence homologies have so far been very simple. It is hardly surprising that they have failed to take account of a number of factors involved in the complex structure of proteins.

It is apparent to the eye that aligned sequences are closer to identity in some regions than in others. When a number of homologous sequences can be considered, it is very clear that the position of identical residues is not random and that there must be a connection with the functional requirements of the protein. This hierarchy of variability, reflecting that discussed in Section III.A (p. 157) has been described in terms of invariant residues, very conservative substitutions, fairly conservative and radical substitutions on the basis of the types of amino acid chains involved. Zuckerkandl and Pauling (1965) have treated "conservatism" in a rather more flexible way than in Epstein's polarity scale, with stress upon the possible function of residues as well as on their chemical kind. This approach has not yielded a method of quantitation of homology but may lead to this later through a better understanding of which properties of a side chain are important in a particular situation.

As was pointed out in Section III.A (p. 158), it is most likely that residues vary in a dependent manner. This phenomenon would contribute to the invalidation of simple quantitative schemes. The compensating pairs of charge changes in ribonuclease, for instance, would add heavily to Δm, based on polarity changes. When sufficient information is available for interdependent changes to be grouped together, a quantitative approach to homology will become more feasible. Homologous groups of proteins should be able to acquire a taxonomy and then, hopefully, a phylogeny.

4. Sequences and Time-scales

As soon as a number of sequences were available for the haemoglobins and cytochrome c, attempts were made to fit the number of differences found on a time-scale derived from the classical approach to evolution. The success of these attempts led to the concept of a steady rate of "evolutionarily effective mutation", characteristic of particular proteins (Zuckerkandl and Pauling, 1962). Taking cytochrome c as an example, since it is a protein which occurs, as far as we know, in all organisms capable of aerobic respiration, let us consider how far general rules can be applied to variability of structure. In cytochrome c proteins of the "mammalian" type (able to transfer electrons to mammalian cytochrome oxidase) there are 104 residues which can be accounted homologous and, of these, 35 are invariant in every organism so far investigated (Smith, 1968). These will form those parts of the molecule where the most stringent structure–function correspondences occur. In the rest of the molecule more or less variability occurs. This may depend on two processes:

(i) Modification with functional significance

Here the amino acid substitutions found are functionally better in the organisms possessing them because of changes in external or in the internal environment, including other macromolecules. Such adaptive changes would seldom concern single residues but would rather be expected to involve numbers of interrelated residues in complex cascade effects of compensating changes.

(ii) Drift in residues with low structure–function correspondence

Such changes would not show interdependence between different sites and might show a wider range of substitutions than adaptive changes.

The total process of change appeared to have a rather simple time dependence (Margoliash and Smith, 1965). If the number of amino acid differences are plotted against time elapsed since a common ancestor of the two organisms lived, an assumed constant rate of effective mutation will yield a curve (Fig. 7) whose divergence from linearity depends upon the number of residues within the molecule which are variable. It can be seen that, for closely related organisms with a small number of differences, the number recorded will equal the number which have occurred. As the number scored increases, it will include more and more repeat changes at the same site, resulting in underestimate of the actual total. Linear extrapolation from a point based on two species of recent divergence (mammals/chicken in Fig. 7) will therefore suggest a too recent divergence for less closely related species.

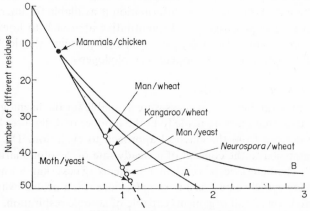

FIG. 7. Relationship between numbers of amino acids different in the sequence of cytochrome *c* and time since the organisms compared had a common ancestor. The figure is constructed by the method of Margoliash and Smith (1965), using the value of 12·4 as the average number of differences between mammals and the chicken and 280 m.y. as the time elapsed since the divergence of avian and mammalian lines. The straight line extrapolated from this point has been used as the abscissa in plotting the two curves representing the observed number of differences against the actual number of changes which would have occurred, including repeat changes at the same position. These values were calculated from probability theory, assuming a total number of variable residues of 70 (curve A) and 50 (curve B). The two curves thus show the time dependence of observed numbers of differences under two conditions of restriction of variation. Numbers of differences in cytochrome *c* between various pairs of distantly related organisms have been indicated on the extrapolation line only

Assuming that a fixed maximum number of residues can be changed and, for simplicity, that all are equally likely to change, the probable number of actual changes corresponding to any number of observed changes can be calculated. The points of divergence for two organisms can then be replotted, using the observed number of changes as the ordinate and the linear extrapolation line as the abscissa from which the actual number of changes is positioned. A new location with respect to the time-scale is then found.

In order to apply such a correction, it is necessary to choose a value for the total number of permitted differences. Curve A in Fig. 7 is based on an assumed maximum of 70. This corresponds to the total number of sites which have so far been observed to vary in any of the organisms for which the protein has been sequenced. On this curve man and yeast (44 differences) or *Neurospora* and wheat (46 differences) had a common ancestor approximately 1·6 b.y. ago. Curve B, based on a maximum number of variable sites of 50, yields a time close to 3 b.y. ago for the divergence of the groups represented by these species, in better agreement with the palaeontological evidence. One might suggest that a total number of variable sites of 50 is, in fact, more realistic than 70, since in no case has this number of differences been exceeded between any two organisms. The actual maximum number would be governed by the actual pathway of variation followed in any particular situation of divergence. That a fixed maximum number cannot be assumed is indicated also by comparisons between the wheat sequence and those of the higher vertebrates, where the number of differences fall in a lower range, between 35 (wheat/man) and 39 (wheat/kangaroo). In passing, it is worthy of comment that the cytochrome *c* sequence data are in accord with the assignment of fungi (*Neurospora* and *Saccharomyces*) to a kingdom equally separate from animals and from higher plants.

From the foregoing, one can conclude that the model wherein a fixed total number of amino acid residues undergo random variation at a fixed average rate is too simple. Dayhoff and Eck (1968) have used an analysis of the number of different amino acids found at each site (in cytochrome *c* of all organisms sequenced to that date) to test a rather more complex model in which three types of amino acid site are recognized. These are immutable, slowly changing and rapidly (three times the slow rate) changing. The model gives a good fit with the total data available. It also predicts that when many more species are sequenced, only 14 residues will be found to be invariant. Dayhoff and Eck suggest that this further variation in sequence should be sought in the untouched parts of the phylogenetic tree. Smith (1968) on the other hand, feels that the number of invariant residues is unlikely to be reduced much below its value of 35 by the collection of more data.

The total number of amino acids invariant in all species is of more interest to the biochemist looking at the mechanism of action of cytochrome *c* than to the phylogeneticist. Of great interest to both must be the number and nature of invariant groups of residues, for on these must be based our knowledge of different functional classes of the protein. More sequences, particularly for cytochromes from lower plants and from the neglected groups of invertebrates, together with a high resolution X-ray picture (which may be available by the time this paper is published) should enable us to formulate a model for the evolution of cytochrome *c* which is more than statistically satisfying.

5. *Sequences and Phylogenetic Trees*

The construction of phylogenetic trees from sequence data has been based on the same kinds of assumption as the derivation of time scale, all residues being considered independently and given equal weight in inferring the length of branches. A "fossil" sequence for the hypothetical organism at a branch-point can be constructed as a first step. Dayhoff and Eck (1968) do this on the basis of three sequences from real organisms. If a particular amino acid appears in at least two of these, it is included in the fossil sequence; otherwise the position is left blank. The fourth real sequence is then positioned in the tree according to its score for similarity to each of the original sequences in those residues in which it differs from the "ancestral" sequence. Another fossil sequence is deduced for the branch-point generated by adding the fourth sequence; and so on. Different phylogenetic trees constructed in this way are compared as to the number of mutational events involved, and the one with least is deemed the most satisfactory. Branch lengths are assigned on the basis of the number of mutations. In this way a phylogenetic tree based on cytochrome c sequences can be made; it compares satisfactorily with phylogenetic trees based on classical data.

The approach to the construction of phylogenetic trees might be refined in the light of further information in the same way as suggested in the previous section for time-scale. It should be possible to separate residues which vary dependently as a group from those showing "random" variability. The former type should be of more use in defining relationships among major groups of organisms. The latter should be more useful in discriminating at the lower levels of phylogeny but will present more problems as a result of repeat or reverse changes.

It is interesting to consider which amino acids occur at very variable sites in the light of the amino acid code (Table I), as this may sometimes suggest the existence of "missing links". For example, position 65 in cytochrome c has the following variations

Tyr	Ser	Phe	Met
Wheat	Fungi	Insects	Vertebrates

The first three amino acids are interconvertible by single nucleotide changes (Table I). Methionine can be reached via a minimum of two changes (from Phe via Leu or Ile). Can a present-day organism be found in which the intermediate occurs? Or did this change require compensating changes and have to be accomplished in a duplicate gene?

Position 33 in cytochrome c shows a similar relationship but with a suggestion of greater variability arising from the new data outlined in chapter 9.

His	Asn	Tyr	Trp
most vertebrates	kangaroo	tuna	bullfrog
insects	turtle		
fungi			
wheat	sunflower	mung bean	

Here, the single-step mutations have occurred both in the vertebrates and the higher plants. Tryptophan, two steps from Tyr (via Cys), (presumably not chain termination), has been reached only in the single amphibian so far sequenced. More information from the Amphibia would certainly be of interest, and this is clearly a site to watch when further plant sequences become available.

C. ENZYMOLOGICAL DATA

When enzymes are chosen for the study of phylogeny, a very large range of techniques is available for use in the comparison of different species. I shall illustrate this with an account of some work on an animal group, the annelid worms. It was the startling pattern of variation in the phosphagen kinase enzymes present in the muscles of these worms which first roused my own interest in the evolution of enzymes.

All animals have, in their muscle, stores of a phosphagen, a guanidine phosphate which can be used to regenerate ATP broken down during muscle contraction, according to the equation

$$\text{Guanidine-P} + \text{ADP} + \text{H}^+ \rightleftharpoons \text{Guanidine} + \text{ATP}$$
$$\text{phosphagen kinase}$$

In vertebrates, the phosphagen is creatine phosphate, in most invertebrates it is arginine phosphate, but the annelids have among them seven different phosphagens, including these two. When the corresponding phosphagen kinases are considered, there is effectively a total of four enzymes, specific for the phosphates of arginine (Arg), creatine (Cre), glycocyamine (Gly) and for the remaining four phosphagens which are various substituted guanidinoethanes, of which taurocyamine (Tau) serves as example (for formulae, see Watts and Watts, 1968a). Crude extracts of muscle from a number of species of worms were tested for ability to form phosphagen (by the reverse reaction) given ATP and one of the four guanidines, Arg, Cre, Gly or Tau. It was found that a particular species might show one, two or three enzyme activities. In this respect, the two major groups of worms showed a sharp difference, the marine Polychaetes being completely variable but the Oligochaetes (the earthworms) showing activity only for Tau.

Besides testing for specificity towards the four natural substrates, some muscle extracts were tested against substrate analogues such as homoarginine and glycylglycocyamine. The arginine kinase of lobster muscle, a much-studied invertebrate enzyme, is completely inactive with these, but several polychaete species showed some phosphorylation. Starch gel electrophoresis of the extracts was also performed. In *Sabella* this indicated two isoenzymes of arginine kinase, the sole activity in this species. In some cases where more than one enzyme activity was found, mixed substrate experiments were done to check, by the additivity of phosphorylation, that the several activities of one low-specificity enzyme were not being measured with the single substrates.

All these findings can be explained if it is assumed that duplication of the gene for arginine kinase, possessed by an ancestral invertebrate, occurred at least twice in the line giving rise to the annelids. Divergent evolution then occurred in the duplicate genes, eventually giving rise to one or more different enzyme activities in some of the descendants. The arginine kinase might be retained alone or in addition to other activities, or might be lost once these were available to serve the needs of muscle metabolism. The various possible outcomes of such a process are shown in column 2 of Table II. The first column

TABLE II

Evolution of phosphagen kinases in polychaete annelids

Organism	Postulated genotype			Phenotype		
Ancestral worm	*AK*	*AK*	*AK*	AK	AK	AK
Modern forms:						
Sabella pavonina	1. *AK*	*AK*	*AK*	AK	(2 isozymes)	
	AK	*AK*	*I*			
	AK	*I*	*I*			
Aphrodite aculeata	2. *AK*	*AK*	*XK*	AK	CK	
Hyalinoecia tubicola	*AK*	*XK*	*XK*	AK	CK	
Sabella alveolata	*AK*	*XK*	*I*	AK		TK
Branchiomma vesiculosum	3. *XK*	*XK*	*XK*			TK
Hermione hystrix	*XK*	*XK*	*I*		CK	
	XK	*I*	*I*			
Arenicola marina	4. *XK*	*XK*	*YK*		GK	TK
Nephthys caeca	*XK*	*YK*	*I*	CK	GK	
Myxicola infundibulum	5. *AK*	*XK*	*YK*	AK	GK	TK
Nereis fucata	6. *XK*	*YK*	*ZK*	CK	GK	TK

Symbols: *AK:* gene for arginine kinase   AK: arginine kinase
          XK, YK, ZK: genes for other kinases   CK: creatine kinase
          I: gene for inactive protein or none GK: glycocyamine kinase
                                                  TK: taurocyamine kinase

shows the polychaete species tested opposite the type of evolution they represent; the third column shows the actual kinases found. The existence of two isoenzymes of arginine kinase in *Sabella* fits into the picture, since this species has not evolved other activities. Lower specificity of the enzymes would also seem to fit, since these are proteins of recent evolution which might not yet be perfect in all their functions.

The Oligochaete annelids all fall into the third group in Table II, where one other enzyme has evolved and arginine kinase has been lost. Why should they all be the same? It has been found by cytological observation (Muldal, 1952) that this group forms a polyploid series. Presumably as a result of interference with sex-determination, hermaphroditism is general and parthenogenesis is common in the group. Organisms with little possibility of recombination would be expected to show a slow rate of evolutionary change. It seems most reasonable to conclude that the polyploidizations occurred after the evolution of lombricine kinase in one line of annelids which thereafter remained distinct from the diploid bisexual polychaetes.

A phosphagen kinase is, in a sense, a special kind of enzyme, particularly suited for evolutionary studies. This is because it is not involved in metabolic pathways with numbers of other enzymes all requiring evolutionary modification if new substrates are to be tackled. Therefore, a phosphagen kinase is able to vary with greater freedom than most enzymes. The new substrates adopted by the annelids could probably all be provided via existing pathways of amino acid metabolism.

Plants possibly do not have phosphagen kinases, and certainly do not have them in high concentration. They must have many other enzymes, homologous between groups, and not involved in pathways of essential metabolism. This is the sphere in which enzymological techniques might rapidly yield phylogenetic information without any need for purification of large quantities of protein such as sequence determination entails.

IV. VARIATION OF PROTEINS IN PLANTS

Although no work appears to have been done on the evolution of plants with the approach used for the annelids and described in the last section, plant proteins and plant enzymes have been used in studies on variation, both within and between species. I shall illustrate this with several examples, all relying upon an electrophoretic separation of protein species. Each of them also involves features of variation not necessarily ever found in the animal kingdom.

A. INTRASPECIFIC VARIATION

A successful species is one which is both well adapted to its environment and reproductively adequate. In the short term, both these conditions can be satisfied with little or no variation, even in the genotype. In the long term, environments are bound to change and the evolutionarily successful organism is one that is versatile. This implies variation of genotype.

In both the animal and plant kingdoms, some species seem to have evolved towards present security rather than long-term adaptability. Present day adaptedness is often very marked in species with genetic systems precluding variation, as in the parthenogenetic earthworms discussed in the last section, or in the well-known plant apomict, the dandelion. Most animals ensure variability through recombination by having two sexes. Most higher plants are hermaphrodite, but many of these ensure the redistribution of genetic material by means of self-incompatibility mechanisms. In these, pollinations made using pollen from a particular infra-specific group upon the stigma of a plant of the same group will fail to produce seed. In many cases, failure of the pollen tube to grow in the style of a self-incompatible plant has been observed.

Pandey (1967) has investigated the basis of incompatibility in *Nicotiana* by electrophoresis of style extracts on acrylamide blocks followed by staining for peroxidase activity. Plants with the same genotype with respect to their incompatibility relations showed the same isoenzyme pattern for peroxidase. When plants which were heterozygous at the incompatibility locus were examined, additive isoenzyme patterns were obtained and no hybrid bands appeared (Fig. 8). In the paper cited, Pandey has investigated only four alleles, in various

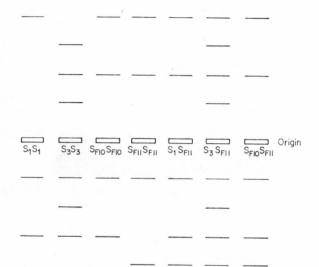

Fig. 8. Electrophoretic patterns of peroxidase activity in style extracts of seven different incompatibility genotypes in *Nicotiana*. S_1, S_3, etc. represent different alleles at the incompatibility locus. (Drawn from Pandey, 1967)

combinations, and has not reported any results for pollen extracts. His interpretation of the connection between the different isoenzyme patterns and incompatibility phenomena would seem to be improbable, but is irrelevant here. The demonstration that intra-specific variation with functional significance can be investigated biochemically by these straightforward procedures is of great interest.

It is essential that techniques to be applied to phylogenetic problems be tested at the level of relationship where cytogenetic analysis is possible. Another example of this kind of verification is provided in a study of peroxidase and polyphenol oxidase isoenzymes in the wheat "Chinese Spring" (Upadhya, 1968). Plants into whose genome had been transferred a chromosomal segment from *Aegilops umbellata* were compared with those of standard genotype and were found to possess an extra isoenzyme band. A great many more experiments of this kind need to be done before enzymes have yielded a new set of characters diagnostic for particular genotypes. It is already possible to have confidence that the approach will prove fruitful.

B. ANALYSIS OF RELATED SPECIES

It has been the feeling of both plant and animal taxonomists that their concepts of the species were fundamentally different, and that this difference of concept arose from the difference in degree of interspecific barriers in the two kingdoms. Stebbins (1950) has discussed this viewpoint very fully and concludes that a common species concept is possible, at least for species with a sexual system of reproduction. "Species," he says, "are separated from each other by gaps . . . in morphological and physiological characteristics which are maintained by the absence or rarity of gene interchange between members of different species." There follows a discussion of the means by which gene interchange is prevented. One of these, of which the higher animals, (insects and warmblooded vertebrates) have a virtual monopoly, is a psychological barrier, the complex of factors which ensures that members of one species will avoid mating with those of another even if the two are sympatric. Lacking only this barrier (more absolute than the most intricate pollination mechanism) among the several that may separate interfertile species, many higher plants are characterized by *rarity* rather than complete *absence* of gene flow between them and hence show reticulate phylogeny rather than a strictly dichotomizing pattern of descent.

Vaughan and his group have made a study of a number of species of *Brassica* and *Sinapis*, and have used acrylamide gel electrophoresis of seed proteins (Fig. 9a) to elucidate the inter-relationships of 10 of these (Vaughan and Denford, 1968). Percentage similarity was assessed on the basis of

$$\frac{\text{Nos. of pairs of similar bands}}{\text{Nos. of different bands} + \text{nos. of pairs of similar bands}} \times 100$$

and reciprocal values used in the construction of a three-dimensional model of relationships between species (Fig. 9b). These values for the distances apart yielded a closed figure, after some adjustment to the positions of the centrally placed *B. juncea* and the two varieties of *B. napus*. As Vaughan and Denford point out, such a three-dimensional structure is a critical test of the validity of the method. The percentage similarity values were also found to agree well with established morphological taxonomy.

As can be seen in Fig. 9b, the two *Sinapis* species are very closely related to each other but only distantly to the *Brassica* species apart from *B. nigra*. The four species located in the centre of the figure are thought to be hybrids. This is borne out by the fact that they showed no specific bands but had various combinations of bands not common between the species on the fringe of the figure. On the basis that *B. campestris* (turnip) possessed only bands appearing with

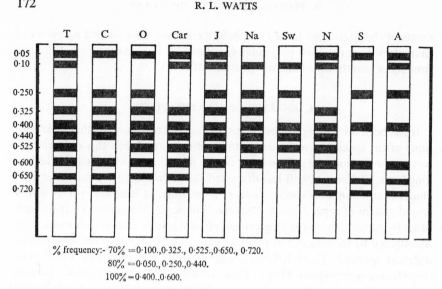

% frequency:- 70% =0·100.,0·325., 0·525.,0·650., 0·720.
                80% =0·050., 0·250.,0·440.
               100% =0·400..0·600.

FIG. 9. (a) Electrophoretic patterns of seed albumins in *Brassica* and *Sinapis* species. Bands with a frequency of 70% or over only are shown

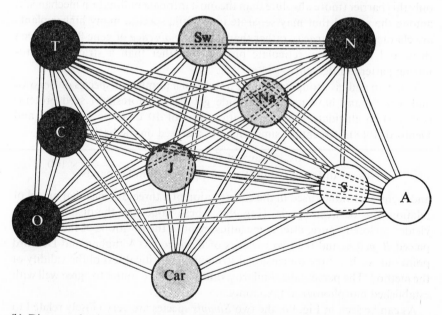

(b) Diagram of a three-dimensional model of relationships among the taxa based on their albumin bands

T: *B. campestris* (turnip), C: *B. campestris* (turnip rape), O: *B. oleracea*, N: *B. nigra*, Car: *B. carinata*, Na: *B. napus* (rape), Sw. *B. napus* (swede), J: *B. juncea*, S: *S. alba*, A: *S. arvensis*

FIG. 9 is reproduced from Vaughan and Denford (1968) with the permission of *The Journal of Experimental Botany* high frequency in other species, Vaughan and Denford have suggested that this may be closest to an archetype of the taxa investigated.

C. ANALYSIS OF HYBRIDS AND POLYPLOID SERIES

Hybridization between close relatives followed by chromosome doubling has been of great importance in the phylogeny of the angiosperms and is of particular interest because many important crop plants are polyploid. Stebbins (1950) has pointed out that chromosome doubling is a most likely method of species separation at a single step, since backcrossing to the parental diploid species will normally give rise to infertile triploids. In addition, he has assembled much evidence that such polyploid species usually only arise in hybrids which are viable but sterile; the two haploid chromosome sets are compatible enough to support vegetative growth but different enough for the complex process of sexual reproduction to be interrupted. Combination of incompatible genes or too little homology between the two sets for chromosome pairing to occur in meiosis may cause failure of gamete production. Once it has occurred, chromosome doubling removes the latter problem for very unlike chromosome sets. Incompatibility or residual homology between the sets may take time to disappear by modification of the polyploid genome. It is not surprising, therefore, that polyploidy is found most often in perennial plants or those with efficient means of vegetative propagation, just as polyploidy in the polychaete worms is associated with a trend away from sexual reproduction. Ohno and Atkin (1966) have suggested the origin of several present day groups of bisexual fish from hermaphrodite ancestors forming a polyploid series. This illustrates another feature of the evolution of polyploids, a trend towards secondary diploidy. Such a trend might be expected on theoretical grounds, since modification of the polyploid genome could occur by mutation in, or by deletion of, genes of homologous function, present in several copies. Hayter and Riley (1967) have found evidence for this kind of modification in a comparison of meiotic behaviour in hexaploid (AABBDD) and tetraploid (AABB) wheats. The control of chromosome pairing, normal in the tetraploid, must be seated in either the A or the B genome. In the hexaploid, this function is disturbed unless chromosome 5 of the D set is present, indicating that it has been lost from the other two sets of chromosomes. Decrease in homology within sets of chromosomes in a polyploid would also be selectively advantageous since it would decrease the likelihood of mispairing in meiosis.

Polyploidy appears not to occur at the organism level among higher animals, although some fully differentiated mammalian cells may contain multiples of the basic chromosome complement. Whether its absence is a consequence of fundamental differences in control of development between these groups and the plants and lower animals or only a secondary effect of differences in their reproductive processes is a question of some interest. Higher animals are unable to hybridize freely because of their complex mating patterns (this, notably, applies far less to the fishes), and they also lack the availability of the processes of apomixis, particularly vegetative propagation, which enables the plants to "weather" the early stages of polyploidy. Yet there is no doubt of the overwhelming efficiency of diploidy at all stages in the life cycle, and of this there is so far no explanation at the molecular level. The answer to this problem may well come from the study of plant polyploids, which not only exist in large numbers in nature but whose formation and subsequent modification can readily be studied under experimental conditions. Elucidation of the phylogeny of polyploid groups via investigation of their proteins is inevitably bound up with the need to understand gene action in diploids and polyploids.

The cultivated wheats and their relatives have been the subject of a study (Johnson and Hall, 1965), which clearly shows the usefulness of comparison of proteins between members of polyploid series. The central point of interest was the derivation of the hexaploid ($2n = 42$) species, *Triticum aestivum*. Morphological evidence indicates that the species arose from a cross between a wheat of the $2n = 28$ (tetraploid) group and *Aegilops squarrosa*, still a common weed of wheatfields in the Middle East, where *T. aestivum* is thought to have originated. Using *T. dicoccoides* as the tetraploid parent, hybrids have been produced artificially which are fertile, closely similar to the natural hybrid and able to cross with it (McFadden and Sears, 1946). Hence the origin of *T. aestivum* might be represented as in Fig. 10.

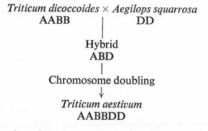

FIG. 10. Postulated derivation of *Triticum aestivum*, based on morphological and cytological data

In prehistoric times, before the origin (or discovery by man) of *T. aestivum*, wheats of both $2n = 14$ (einkorn) and $2n = 28$ (emmer) were in cultivation. Johnson and Hall suggest that emmer has the same genomic constitution (AABB) as *T. dicoccoides* and was in turn derived from a hybrid between einkorn (*T. monococcum*) and a species of *Aegilops* other than *A. squarrosa*. In order to establish the pattern of gene homology between related species more directly than can be achieved by morphological or cytological investigation, they compared the electrophoretic behaviour of seed protein extracts in blocks of polyacrylamide gel. Some of the resulting patterns are shown in Fig. 11.

The comparison of *T. durum* and *T. dicoccum* demonstrates the similarity in protein components between two species thought to have the same genomic constitution, while that between *Secale cereale* and *T. monococcum* shows the sharp difference for two species whose chromosomes are not thought to be closely homologous. In the top section of the figure, it can be seen that *T. aestivum* has protein components in common with both *T. dicoccum* and *T. monococcum*. The similarity between the two latter species is less clear. One major and two minor bands of intermediate mobility and one slower band common to *T. monococcum* and *T. aestivum*, are absent from *T. dicoccum*.

There are two possible phylogenetic explanations of these findings. (i) The genome homologies are as indicated in Fig. 11. The D chromosomes of *T.*

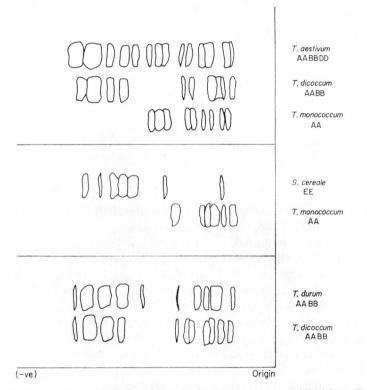

FIG. 11. Electrophoretic patterns of seed proteins from various species of Triticinae. (Drawn from Johnson and Hall, 1965)

aestivum were derived from an *Aegilops* parent, the A and B groups from a tetraploid precursor of the emmer, possibly *T. dicoccoides*, which possessed the protein components subsequently lost in *T. dicoccum* and derived from *T. monococcum* (Fig. 12a). (ii) The D chromosomes of *T. aestivum* came from *T. monococcum* which has been mis-assigned the A genome and may possibly possess homology with *Aegilops*. In this case, the AABB wheats were not derived directly from *T. monococcum* (Fig. 12b). It might prove possible to distinguish which of these two alternatives is correct if further species, particularly *T. dicoccoides* and *Aegilops squarrosa* were analysed in the same way. The protein complement of the hexaploid derived from their artificial hybrid should be of great interest, and also its possible modification from generation to generation.

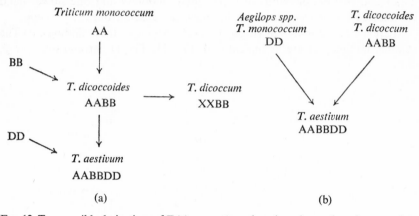

FIG. 12. Two possible derivations of *Triticum aestivum*, based on electrophoretic comparison of its proteins with those of its relations

V. CONCLUSIONS

It may seem that this paper has dealt almost entirely in animal proteins and mere prospects for plant proteins. Plant scientists, accustomed as they are to be first or even alone in the field with chemical investigations, should not be discouraged by this. Plant proteins may be harder to come by, and may show less versatility of function than those of animals, which include muscle and blood among their tissues, but, as in any organism, they represent the first phenotypic realization of the potential of the genotype. Because there are firm chemical and biological restrictions applying to the way proteins may vary and we are now technically able to gain a proper understanding of these, comparison of proteins between organisms should be able not only to tell us how alike are present day forms, but to give us some idea of the macromolecular constitution of ancestors which we shall never be able to study directly.

A protein is not just an amino acid sequence. Sequence data will be of indisputable use in establishing the phylogeny of organisms only when the relationship between primary and higher structures of the protein molecule is better understood. Meanwhile, comparative studies of proteins by as many techniques as possible in addition to sequence determination will add to knowledge of both structure–function relations in the molecules themselves and phylogenetic relations between the organisms from which they come. Work upon plant proteins can make a unique contribution to such a fundamental understanding because plants, with their variety of genetic systems, lend themselves to the experimental study of hybridization and its consequences. Here it should be possible to follow the interaction of related genes and their control systems in terms of their primary products, the protein molecules, and to understand how these effects are modified in successive generations.

Amid the many chemical riches of the plant world, proteins occupy a special position by virtue of their relationship to the individuality of organisms and the characterization of species. The door to their study now only stands ajar. It is up to us to see that it is shortly thrust wide open.

REFERENCES

Allewell, N. (1968). *In* "Structural Aspects of Enzymatic Activity." British Biophysical Society Symposium.

Anfinsen, C. B. (1959). "The Molecular Basis of Evolution." Wiley, New York.

Baglioni, C. (1962). *Proc. natn. Acad. Sci. U.S.A.* **48**, 1880–1886.

Blake, C. C. F., Mair, G. A., North, A. C. T., Phillips, D. C., and Sarma, V. R. (1967). *Proc. R. Soc. Ser. B*, **167**, 365–377.

Bridges, C. B. (1936). *Science, N.Y.* **83**, 210–211.

Dayhoff, M. O., and Eck, R. V. (1968). "Atlas of Protein Sequence and Structure, 1967–1968." National Biomedical Research Foundation.

Dickerson, R. E., Kopfa, M. L., Weinzierl, J., Varnum, J., Eisenberg, D., and Margoliash, E. (1967). *J. biol. Chem.* **242**, 3015–3018.

Edsall, J. T., and Wyman, J. (1958). "Biophysical Chemistry," Vol. 1. Academic, New York and London.

Epstein, C. (1967). *Nature, Lond.* **215**, 355–359.

Epstein, C. (1968). *In* "Homologous Enzymes and Biochemical Evolution" (N.v. Thoai and J. Roche, eds.), pp. xxi–xlix. Gordon and Breach, New York, London, Paris.

Ford, E. B. (1965). "Genetic Polymorphism." Faber, London.

Hayter, A. M., and Riley, R. (1967). *Nature, Lond.* **216**, 1028–1029.

Ingram, V. M. (1961). *Nature, Lond.* **189**, 704–708.

Johnson, L., and Hall, O. (1965). *Am. J. Bot.* **52**, 506–513.

Karlson, P. (1965). "Introduction to Modern Biochemistry." Academic, New York and London.

Kendrew, J. C., Dickerson, R. E., Strandberg, B. E., Hart, R. G., and Davies, D. R. (1960). *Nature, Lond.* **185**, 422–427.

Lehmann, H., and Carrell, R. W. (1969). *Br. med. Bull.* **25**, 14–23.

Linderstrøm-Lang, K. (1937). *Collegium*, **10**, 561–569.

McFadden, E. S., and Sears, E. R. (1946). *J. Hered.* **37**, 81–89.

Mahler, H. R., and Cordes, E. H. (1968). "Basic Biological Chemistry." Hooper International Editions.

Margoliash, E., and Smith, E. L. (1965). *In* "Evolving Genes and Proteins" (V. Bryson and H. J. Vogel, eds.), pp. 221–242. Academic, London and New York.

Muldal, S. (1952). *Heredity, Lond.* **6**, 55–76.

Ohno, S., and Atkin, N. B. (1966). *Chromosoma*, **18**, 455–466.

Pandey, K. K. (1967). *Nature, Lond.* **213**, 669–672.

Pauling, L., and Corey, R. B. (1951). *Proc. natn. Acad. Sci. U.S.A.* **37**, 251–256.

Pauling, L., Corey, R. B., and Branson, H. R. (1951). *Proc. natn. Acad. Sci .U.S.A.* **37**, 205–211.

Perutz, M. F., Rossman, M. G., Cullis, A. F., Muirhead, H., Will, G., and North, A. C. T. (1960). *Nature, Lond.* **185**, 416–422.

Sanger, F. (1955). *Society for Experimental Biology Symposium*, **9**, 10–30.

Smith, E. L. (1968). *In* "Homologous Enzymes and Biochemical Evolution." (N.v. Thoai and J. Roche, eds.), pp. 43–67. Gordon and Breach, New York, London, Paris.

Smithies, O. (1964). *Cold Spring Harb. Symp. quant. Biol.* **29**, 309–319.

Stebbins, G. L. (1950). "Variation and Evolution in Plants." Columbia U.P., New York.

Upadhya, M. D. (1968). *Experientia*, **24**, 613–614.

Vaughan, J. G., and Denford, K. E. (1968). *J. exp. Bot.* **19**, 724–732.

Watts, D. C. (1967). *Guy's Hosp. Rep.* **116**, 277–302.

Watts, D. C. (1968). *In* "Homologous Enzymes and Biochemical Evolution" (N.v. Thoai and J. Roche, eds.), pp. 279–296. Gordon and Breach, New York, London, Paris.

Watts, R. L., and Watts, D. C. (1968a). *Nature, Lond.* **217**, 1125–1130.

Watts, R. L., and Watts, D. C. (1968b). *J. theor. Biol.* **20**, 227–244.

Wyckoff, H. W. (1969). *Brookhaven Symp. Biol.* **21**, 252–257.

Yanofsky, C. (1963). *In* "Cytodifferentiation and Macromolecular Synthesis" (M. Locke, ed.), pp. 15–29. Academic, New York and London.

Zuckerkandl, E., and Pauling, L. (1962). *In* "Horizons in Biochemistry" (M. Kasha and B. Pullman, eds.), p. 189. Academic, New York and London.

Zuckerkandl, E., and Pauling, L. (1965). *In* "Evolving Genes and Proteins" (V. Bryson and H. J. Vogel, eds.), pp. 97–166. Academic, London and New York.

CHAPTER 9

Amino Acid Sequence Studies of Plant Cytochrome *c*, with Particular Reference to Mung Bean Cytochrome *c*

D. BOULTER, M. V. LAYCOCK,* J. RAMSHAW AND E. W. THOMPSON

Department of Botany, University of Durham, England

I. INTRODUCTION

One of the major generalizations of biochemistry is the overall biochemical unity of living organisms. By and large, the same basic processes occur in most organisms and are catalysed by homologous enzymes. Thus, an organism will contain an enzyme for a particular function which will resemble in three-dimensional structures its counterpart in many other organisms.

It became clear when DNA was shown to be the genetic material and to act by the production of messenger RNA molecules, that the sequence of base pairs in part of the DNA determined the sequence of amino acids in particular proteins. As pointed out by Zuckerkandl and Pauling (1965) therefore, a considerable amount of the evolutionary history of a species was encoded in the structure of its macromolecules, and that by comparing the base sequences of DNA and/or the amino acid sequences of the proteins of one organism with those of another, an estimate could be made of the relationship of the two organisms. Zuckerkandl and Pauling (1965) distinguished between the semantides, DNA, RNA and proteins, and the episemantides, the intermediary compounds of metabolism, the latter being the products of the former and containing smaller amounts of information.

At present it is not possible to sequence large DNA and RNA molecules, although hybridization techniques can be used to gain some insight into relationships (Kohne, 1968). At the moment however, there are difficulties both of technique and in interpretation using this method; particularly difficult is the fact that in evolution inversions, translocations and repetitions of the DNA

---

* Present address: Atlantic Regional Laboratory, Halifax, Nova Scotia, Canada.

sequences have occurred and these may obscure evolutionary relationships when hybridization techniques are used. Proteins, on the other hand, can be readily sequenced and their homologies established. There are a large number of different proteins which could be used for this purpose; cytochrome c was chosen for the following reasons. (1) It is ubiquitous in aerobic organisms, where it is found on the mitochondria as part of the electron transport chain. (2) It is relatively easy to purify since it is a small molecule of about 100 amino acid residues, is highly basic (isoelectric point about 10), relatively stable, and is a coloured protein. Its relatively small size is very important since although present techniques can be applied to proteins of molecular weight of up to 40,000, progress is more rapid, the smaller the molecule. (3) The sequence of cytochrome c has already been established for various animals, bacteria, fungi and for wheat germ (Dayhoff and Eck, 1968), and these results have shown it to be a representative protein for the purpose in hand, i.e. except in minor details, the sequence data fit the phylogenetic tree already established from the fossil record and by other means.

Table I gives some of the results obtained with animal cytochrome c, and it can be clearly seen that the differences found there are in agreement with existing ideas on the phylogenesis of animals. Figure 1 is a phylogenetic tree which has been established from the existing data (Dayhoff and Eck, 1967). Minor

TABLE I

Evolution of cytochromes c

Species comparison	Variations	Divergence of lines in millions of years
Horse–Chicken	11 ⎤	
Man–Chicken	13 ⎦	280 (assumed)
Horse–Tuna	19 ⎤	
Man–Tuna	21 ⎬	490 (calculated)
Chicken–Tuna	17 ⎦	
Horse–Moth	19 ⎤	
Man–Moth	31 ⎪	
Chicken–Moth	27 ⎬	750 (calculated)
Tuna–Moth	33 ⎦	
Horse–Yeast	45 ⎤	
Man–Yeast	44 ⎪	
Chicken–Yeast	46 ⎬	1180 (calculated)
Tuna–Yeast	48 ⎪	
Moth–Yeast	48 ⎦	

Data taken from Margoliash and Schejter (1966).

differences between the cytochrome data and existing evidence occurs only with turtle, chicken and kangaroo (Fitch and Margoliash, 1967). The rate of change of cytochrome *c* during animal evolution indicates that it is very suitable for ordination studies of major groupings.

Unlike the situation with animals, where there is an adequate fossil record to establish the phylogenetic tree, no comparable record exists for the plant kingdom. By the time of the Cretaceous period, when flowering plant fossils became abundant, all the major existing orders are represented, and it is not possible therefore to construct a phylogenetic tree from this source. For this reason,

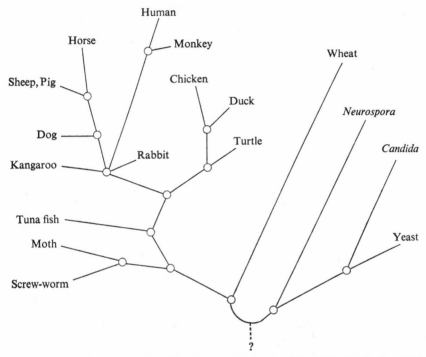

FIG. 1. Redrawn from the "Atlas of protein sequence and structure," by Dayhoff and Eck (1967)

botanists have had to use indirect methods and as a result there are several conflicting schemes; the use of these more objective criteria could therefore prove to be extremely important.

One of the major difficulties encountered has been the fact that cytochrome *c* occurs in very small amounts in plants; thus, 1 kg of horse heart will yield 250 mg of pure cytochrome *c*, whereas 1 kg of plant material yields about 4 mg of pure cytochrome *c*. Choice of material is important therefore, and dark-grown germinating seeds have been used since seeds are readily available, can be stored, and are biochemically conservative, i.e. their constituents do not change radically in seeds grown under different conditions; on germination they produce metabolically active seedlings. The use of dark-grown material avoids the complication of separating cytochrome *c* from the photosynthetic cytochromes. The difficulty of obtaining sufficient pure cytochrome *c* has also influenced the analytical methods used, and thus paper electrophoresis rather than ion-exchange chromatography has been used to separate peptides which have then been sequenced by Dansyl/Edman rather than by subtractive Edman methods, in contrast to the usual procedures with animal cytochromes *c*.

II. EXPERIMENTAL METHODS

Figure 2 is a block diagram outlining the major steps in the purification procedure for cytochrome *c*. The purity of the final product has been tested: (1) by electrophoresis on acrylamide gels, which showed it to consist of a single major component together with a very minor impurity; and (2) by the spectral

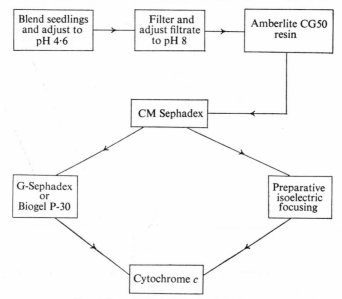

FIG. 2. Preparation of plant cytochrome *c*

characteristics at 550, 410 and 280 nm. The purified cytochrome *c* was separately digested with chymotrypsin and with trypsin, giving mixtures of peptides which were separated one from another, and their amino acid sequences determined by the Dansyl/Edman method (Gray and Hartley, 1963). *N*-terminal dansyl-ated amino acids were identified by three-dimensional polyamide sheet chromatography (Woods and Wang, 1967), and the amino acid sequence of each of the peptides was determined. The sequences were checked: (1) by total amino acid composition, (2) by calculating the mobility under the electrophoretic conditions used with the known mobility of similar peptides (Offord,

1966) and (3) by various colour reactions. The peptides were then fitted into the sequences given in Fig. 3 from a knowledge of the wheat germ sequence (Stevens *et al.* 1967), and from the known specificity of the chymotryptic and tryptic reactions.

III. AMINO ACID SEQUENCES OF HIGHER PLANT CYTOCHROMES *c*

Figure 3 gives the amino acid sequences as far as they are known for higher plant cytochrome *c*. These sequences have residues to the left of residue 1 (see legend of Fig. 3), i.e. they have a tail and in this respect they resemble all the known sequences in the fungi and also those of screw-worm fly and silkworm

```
Wheat:      Acetyl.Ala.Ser.Phe.Ser .Glu.Ala.Pro.Pro.
Mung Bean:  Acetyl.Ala.Ser.Phe.Ala.Glu.Ala.Pro.Pro.
Sunflower:  Acetyl.Ala.Ser.Phe.Ala.Glu.Ala.Pro.Ala.
              (−8)          (−5)          (−1)
```

Gly.Asn.Pro.Asp.Ala.*Gly*.Ala .Lys.Ile.*Phe*.Lys.Thr.Lys.*Cys*.
Gly.Asn.Ser .Asp.Lys.*Gly*.Met.Lys.Ile.*Phe*.Lys.Thr.Lys.*Cys*.
Gly.Asn.Pro.Thr.Thr.*Gly*.Ala .Lys.Ile.*Phe*.Lys.Thr.Lys.*Cys*.
(1)   (5)   (10)
Ala.Gln.*Cys*.*His*.Thr.Val.Asp.Ala.Gly.Ala.Gly.His.*Lys*.Gln.*Gly*.
Ala.Gln.*Cys*.*His*.Thr.Val.Asp.Lys.Gly.Ala.Gly.His.*Lys*.Gln.*Gly*
Ala.Gln.*Cys*.*His*.Thr.Val.Glu.Lys.Gly.Ala.Gly.His.*Lys*.Gln.*Gly*.
              (20)
Pro.Asn.*Leu*.His .*Gly*.Leu.Phe.Gly.*Arg*.Gln.Ser.*Gly*.*Thr*.Thr.Ala.
Pro.Asn.*Leu*.Asn.*Gly*.Leu.Phe.Gly.*Arg*.Gln.Ser.*Gly*.*Thr*.Thr.Ala.
Pro.Asn.*Leu*.Asn.*Gly*.Leu.Phe.Gly.*Arg*.Gln.Ser.*Gly*.*Thr*.Thr.Ala.
(30)   (40)

Gly.Tyr.Ser.*Tyr*.Ser.Ala .*Ala*.*Asn*.Lys.Asn.Lys .Ala.Val.Glu.*Trp*.
Gly.Tyr.Ser.*Tyr*.Ser.Thr.*Ala*.*Asn*.Lys.Asn.Met.Ala.Val. Ile .*Trp*.
Gly.Tyr.Ser.*Tyr*.Ser.Ala .*Ala*.*Asn*.Lys.Asn.Met.Ala.Val. Ile .*Trp*.
              (50)

Glu.Glu.Asn.Thr.Leu.Tyr.Asp.*Tyr*.*Leu*.Leu.*Asn*.*Pro*.TML.*Lys*.*Tyr*.
Glu.Glu.Lys .Thr.Leu.Tyr.Asp.*Tyr*.*Leu*.Leu.*Asn*.*Pro*.TML.*Lys*.*Tyr*.
Glu.Glu.Asn.Thr.Leu.Tyr.Asp.*Tyr*.*Leu*.Leu.*Asn*.*Pro*.TML.*Lys*.*Tyr*.
(60)   (70)

Ile.*Pro*.*Gly*.*Thr*.*Lys*.*Met*.Val.*Phe*.Pro.*Gly*.Leu.TML.*Lys*.Pro.Gln.
Ile.*Pro*.*Gly*.*Thr*.*Lys*.*Met*.Val.*Phe*.Pro.*Gly*.Leu.TML.*Lys*.Pro.Gln.
Ile.*Pro*.*Gly*.*Thr*.*Lys*.*Met*.Val.*Phe*.Pro.*Gly*.Leu.TML.*Lys*.Pro.Gln.
              (80)

Asp.*Arg*.Ala.Asp.Leu.Ile.Ala.Tyr.Leu.Lys.Lys .Ala.Thr.Ser .Ser.
Asp.*Arg*.Ala.Asp.Leu.Ile.Ala.Tyr.Leu.Lys.Glu.Ser .Thr.Ala.***
Glu.*Arg*.Ala.Asp.Leu.Ile.Ala.Tyr.Leu.Lys .Thr.Ser .Thr.Ala.Ala.
(90)   (100)   (104)

FIG. 3. Wheat cytochrome *c* sequence from Stevens *et al.* (1967). Data for mung bean and sunflower from the present study. TML = ε-*N*-trimethyllysine. Key to other amino acid abbreviations is given in Chapter 8, p. 152. Numbering as with animal cytochrome.

moth. Recently it became known (Stevens *et al.* 1967) that wheat germ was the first sequence with both a tail and an acetylated *N*-terminus. Stevens *et al.* (1967) pointed out that this finding destroyed the idea that the function of the tail was the same as that of acetylation of the *N*-terminus; i.e. to eliminate any effect of the positive charge of the *N*-terminus on the similar charge on the haem His. Wheat germ is not unique in having both a tail and *N*-acetylation of its *N*-terminus since this occurs in sunflower and mung bean (Fig. 3).

In bacteria, where *N*-formyl-methionyl *t*RNA is involved in protein chain initiation, proteins occur *in vivo* which are not formylated, and enzymes exist which can remove the formyl group, and in some cases possibly several of the amino acids at the *N*-terminal end as well (Adams and Capecchi, 1966). Recently, Laycock and Hunt (1969) have implicated *N*-acetyl-valyl *t*RNA in the initiation of haemoglobin synthesis and it is possible that protein synthesis on the 80 s ribosomes could occur by a blocked *N*-terminus. Since animals and fungi have 80 s cytoplasmic ribosomes and cytochrome *c* is known to be synthesized in the cytoplasm (Gonzalez-Cadavid and Campbell, 1967) it could be, by analogy with bacteria, that the cytochromes of fungi, fly and moth have additional amino acids and an *N*-acetylated *N*-terminus when first synthesized.

Examination of the sequences given in Fig. 3 strongly suggest that fungi, higher plants and animals, are all derived from a common ancestor since almost one third of the residues in all species so far examined are invariant (in italics, Fig. 3).

In addition it is likely that the higher plants have a common ancestor of their own. As discussed above, wheat cytochrome *c* has an unusually long tail, which begins *N*-acetyl-Ala-Ser-Phe, and this has also been found to be the starting sequence of mung bean and sunflower cytochrome *c*. Comparisons of the wheat sequence with that of all others shows that in twenty-two positions of wheat there are amino acid residues not previously encountered in those positions in other cytochromes (Fig. 3). These residues are: Ala (position −8), Ser (−7), Phe (−6), Ser (−5), Ala (−3), Pro (−1), Pro (3), Ala (5), Asp (21), Ala (24), Gly (25), Gln (28), Gln (39), Thr (42), Thr (43), Ala (56), Glu (60), Tyr (65), Leu (69), Pro (83), ϵ-*N*-trimethyllysine (TML) (86) and Gln (89). In mung bean nineteen are the same, the differences being: Ala (−5), Ser (3) and Lys (5). In sunflower eighteen are the same, the differences being: Ala (−5), Ala (−1), Thr (5) and Glu (21). Of particular interest is the report by Delange *et al.* (1969) of TML in positions 72 and 86 in wheat cytochrome *c*. This unusual amino acid does not occur in animal cytochrome *c*, but is known to occupy position 72 in *Neurospora crassa*. Analysis of mung bean and sunflower cytochrome *c* gave TML in positions 72 and 86. The presence of this amino acid is strong support for a higher plant common ancestor.

In the mung bean and sunflower sequences, seven positions only have an amino acid residue not previously encountered in these positions, i.e. Met (7), Thr (50), Met (55) and Lys (62) in mung bean; Thr (4), Thr (5), Met (55) and Thr (100) in sunflower.

IV TAXONOMIC IMPLICATIONS OF THE CYTOCHROME *c* DATA

The results described above show the close similarities of the three sequences, but of equal importance, particularly for taxonomic studies, are the differences between the species. Fourteen differences have been found between wheat and mung bean. These are: Ala for Ser (−5), Ser for Pro (3), Lys for Ala (5), Met for Ala (7), Lys for Ala (22), Asn for His (33), Thr for Ala (50), Met for Lys (55), Ile for Glu (58), Lys for Asn (62), Glu for Lys (100), Ser for Ala (101), Ala for Ser (103) and deletion for Ser (104). Fourteen differences have also been found between sunflower and wheat, which are: Ala for Ser (−5), Ala for Pro (−1), Thr for Asp (4), Thr for Ala (5), Glu for Asp (21), Lys for Ala (24), Asn for His (33), Met for Lys (55), Ile for Glu (58), Glu for Asp (90), Thr for Lys (100), Ser for Ala (101), Ala for Ser (103) and Ala for Ser (104). There are eleven differences between sunflower and mung bean.

The phylogenetic position of the fungi is at present uncertain. The 40–50 differences between fungi and mammals and between fungi and higher plants, places them fairly well on in the evolutionary plan of Margoliash and Schejter (Fig. 1). The 41 differences between *Neurospora* and Baker's yeast, and the 27 between Baker's yeast and *Candida* are greater than might be expected.

When sufficient data is available, statistical analysis similar to that of Fitch and Margoliash (1967) will have to be used. In addition each position may have to be weighted. The present results indicate that differences of the order of 10–15 residues could occur between plants, which were distantly related, and that cytochrome *c* therefore will be very useful in ordinating the major plant groups. Finer delimitations may require the use of a more rapidly evolving protein than cytochrome *c*. Recent work with *Leucaena glauca* ferredoxin (Benson and Yasunobu, 1969) indicates polymorphism for certain amino acid residues and suggests therefore the possibility of using such differences at the level of population studies. The present investigation has so far not given any evidence of polymorphism. However, many pooled seeds of highly cultivated species have been used as starting materials, and these results may not be typical of the situation in natural populations.

Although the sequencing techniques require some sophisticated equipment and experience, it would appear that the methods are sufficiently rapid to accumulate sufficient data to allow the determination of the major branches of the phylogenetic tree of the plant kingdom over the next decade.

REFERENCES

Adams, J. M., and Capecchi, M. R. (1966). *Proc. natn. Acad. Sci. U.S.A.* **55**, 147–155.

Benson, A. M., and Yasunoba, K. T. (1969). *J. biol. Chem.* **244**, 955–981.

Dayhoff, M. O., and Eck, E. (1967). "Atlas of protein sequence and structure 1966–67." National Biomedical Research Foundation, Maryland, U.S.A.

Dayhoff, M. O., and Eck, E. (1968). "Atlas of protein sequence and structure 1967–68." National Biomedical Research Foundation, Maryland, U.S.A.

Delange, R. J., Glazer, A. N., and Smith, E. L. (1969). *J. biol. Chem.* **244**, 1385–1388.

Fitch, W. M., and Margoliash, E. (1967). *Science*, **155**, 279–284.
Gonzalez-Cadavid, N. F., and Campbell, P. N. (1967). *Biochem. J.* **105**, 443–450.
Gray, W. R., and Hartley, B. S. (1963). *Biochem. J.* **89**, 379–380.
Kohne, D. E. (1968). *In* "Chemotaxonomy and Serotaxonomy" (J. G. Hawkes, ed.), pp. 117–130. Academic, London and New York.
Laycock, D. G., and Hunt, J. A. (1969). *Nature*, **221**, 1118–1122.
Margoliash, E., and Schejter, A. (1966). *Adv. Protein Chem.* **21**, 113–286.
Offord, R. E. (1966). *Nature*, **211**, 591–593.
Stevens, F. C., Glazer, A. N., and Smith, E. L. (1967). *J. biol. Chem.* **242**, 2764–2779.
Woods, K. R., and Wang, K. T. (1967). *Biochem. Biophys. Acta*, **133**, 369–370.
Zuckerkandl, F., and Pauling, L. (1965). *J. theor. Biol.* **8**, 357–366.

CHAPTER 10

Molecular Approaches to Populational Problems at the Infraspecific Level

B. L. TURNER

*The Cell Research Institute and Department of Botany,
The University of Texas at Austin, Austin, Texas, USA*

I. INTRODUCTION

Numerous workers have contributed to the ever mounting list of chemical components found among higher plant species. Most such work has been based upon relatively few samples, usually one or two individuals from a single population and even then little effort has been made to compare whole plants at comparable stages of development (to say nothing of the failure to document investigations through the preservation of voucher specimens). Alston (1967) has reviewed critically many of these studies and has predicted that in future chemosystematic work we can expect "a gradual shift in emphasis from the problem-exposing or the data-exposing phase to the problem-attacking phase." As will be indicated below, some of the most intriguing evolutionary problems exist at the infraspecific or populational level and, because of the difficulty in perceiving and/or gathering morphological data for their resolution, it would appear that such problems will prove especially attractive to the chemosystematist.

While a broad conceptual framework now exists for the evaluation of chemical data at the specific level or lower (Alston, 1967; Turner, 1967, 1969) relatively few studies have attempted to accumulate sufficient chemical data for meaningful systematic interpretations. Plant systematists in general, accustomed as they are to working with numerous, easily observed, usually quite variable morphological characters, are rightfully suspicious of the sweeping generalizations and extrapolations often made by chemosystematists using blatantly inadequate populational data. Indeed, until the comparative phytochemist becomes as much aware of inter- and intra-populational variability in chemical characters as the classically oriented systematist has been to morphological characters, most plant systematists will continue to play down the significance of chemical data.

A few early workers in the area of chemosystematics, notably Mirov (1956) working with *Pinus*, were very much aware of the kinds of data and sampling needed for adequate populational study but, largely because of inadequate instrumentation, the time involved in merely accumulating relatively simple listings of data precluded the intensive survey for numerous compounds from a large assemblage of individuals and populations. In addition, numerical methods using high speed computers, essential to the evaluation of such large masses of data, have been relatively recent developments (Sokal and Sneath, 1963; Flake and Turner, 1968).

II. Infraspecific Structure of Plant Populations

No attempt will be made here to discuss the many facets of this topic. Rather I would like to refer the reader to an excellent article on the subject by Ehrendorfer (1968). He neatly summarizes and evaluates much of the recent work in the area of infraspecific differentiation, showing that populational variability in morphological, chromosomal, as well as chemical characters can be exceedingly complex. Most variation patterns are correlated with ecological or geographical patterns or gradients. Factors affecting such patterning include (1) size of population, (2) environmental position, (3) migration routes, (4) variation and selection and (5) reproductive mechanisms.

Ehrendorfer recognized two extreme types of geographical and ecological differentiation in populations which he referred to as allopatric and partially sympatric. *Allopatric differentiation* occurs when subpopulations become divergent at the periphery of an already variable gene pool so that differentiation takes place without the build-up of strong barriers to gene exchange among individuals of the parental population. Divergence of subpopulations is usually dependent upon isolation of the latter from the parental gene pool. Any new contact of isolated peripheral populations with the parental gene pool, if gene flow is relatively unrestricted, will produce clinal or stepwise intergradation (depending upon ecogeographical factors) or at least regional differentiation. *Partially sympatric differentiation* occurs when inbreeding subpopulations are formed within the geographical confines of the parental population, usually because of abrupt intrinsic barriers to gene exchange among at least some of the subpopulations. This produces a populational structure or pattern which is essentially a mosaic, the barriers to gene exchange having produced a sorting of variation and divergence among the populations.

These two phyletic models for populational differentiation are illustrated in

Fig. 1. They are at best highly simplified and all manner of variations upon and between the models can be anticipated. What matters is that the systematist be made aware of the kind and extent of chemical variation likely to be encountered in natural populations, always keeping in mind that the variation is

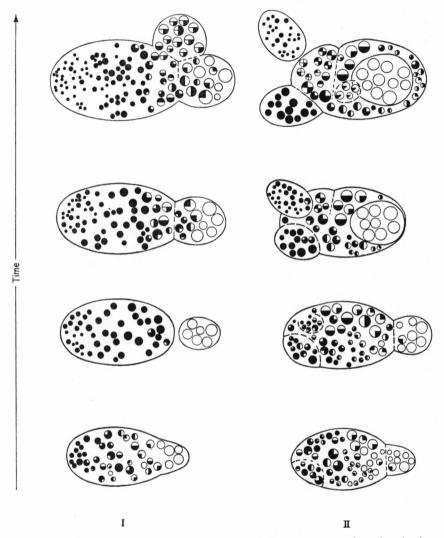

I   II

FIG. 1. Evolutionary scheme in two idealized population groups over four time levels: I, allopatric; II, partly sympatric differentiation. Further explanation in the text. (After Ehrendorfer, 1968)

likely to reflect, to some extent, the influence of all the factors mentioned above.

Most of our knowledge of differentiation in natural populations has come out of the study of morphological or chromosomal characters. Chemosystematic investigations, especially those involving the use of micromolecular data, have largely been concerned with taxonomic problems at the specific level, particularly where hybridization among the taxa concerned has been a factor in the variation (Turner, 1969). There has, however, been a recent trend toward more intensive studies at the infraspecific level using chemical data and, as will be indicated below, these have paid off handsomely.

III. CHEMICAL INVESTIGATIONS AT THE INFRASPECIFIC LEVEL

Several interesting publications have appeared on the significance of variation in chemical constituents in and among infraspecific populations. Many of the earlier studies, including those of his colleagues and students working with *Baptisia*, have been reviewed by Alston (1967). More recently I have reviewed several interesting papers in this area (Turner, 1969), and Mabry (Chapter 13) has given a detailed account of his very extensive populational studies in *Ambrosia*.

In my treatment of infraspecific variation in chemical constituents in the present paper, I have selected for review four studies which seem to me especially instructive. Each of these adds a significant dimension to the field and taken together they provide a historical perspective which, I hope, will compensate for my inability, because of space limitations, to cover all those published reports relating to the subject.

The alkaloid investigations on *Cinchona* by the indefatigable W. H. Camp is important because, to my knowledge, it is one of the earliest populational studies in which chemical data have been brought to bear on a systematic problem. The study would not have been possible at the time (undertaken during World War II, 1942–1945) were it not for governmental support of the project, for it engaged the services of several first-rate taxonomists and organic chemists, the purpose being to locate high yielding quinine trees. To judge from Camp's paper, very little planning went into the systematic side of the "team effort", especially as concerns populational sampling and what might be needed for adequate statistical treatment. Still, the study is important because it showed the existence of chemical clines across a broad region and this only a few years after the term *cline* was proposed for such character gradients (Huxley, 1942).

The study of chemical variation in populations of lichens by Culberson and Culberson (1967) is reviewed because it illustrates the value of intensive infrapopulational studies at the local level, even when only a few compounds are available for investigation.

The work on contour mapping of chemical characters in *Juniperus pinchotii* by Adams (1969) is significant in that it points to the direction that future studies in this area might take. It also provides a very convenient method for the presentation of chemical data since variation-trends across broad geographical regions for single compounds or those in combination can be visualized at a glance.

Finally, the study by Flake, von Rudloff and Turner (1969) in *Juniperus virginiana* is included because chemical and numerical approaches have been used to help resolve long-standing systematic problems in that species. Comparison of this study on clinal variation with that of Camp's on *Cinchona* should also prove instructive.

A. POPULATIONAL VARIABILITY OF ALKALOIDS IN *Cinchona*

Camp (1949), in an often overlooked but significant chemosystematic paper (modestly entitled "Cinchona at High Elevations in Ecuador"), presented populational data for four alkaloids (quinidine, cinchonidine, quinine and cinchonine) found in the bark of a group of taxa belonging to the genus *Cin-*

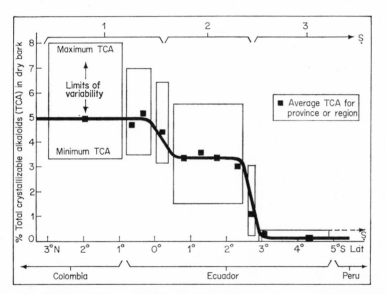

FIG. 2. The cline in total crystallizable alkaloids in three races of a species-complex in *Cinchona* from along the crest of the Andes from about 3° N Lat in Colombia, to 5° S Lat, near the Ecuador–Peru border. (Modified from Camp, 1948). Additional explanation in text

chona. Quantitative variation in these compounds was obtained from over 120 trees representing eleven populations, the latter sampled along a north–south transect of 500 miles from Colombia to Peru (Fig. 2).

The study suffers in that the populational samples are of varying sizes (from 3 to 30 individuals) and statistical data are not given for all of the populations studied. Also, little information is presented as to the age of the trees, season collected, or how the bark samples were subsequently stored or treated. While not evaluated statistically, the data available to him suggested that the taxon under investigation was composed of a chemical cline divisible into "steps."

The *Cinchona* study is significant in that it is one of the very earliest attempts to apply chemical data to populational problems, concluding in this instance, that what had been treated variously as three or more specific taxa was in reality a clinal complex composed of three intergrading populational units. Camp also concluded that the cline was largely a result of gene-flow into the taxon from adjacent species and proposed the term *exogenous cline* to describe such clinal types. He distinguished this from the usual situation in which genetic differentiation *within* the population (i.e. mutation and selection among local or intergrading habitats) accounts for the pattern of variation disclosed. He termed the latter an *endogenous cline*. These are important concepts since both types undoubtedly occur in nature, both singly and in combination.

In the *Juniperus* studies outlined below, it will be shown that what was previously thought to be an exogenous cline in *J. virginiana* is in fact an endogenous cline, the crucial evidence coming from gas chromatographic studies and numerical treatments of the resulting data. Finally, an attempt will be made to indicate the direction in which infraspecific studies will move, at least from the standpoint of presentation of data (contour mapping) and numerical evaluation (character weighting according to variance within and between populations).

B. POPULATIONAL VARIABILITY OF LICHENIC ACIDS IN *Ramalina*

Most of the populational studies of plant taxa published to date have been relatively broad regional studies. A notable exception has been that of Culberson and Culberson (1967) who studied the variation in lichenic acids among a group of lichens belonging to the genus *Ramalina*. Their study of 980 plants (10 from each of 98 quadrats) is effectively summarized in Figs. 3 and 4; briefly, it shows how much information can be obtained from detailed populational work, even when examining for only a few compounds in a very local area. As a result of this work and subsequent studies, Culberson (1969) concluded that the *Ramalina* complex at this locality was composed of three chemically differentiated "races" which were acting like species (i.e. each occurred in different but adjacent habitats, and in spite of prolific sexual reproduction, they retained their identities when grown side by side, there being no genetic recombination of the chemical characters; in short, they appeared to have their own populational "gene pool" in spite of the opportunities for outcrossing with the other "races").

C. CONTOUR MAPPING OF CHEMICAL DATA IN *Juniperus pinchotii*

Most systematists working at the infraspecific level are usually interested in evolutionary processes. That is, they are concerned in establishing the extent of variation among characters, the correlation of this variation with geographical or environmental factors, why such correlations exist, and the source of such

Fig. 3. The promontory in North Wales studied for the behaviour of the chemical races of the *Ramalina siliquosa* lichens. All of the cliff faces support a *Ramalina* vegetation. To the south and west, the *Ramalina* zone faces the sea and is directly above a *Verrucaria maura* zone (V), which in turn is above the algae of the *Fucus-Ascophyllum* zone (F), here seen exposed at low tide. Toward progressively more sheltered conditions around the headland, the *Ramalina* zone on the northwest side faces a rocky beach at the left of the photograph; on the northeast, the most protected place of all, it faces a grassy slope (G). The location of the six line-transects, indicated by numbers, is approximate and the distance between transects 1 and 2, 2 and 3, and 3 and 4 (actually 2·5, 3·4 and 4·1 m respectively) is distorted by perspective. Fig. 4 has a breakdown of the chemical types found in the six transects. (After Culberson and Culberson, 1967)

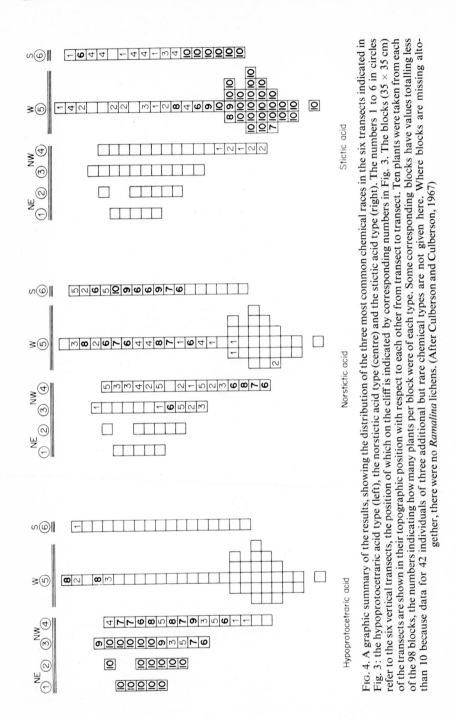

Hypoprotocetraric acid   Norstictic acid   Stictic acid

FIG. 4. A graphic summary of the results, showing the distribution of the three most common chemical races in the six transects indicated in Fig. 3: the hypoprotocetraric acid type (left), the norstictic acid type (centre) and the stictic acid type (right). The numbers 1 to 6 in circles refer to the six vertical transects, the position of which on the cliff is indicated by corresponding numbers in Fig. 3. The blocks (35 × 35 cm) of the transects are shown in their topographic position with respect to each other from transect to transect. Ten plants were taken from each of the 98 blocks, the numbers indicating how many plants per block were of each type. Some corresponding blocks have values totalling less than 10 because data for 42 individuals of three additional but rare chemical types are not given here. Where blocks are missing altogether, there were no *Ramalina* lichens. (After Culberson and Culberson, 1967)

variation. Variation among individuals in a given natural population is the result of the interplay of genetic and environmental factors. In fact the first question that most non-systematists ask about character variations found in nature is whether or not these have a genetic basis. This is not the place for a prolonged digression on this subject but suffice it to say that nearly all experimental work to date has indicated that variations in natural populations have a genetic basis, however much their expression might be influenced by environmental factors (Clausen *et al.* 1940). Because of this well-established fact, most systematists feel that they do not have to prove each time anew that the populational variation found in a given character is or is not the result of environmental factors. Rather it is taken as axiomatic that most, if not all, characters which are relatively well established in natural populations have been selected for genetically because of adaptational fitness, either singly or in combination (i.e. coherence; cf. Clausen and Hiesey, 1958). To ask that a systematist prove *each time anew* that the characters under investigation vary because of genetic factors is like asking that a chicken be made to lay an egg in order to prove its capacity. The fact that an occasional chicken doesn't lay eggs, doesn't prohibit the commercial egg producer from purchasing unproved layers by the thousands. I hope this crude analogy makes its point; too many non-systematists have laboured this question, as if the science of systematics dealt with absolutes and not probabilities.

In fact, systematics at the infraspecific level, in an operational sense, might best be defined as the application of statistical methods (or probability theory) to natural populations. It is perhaps for this reason that many if not most plant systematists prefer to lump plus-or-minus "recognizable" races into "aggregate species": the statistical data needed to justify their recognition are too difficult to assemble or else, where the genus is unusually large, the interpretation of even relatively simple statistical data can often times be as subjective and difficult as that process involved in sorting an assemblage of variable plants. It is for this reason, among others, that numerical treatments of taxonomic data have proved so uninspiring to plant systematists generally. For infraspecific populational work in particular, what has been needed is a pictorialized presentation of these kinds of data such that a worker might visualize how the characters in question are varying singly or in combination across the geographic range of a species. Of the numerical techniques developed for this purpose the most promising appears to be that of contour mapping. I would like, therefore, to discuss very briefly the interesting work of Adams (1969) who has applied these techniques in his populational studies of *Juniperus*.

Juniperus is a genus of about 60 species belonging to the conifer family Cupressaceae (Dallimore and Jackson, 1966). A number of species are native to North America where they often dominate the more xeric tree communities, especially in the southwestern United States. The taxa are often difficult to distinguish and much disagreement has been expressed as to the relationship and status of certain populations or taxa. One such complex, *J. pinchotii*, is centred in western Texas. Adams (1969), using gas–liquid chromatography, has studied in considerable detail the volatile terpenoids and selected morphological features in this group and these data have been brought to bear on some of the taxonomic questions raised by various workers. Briefly, some contend that the species is heavily hybridized throughout its range with the often sympatric *J. ashei*, while others treat the group as relatively uniform, albeit variable,

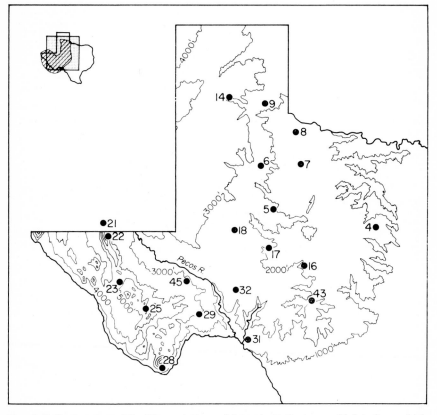

FIG. 5. Collection sites of the 20 populations of *J. pinchotii*, the terpenoid data from which were used to construct Figs. 6–10. Note that sites 25 and 28 are in the mountainous regions of trans-Pecos, Texas and the two populations from this area possess divergent accumulations of terpenoids (Fig. 10). Additional explanation in text

except for some of the more western mountainous populations which seem sufficiently divergent to be considered a separate species.

After the accumulation of terpenoid data (samples of 4 branches from each of 5 trees from the 20 populations shown in Fig. 5) Adams was faced with the problem of presenting his data so that these might be readily comprehended by systematic workers generally. To this end he settled upon a numerical treatment and algorithm that is too complex to outline here, but briefly expressed the chemical data are placed on punch cards and fed into a computer such that the data are expressed automatically on a grid system in the form of contours which can be read essentially as one might read a topographic map.

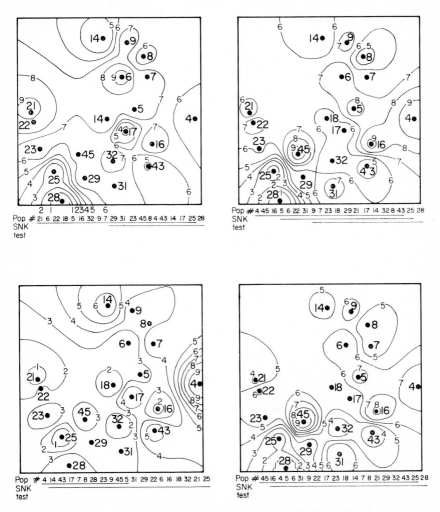

FIGS. 6–9. Contour map showing percentage distribution of camphor (Fig. 6, top left), camphene (Fig. 7, top right), myrcene (Fig. 8, bottom left) and tricyclene (Fig. 9, bottom right) in natural populations of *J. pinchotii*. The area concerned and range of the species is indicated by small inserts in the upper left hand corner of Fig. 5. Populations are numbered and are indicated by dots. Additional explanation in text. (From Adams, 1969)

For example, Figs. 6–9 show the distribution of tricyclene, camphene, myrcene and citronellal across the range of *J. pinchotii* as expressed in the form of contours showing the relative percent concentration of those compounds in natural populations. Approximately 70 terpenoids were studied in

J. pinchotii. Of these, 19 showed highly significant differences between populations. Adams then treated the latter data numerically and statistically (according to highly significant differences at the 1 % level among the populations, indicated by SNK tests shown by bar-groupings beneath the figures—the significantly different populations being indication by inclusion in a given

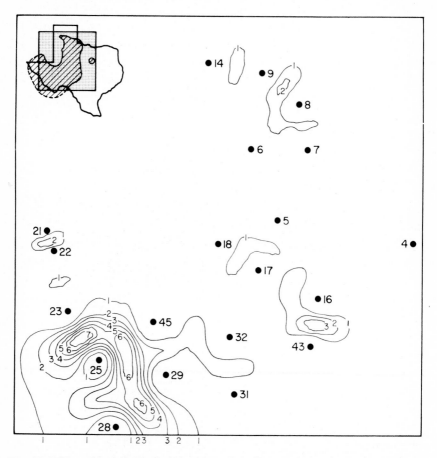

Fɪɢ. 10. Differential of 19 chemical characters (i.e. taken together and treated numerically according to those populations showing infraspecific differentiation or divergence) found in *J. pinchotii*. Note that the divergent populations ("topographic highs") are found in the mountainous regions of trans-Pecos, Texas (cf. Fig. 5). See text for additional discussion. (After Adams, 1969)

bar-group). Figure 10 shows the composite differential of these 19 compounds taken together so as to expose the divergent populations; i.e., as expressed in a contour map showing the relative rates of change (differential) among the 20 populations, according to the combined distributional significance of the 19 terpenoid compounds for which data were available.

It will be noted that populations 25 and 28 diverge considerably from the other populations, the former occupying the mountainous regions of westernmost Texas (Fig. 5). In fact, these are the very populations which have been recognized as either (1) belonging to the species *J. monosperma*; (2) being hybridized; or (3) a species in its own right. The work by Adams negates hypotheses (1) and (2) in that the populations concerned, on chemical and numerical grounds, clearly cluster *within* the *J. pinchotii* complex and not with *J. monosperma*. Also, no evidence was assembled to support the view that extant hybridization was a factor in the variation in the two populations. While his data do not resolve the question of specific status, assignment of rank to a taxon being subjective, he did show that these populations are, phyletically speaking, most closely related to *J. pinchotii*.

Presentation of data in this fashion makes possible the rapid and easy assimilation of considerable information at a glance. Of course, any contour presentation will be only as reliable as the data on which it is based. For example, the number of trees (5) from each of the 20 populations sampled in *J. pinchotii* was probably not sufficiently large to be representative of the populations. Also, a larger number of populations, perhaps 40, would have given somewhat more meaningful contours. But these are problems of sampling inherent in all such populational studies: one can only approach the absolute.

Contour mapping of the sort discussed above is almost certain to have a wide appeal among systematists, comparative phytochemists and ecologists, especially those interested in evolutionary processes. This approach might appropriately be referred to as "character ecology" or "adaptational ecology of characters" since it makes possible the formulation of evolutionary models that can be subjected to experimental analysis both through transplant studies and growth chamber experiments.

D. CLINAL STUDIES OF TERPENOIDS IN *Juniperus virginiana*

Several North American species of *Juniperus* have been studied in considerable detail as concerns their participation in situations involving hybridization. Hall (1952) analysed the morphological variation found in six morphological characters of the allopatric species *J. ashei* and *J. virginiana*, concluding that the former has influenced the latter by hybridization and subsequent introgression throughout most of the southeastern United States (Fig. 13). Indeed, the study of hybridization between *J. ashei* and *J. virginiana* has been repeatedly referred to as one of the most detailed, best documented examples of allopatric introgression in the literature (Davis and Heywood, 1966).

Hall's study was primarily a morphological one and, because of the considerable subjectivity inherent in the evaluation of morphological characters, particularly those involving habital features of the plant, we selected terpenoid characters for intensive study using gas chromatographic techniques. Gas chromatographic techniques are unusually well suited to the study of infraspecific variability, especially where the data processing capabilities of modern computers are used in the evaluation of populational data. Modern chromatographic units with automatic print-out devices, accurate to the second decimal place, enable the chemosystematist to procure large sets of quantitized data of an unbiased nature in a matter of minutes. This is usually not possible in the case of morphological features.

Using this approach we ascertained the quantitative variation in terpenoid compounds for individuals both within and between populations; these data were then used in combination to test for hybridization, introgression and/or clinal variation within the *J. virginiana* complex.

Several techniques have been developed for the detection of hybridization and introgression in natural populations. One of the most frequently used has

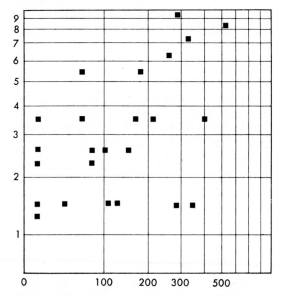

FIG. 11. Scatter diagram of 24 specimens from a purportedly introgressed *J. virginiana* population from Austin, Texas. Leaf gland length–width ratio is plotted on the vertical axis; length of lateral whip on the horizontal axis. (Data from locality 9, Fig. 10 of Hall, 1952)

been the pictorialized scatter diagram. Anderson (1949) has discussed in detail the theoretical basis of its applicability in studies involving introgression, and numerous plant workers have used this approach, presumably with considerable success (Davis and Heywood, 1966).

Basically, scatter diagrams are just another way of presenting populationally data in some form other than tabular listings. Variability data are assembled for each plant of a given population and this is presented in the form of a pictorialized spot on a two-dimensional grid, the coordinates being determined by the intersection of lines from the variables plotted along the ordinate and abscissa. The selection of characters for the latter is most important in constructing scatter diagrams for it is these axes that permit the recognition of the so-called "recombination spindle." That is, the visual detection of aggregations of spots near the ends of the opposed diagonals on such a two-dimensional grid suggests the occurrence of two gene pools or populations at the locality sampled, any hybrid individuals and/or backcrosses would occupy appropriate loci between those aggregations. Fig. 11 shows a scatter diagram for a population of *Juniperus* from Austin, Texas. This is based on data assembled by Hall (1952). He interpreted this (and numerous similar diagrams) as indicative of introgression from *J. ashei* into *J. virginiana*. Using chemical characters for the

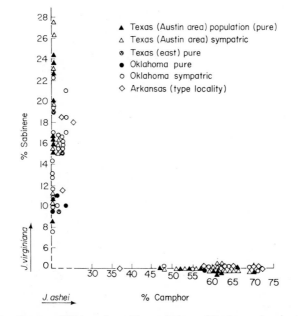

Fig. 12. Scatter diagram of 88 trees from 11 populations of *Juniperus* showing the absence of chemical recombinations (for camphor and sabinine) in purportedly hybrid and/or introgressed populations. Additional explanation in text

construction of scatter diagrams, however, from populations also from Austin and other localities sampled by Hall in his studies, von Rudloff, Irving and Turner (1967, 1969) could detect no evidence of hybridization between these two species. Figure 12 shows a scatter diagram constructed from 88 trees collected from 11 populations, *most of the specimens having been selected as putative hybrids and backcrosses*. It will be noted that individuals intermediate for the chemical characters concerned are strangely absent. Indeed, *J. virginiana* and *J. ashei* are remarkably different in their terpenoids (von Rudloff, 1968; Vinutha and von Rudloff, 1969), of the approximately 90 compounds found in the two species only 11 are common to both taxa. Finally, it should be noted that none of the approximately 79 compounds which are confined to one or the other species was ever detected outside of the taxon to which it normally belonged. It was therefore concluded that hybridization and introgression were not factors in the obvious morphological variation found in populations of *J. virginiana*.

Barber and Jackson (1957) have been the only other workers to question allopatric introgression as an explanation for the variation found in *J. virginiana*. They suggested rather that the variation found in the latter species was

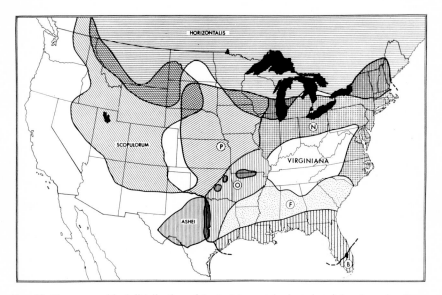

FIG. 13. The geographical distribution of *J. virginiana* and selected peripheral species. Races of *J. virginiana* are indicated by letters as follows: (P) Platter River Race; (N) Northern Race; (F) Florida Race; and (O) Ozark Race. Populations typical of the species occur in the Appalachian Region of the Eastern United States (labelled "virginiana"). Vertical diagonals (B) in the southeastern-most U. S. show the range of the closely related species, *J. barbadense*. *Juniperus ashei* in central Texas (with outliers in Oklahoma, Arkansas and Missouri) has purportedly introgressed with *J. virginiana* producing populations (e.g. Ozark Race) approaching the former in habital features. Chemosystematic data do not support this interpretation. Additional explanation in text. (After Hall, 1952)

due to differential selection along an ecological gradient. The so-called introgressed populations of *J. virginiana* (consisting of trees with wider aprons, shorter terminal shoots, etc.) were believed to be due to selective influences working on a variable gene pool in the western part of its range so as to produce populations superficially like *J. ashei*. In short, these authors suggest that the morphological variation found in *J. virginiana* is clinal; i.e. the species has formed, or is in the process of forming, regional races as a result of adaptational mechanisms arising out of its own gene pool, this being unrelated to the possible influx of genes from the largely allopatric *J. ashei*.

Confirmation of a clinal structure in the populations comprising *J. virginiana* was made possible through a clump-searching algorithm recently developed to assist in problems of biological classification, particularly at the infraspecific level (Flake and Turner, 1968). Terpenoid data were collected in the form of

FIG. 14. Map showing clinal structure of *J. virginiana* populations along a transect from Washington, D.C. (9) to eastern Texas (1). Contours identify those populations with statistically significant aggregational differences. Thus, populations 6, 7, 8 and 9 are relatively homogeneous and cluster successively at high similarity levels (·9574 per cent similarity or better). Note that populations 10 and 11, *sampled* at a different season, do not cluster with their nearest neighbours, indicating the importance of seasonal influences in the chemical composition of populations. Additional explanation in text. (After Flake *et al.* 1969)

samples of about 10 plants from 9 population sites at approximately 150-mile intervals on a transect from near Texarkana, Texas, to Washington, D.C. Gas chromatographic analysis yielded percentage distributions of 37 distinct compounds that were detected in sufficient quantities to be well above instrumental error as well as having the appearance of being present in reasonably consistent amounts within the individual populations.

A contour map showing the hierarchy of aggregations for eight of the *J. virginiana* populations, using characters weighted according to variance, is shown in Fig. 14. The aggregation at the highest level contains populations 8 and 9, located near Durham, North Carolina and Washington, D.C. At lower levels, the aggregations tend to enclose successive populations along the transect from Washington, D.C., to the vicinity of Texarkana, Texas, indicating a northeast to southwest clinal variation in *J. virginiana*.

The terpenoid data were treated statistically by Flake *et al.* (1969). The results of their statistical analysis showing the aggregations with statistically significant level differences may be summarized by listing the population identification numbers in the order in which they appear in the hierarchy of aggregations in Fig. 14 as follows:

$$\underline{1} \quad \underline{2} \quad \underline{\underline{4} \quad 3 \quad \underline{\underline{6} \quad 7 \quad 8} \quad 9}$$

These population numbers may be used to specify the various aggregations by associating each aggregation with the population that entered the hierarchy at the aggregation's level. Those population numbers that are underscored by the same line identify aggregations with insignificant level differences. Those population numbers not underscored by the same line identify aggregations with significantly different levels. These results are illustrated by the aggregation contours in Fig. 14. The dotted contour section in the vicinity of population 6 signifies the corresponding overlap of the significant ranges.

It seems pertinent to compare the hierarchical structure produced by these approaches with phytogeographical theory. The clinal structure from northeast to southwest with the more northeastern populations (6, 7, 8 and 9) clustering more or less homogeneously at the highest significant level of similarity, the geographically more distant populations clustering at successively lower levels of similarity, is consistent with what is believed to have been the phyletic history of the extant populations that comprise the species; i.e. that *J. virginiana* was originally a species belonging to an essentially northeastern flora, having occupied the ancient land mass of Appalachia in remote times but subsequently extending its range to peripheral western areas as previously submerged land became newly exposed for occupancy.

REFERENCES

Adams, R. P. (1969). "Chemosystematic and Numerical Studies in Natural Populations of *Juniperus*." Doctoral Thesis, University of Texas.

Alston, R. E. (1967). *In* "Evolutionary Biology I" (T. Dobzhansky, M. K. Hecht and W. C. Steeve, eds.). Meredith, New York.

Anderson, E. (1949). "Introgressive Hybridization". Wiley, New York.

Barber, H. N., and Jackson, W. D. (1957). *Nature*, **179**, 1267.

Camp, W. H. (1949). *Brittonia*, **6**, 394.

Clausen, J., Keck, D. D., and Hiesey, W. M. (1940). *Carnegie Inst. Wash. Publ.* **520**, 1.
Clausen, J., and Hiesey, W. M. (1958). *Carnegie Inst. Wash. Publ.* **615**, 266.
Culberson, W. L. (1969). *Taxon*, **18**, 152.
Culberson, W. L., and Culberson, C. F. (1967). *Science*, **158**, 1195.
Dallimore, W., and Jackson, A. B. (1966). "A Handbook of Coniferae and Gingoaceae." Arnold, London.
Davis, P. H., and Heywood, V. H. (1966). "Principles of Angiosperm Taxonomy." Oliver and Boyd, London.
Ehrendorfer, F. (1968). *In* "Modern Methods In Plant Taxonomy." Academic, London and New York.
Flake, R. H., and Turner, B. L. (1968). *J. Theor. Biol.* **20**, 260.
Flake, R. H., von Rudloff, E., and Turner, B. L. (1969). *Proc. natn. Acad. Sci. U.S.A.* (In press.)
Hall, M. T. (1952). *Ann. Mo. Bot. Gdn.* **39**, 1.
Huxley, J. (1942). "Evolution: the Modern Synthesis." Harper, New York.
Mirov, N. T. (1956). *Can. J. Bot.* **34**, 443.
Sokal, R. R., and Sneath, P. H. A. (1963). "Numerical Systematics." Freeman, San Francisco.
Turner, B. L. (1967). *J. Pure Appl. Chem.* **14**, 189.
Turner, B. L. (1969). *Taxon*, **18**, 134.
von Rudloff, E., Irving, R., and Turner, B. L. (1967). *Am. J. Bot.* **54**, 660.
von Rudloff, E. (1968). *Can. J. Chem.* **46**, 679.
von Rudloff, E., Irving, R., and Turner, B. L. (In preparation).
Vinutha, A. R., and von Rudloff, E. (1969). *Can. J. Chem.* **46**, 3743.

CHAPTER 11

Environment and Enzyme Evolution in Plants

H. W. WOOLHOUSE

Department of Botany, University of Sheffield, Sheffield, England*

I. INTRODUCTION

It is now apparent that the metabolic pathways mediating the major physiological processes such as respiration, nucleic acid and protein synthesis are common to all living organisms. The enzymes catalysing the individual steps in these pathways are being characterized and, as may be seen from the "Atlas of Protein Structures" (Dayhoff and Eck, 1968) the amino acid sequences of a few particular enzymes are known for a number of different organisms. Comparison of particular enzymes shows that there are a large number of possible variations in the amino acid sequences whilst conserving the overall specific catalytic properties of the enzyme molecule. As an example, Table I (after Margoliash and Fitch, 1968) summarizes the extent of these variations in the case of cytochrome *c*. Partial degradation and inhibitor studies show that the active centres of many enzymes occupy a relatively small amino acid sequence in the molecule, the remaining portions being concerned in varying degrees

* Present address: Department of Botany, University of Leeds, Leeds, England.

with such properties as the degree of substrate specificity, thermal stability and the attachment of the enzyme to other components of the cell. Thus it must be accepted that any attempt to establish phyletic relationships from the extent of the similarity of the amino acid sequences of enzymes can only become convincing if one has a knowledge of the extent to which they may have been secondarily selected as a result of particular internal and environmental demands upon the organism. Using a statistical procedure, Margoliash and Fitch (1968) have compared the average number of variant residues between the cytochromes *c* of species from a number of different phyla (Table I) and as will be readily seen the amount of variation differs, the more widely separate the phyla.

TABLE I

Average numbers of variant residues and minimal mutation distances between cytochromes *c* of various species (comparison by taxonomic groupings)

Groups compared	Average number of variant residues	Average of the minimal mutation distance
Mammals to birds	$10·6 \pm 1·5$	$13·6 \pm 1·9$
Mammals and birds to reptiles	$14·5 \pm 4·5$	$18·6 \pm 6·8$
Mammals, birds and reptiles to fish	$17·9 \pm 1·8$	$26·5 \pm 2·4$
Vertebrates to invertebrates	$22·1 \pm 2·0$	$30·4 \pm 3·2$
Vertebrates and invertebrates to fungi	$43·1 \pm 2·3$	$59·9 \pm 2·8$

When we come to consider enzyme structure and amino acid sequences in a narrower phylogenetic grouping however, let us say within the angiosperms, what is the likelihood of phyletically significant amino acid sequences being preserved in an enzyme within this group? Elsewhere in this book (Chapter 9), Boulter describes the beginnings of the imposing task of amino acid sequence determinations on cytochrome *c* from a number of different angiosperm families. If, however, one considers two of the larger families representing very different levels of evolution, e.g. the Ranunculaceae and Compositae, one has within each family a range of different life forms, herbaceous, climbing, shrubby and arborescent. Moreover, the different life forms within each family are each dispersed over a number of widely different environments, aquatic, xeric, arctic, temperate and tropical. Thus the question arises as to whether the major constraints determining the amino acid sequences of the cytochromes in plants of these families are more likely to be determined by the demands inherent in these life forms and environments than by the families to which they happen to belong. *A priori* it is at least probable that similar amino acid sequences will be found in cytochrome *c* from the two families, either because they have never differed or have variously converged under common selection pressures. The various possibilities in this situation are summarized in the well-known isology diagram (Florkin 1966) (Fig. 1).

If this possibility is accepted, the question of course remains: what are the unique features at the chemical level which determine the organizational characteristics which delimit a particular order, family or genus of plants? In order to answer this question, attention would need to focus on the factors which control the extent and timing of cell division, growth, cessation of growth and differentiation in cells. Although we know a great deal about the hormonal and other factors controlling these processes in plants, the quantitative aspects of hormone production, transport and mode of action are not sufficiently known for it to be possible to say which of these factors developed as

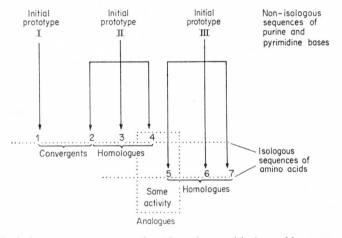

FIG. 1. Evolutionary convergence, analogy, homology and isology of base sequences and amino acid sequences. The arabic figures designate sequences of amino acids and the roman figures of purine pyrimidine bases

important innovations at a particular level of evolution, or are crucial to the distinctive morphology of a particular group or species.

Despite this view that one must look for innovations in major networks of "switches" controlling activation and repression of the genetic programme in order to understand many of the major events of evolution of organizational complexity, the fact remains that much evolutionary change can result from the accumulation of individually relatively minor modifications of the genetic programme, accelerated where there is the development of isolation mechanisms. It is useful to examine briefly at the outset how this process takes place.

II. NATURAL SELECTION AND ENZYME STRUCTURE

Changes in a matter of biochemical detail such as an amino acid sequence in an enzyme protein will only come about if they confer a selective advantage on the organism or are tightly linked to some other genetic factor which confers

9

such an advantage. Selective advantage is usually measured in terms of the selection coefficient (Haldane, 1932) and in the case of an enzyme, a high selection coefficient would imply that the new form of the protein functions in some way to render the organism able to contribute a larger proportion of its progeny to succeeding generations in the population. The following discussion concerns two enzymes from the plant kingdom which ecological and physiological considerations suggest may be undergoing frequent evolutionary changes in relation to the nature of the environment. The enzymes are (1) carboxydismutase, catalysing the carboxylation of D-ribulose-1,5-diphosphate to form phosphoglyceric acid and (2) the phosphatase enzymes associated with the cell walls of plant roots.

I have deliberately chosen examples from the aerial and edaphic environments in order to illustrate some of the complexities which may arise in unravelling the nature of the selective forces and the relevance of particular adaptive changes in a molecule in any particular ecological situation. It will be apparent that we have very little detailed information concerning the molecular architecture of either of these groups of enzymes; it will be my purpose to show why they constitute a particularly exciting prospect for such work in the future, and may provide exacting tests of the potential phyletic versus biological significance of enzyme structures.

III. Carbon Dioxide Fixation and the Aerial Environment

The rate of photosynthesis in higher plants may be controlled by both physical and chemical factors. The importance of an adequate light intensity of appropriate wavelength has been apparent for a long time and the quantitative significance of factors governing the entry of CO_2 molecules into the leaf, notably CO_2 concentration, air movement and stomatal aperture have been considered theoretically and measured experimentally (Gradmann, 1928; Penman and Schofield, 1951; Gaastra, 1959). Once CO_2 molecules have entered a leaf they must traverse the wet cell walls of the mesophyll cells, a layer of cytoplasm and the chloroplast membranes; even at this point, the entry of the carbon atom into organic combination requires that the assimilating enzyme should have a sufficient affinity for the substrate (CO_2 or HCO_3^-) molecule and an adequate supply of an appropriate acceptor molecule. Collectively, this complex of resistances within the leaf is referred to as the mesophyll resistance (r_m) and it has been shown that this component may frequently be of a comparable magnitude to the external and stomatal diffusion resistances limiting the rate of entry of CO_2 into the leaf (Gaastra, 1959). Rackham (1966) analysed the physical aspects of the mesophyll resistance and concluded that they were unlikely to constitute a major component, despite the several orders of magnitude slower rate of diffusion of CO_2 in water than in air. Studies showing an increasing r_m in ageing leaves, in which the rate of photosynthesis declined in parallel with a decrease in the activity of the CO_2 fixing enzyme carboxydismutase, led to the conclusion that the amount or affinity of the carboxylating enzyme for CO_2 may be a major component of the mesophyll resistance and hence quantitatively important in limiting the overall rate of photosynthesis (Woolhouse, 1968). This conclusion was supported by the finding that the light reactions of photosynthesis did not show a similar decline in the ageing leaves studied, so that the supply of ATP, $NADH_2$ and hence acceptor molecules for the fixation of CO_2 was not a rate-limiting factor. Moreover species such as sorghum and maize which are known to fix CO_2 through phosphoenol pyruvate (PEP) carboxylase, an enzyme with a higher affinity for CO_2 than carboxydismutase, were found to have lower mesophyll resistances (Woolhouse, 1968).

IV. CARBOXYDISMUTASE (D-RIBULOSE-,1,5-DIPHOSPHATE CARBOXYLASE E.C. 1.1.39)

A. GENERAL PROPERTIES

This enzyme is the major component of the soluble leaf proteins designated Fraction I (Wildman and Bonner, 1947) and shown to be relatively homogeneous during electrophoresis and ultracentrifugation (Eggman et al. 1953) with a sedimentation coefficient of 17–19s. Using carefully prepared "unleached" chloroplasts, Park and Pon (1961) and Heber et al. (1963) provided strong evidence that this protein fraction is normally a chloroplast constituent. Dorner et al. (1957) showed carboxydismutase to be associated with this fraction. Subsequently ribose phosphate isomerase and phosphoribulokinase were also found to be associated with Fraction I. These enzymes may however be separated from the carboxydismutase by Sephadex G200 gel filtration (Trown, 1965); even when this is done, the carboxydismutase may still account for up to 50% of the total soluble protein in a mature leaf. The purified enzyme has a molecular weight of approximately 5×10^5 although dimers and trimers have frequently been found to occur; under the electron microscope, it appears as a cuboid structure of side 120 Å, each cube containing 24 sub-units (Hesselkorn et al. 1965). Amino acid analysis of purified carboxydismutase indicates $96\frac{1}{2}$ cystine residues per molecule, that is an average of $4\frac{1}{2}$ cystines per sub-unit. Dissociation of the molecule to monomeric sub-units has not yet been achieved with certainty, although Sugiyama and Azakawa (1967) have obtained tentative evidence for two different monomeric sub-units by treating the protein with 5×10^{-3} M sodium dodecyl sulphate followed by polyacrylamide gel electrophoresis. Unfortunately, this reagent binds irreversibly to the protein and the products have proved difficult to fractionate by other preparative techniques. Carboxydismutase is inhibited by parachlormercuribenzoate (PCMB) and the activity of the enzyme over a range of PCMB concentrations suggests a minimum of 20 SH groups necessary for activity. However pretreatment of the enzyme with ribulose 1,5-diphosphate before addition of PCMB prevents inhibition but only slightly reduces PCMB binding by the enzyme, suggesting that only one or two SH sites on the enzyme may be directly involved in CO_2 binding (Sugiyama and Azakawa 1967; Rabin and Trown, 1964). Other studies of carboxydismutase from higher plants show that it may be light activated and the activity is also influenced by magnesium and bicarbonate ion concentrations (Sugiyama *et al.* 1968).

It is apparent that an enzyme molecule of this complexity may undergo a wide variety of changes influencing its catalytic activity. The various known types of enzyme variation have recently been summarized by Epstein and Schechter (1968) (Table II). Although investigations of the properties of

TABLE II

Origins of multiple forms of proteins

 I. Evolutionarily unrelated: "convergent" evolution
II. Evolutionarily related
   A. Genetically unrelated: "divergent" evolution of duplicated genes
   B. Genetically related
      a. Covalent differences

         1. Introduced during translation
         2. Introduced after translation

            (i) Deamination
            (ii) Attachment of carbohydrate
            (iii) Phosphorylation, sulphation
            (iv) α- and ϵ-NH_2 acetyls, formyls, Schiff's bases
            (v) Oxidation of sulphydryl groups
            (vi) Oxidation and reduction of prosthetic group
            (vii) Cleavage of peptide chain

      b. Mixed multimers
      c. Non-covalent differences

         1. Aggregation
         2. Binding of small molecules
         3. "Stable" conformational variants

---

carboxydismutase from different species, plants of different habitats and plants grown under a variety of conditions is only just beginning, it is of interest to consider the available information in relation to this summary of known enzyme changes.

B. EVOLUTION

It is not known whether the carboxydismutases found in different organisms are related or result from convergent evolution. The enzyme is found in many microorganisms; for example, bacteria using methane as a source of carbon, do so by first oxidizing it to CO_2 and then fixing the CO_2 by means of a carboxydismutase. The enzyme is also present in the chemoautotrophs *Thiobacillus thioparus* and *Thiobacillus neopolitanus* (MacElroy *et al.* 1968). Here also the enzyme has a molecular weight of 5×10^5 but interestingly it has not proved possible in this case to separate the associated phosphoribulokinase enzyme, and the suggestion has been made that this may add a further control point for the fixation of CO_2, since factors affecting the activity of the whole complex would thereby regulate the amount of ribulose diphosphate available as a CO_2 acceptor (Johnson 1966a, 1966b). Allosteric regulation of the phosphoribulo-kinase has been demonstrated (MacElroy *et al.* 1968).

C. COVALENT DIFFERENCES

Of the covalent modifications which may be introduced into the carboxy-dismutase enzyme thereby regulating activity, the oxidation of sulphydryls is particularly attractive, for example the photoactivation of the enzyme may operate in this way. Thus, it has recently been shown that in species of *Amaranthus* which fix CO_2 by means of a PEP carboxylase the enzyme PEP synthetase is inactive in extracts of leaves placed in darkness but becomes fully active within 20 minutes of transfer to light. The enzyme may be activated in cell-free extracts of darkened leaves by treatment with sulphydryl reagents, suggesting that *in vivo* a photo-reduction of sulphydryls is involved in regulating the activity of the enzyme and hence the rate of CO_2 fixation (Slack, 1968). Clearly if the abundant sulphydryls in the carboxydismutase molecule prove to be capable of reversible oxidation/reduction in this way, they could function in modifying the structure and hence regulating the catalytic activity of the molecule.

D. MIXED MULTIMERS

The finding that there are probably two different monomeric sub-units in the carboxydismutase molecule offers the possibility of a wide range of variation in the enzyme by changes in the relative proportions of the two sub-units in the active enzyme. The studies of Markert (1963, 1968) with mammalian lactic dehydrogenase have clearly shown that the active enzyme is a tetramer; there are two different sub-units designated A and B, the synthesis of which is under independent genetic control (Shaw and Barto, 1963) and these may aggregate in varying proportions to yield all five possible isozymes, A_4, A_3B_1, A_2B_2, A_1B_3 and B_4. It will be immediately clear that if such a situation exists with respect to carboxydismutase there would be 25 possible isozymes which could be formed. Moreover if a plant was heterozygous for only one of the sub-unit loci the number of possible isozymes would be increased to 325, heterozygosity at both loci would yield the possibility of 2,925 different isozymes being formed! If the dimers and trimers of the carboxydismutase molecule which have been detected in fractionation studies are able to exist *in vivo* the number of possible polymers clearly becomes very large indeed. In one respect this situation offers some hope with respect to the potential phyletic value of amino acid sequence studies of the enzyme, for it may be argued that if such a wide range of potential variation exists at the level of sub-unit polymerization, the requirements of local ecological pressures may be satisfied without the need for a high degree of variation at the amino acid sequence level.

Studies of the lactic dehydrogenase system in mammals have shown that the polymeric variants of the enzyme differ with respect to such properties as substrate specificity, K_ms, thermostability, denaturation characteristics and extent of substrate inhibition. The latter characteristic has been suggested as being particularly important in mammals in the light of the finding that tissues such as skeletal muscle, which may be subject to periods of relative anaerobiosis, the A_4 tetramer predominates; this form of the enzyme is relatively unaffected by high lactate or pyruvate concentrations. In well-oxygenated tissues such as heart muscle on the other hand, where lactate accumulation is unlikely to occur, there is a preponderance of the B_3 and B_4 polymers which show marked inhibition by high concentrations of lactate.

In the case of carboxydismutase the main selection pressures in the aerial environment affecting the enzyme are likely to be light intensity and temperature. Many plants when grown under conditions of shade show an increased leaf area; in this way the leaf area ratio of the plant is increased thus compensating for the declining net assimilation rate and so maintaining the relative growth of the plant. This increase in leaf area under shade is usually a result of lateral cell expansion; in consequence the stomata come to be further apart, the diffusion paths are lengthened and so the diffusional resistances to CO_2 uptake are increased (Holmgren et al. 1965). If we adopt an electrical analogue (Gradmann, 1928) and consider the constraints upon CO_2 uptake by the leaf as a series of resistances connected in series (Fig. 2) it will be seen that if r_a and r_s are increased so that they become large when compared to r_m, there will, in effect be a decreased selection pressure on the components of r_m (e.g. the affinity of carboxydismutase for CO_2) since changes in the enzyme will now count for very little in the overall resistance to CO_2 uptake. Moreover it may be anticipated that at low light intensities factors affecting the photochemical activity of the chloroplasts would become rate-limiting components of r_m. Experimental evidence for modifications of this type in the electron transport system of the chloroplasts of sun and shade ecotypes of *Solidago virgaurea*, has been presented by Bjorkman (1968a).

That there is a lower carboxydismutase activity in plants growing under conditions of shade is readily demonstrated (Bjorkman, 1968b; Woolhouse, unpublished) but there are several points which have not been unequivocally resolved to date. Firstly, it may be that the races really do differ simply in the amount of carboxydismutase protein produced. If this proves to be the case, it would seem likely that the hormonal control of enzyme production is what distinguishes the two races. Thus, if plants of *Perilla frutescens* are detopped so that only two leaves remain, the rate of photosynthesis in these remaining

leaves rises over a period of about seven days with a concomitant increase in
Fraction I protein and carboxydismutase activity; evidence was adduced for
the role of root hormones in controlling Fraction I synthesis under these conditions (Woolhouse, 1967). Further evidence for the role of cytokinins in controlling carboxydismutase activity is provided by Wareing *et al.* (1968) and
Lips and Ben-Zioni (1968). Secondly although when shade adapted ecotypes of
S. virgaurea are grown alongside plants from open habitats under conditions
of high light intensity they develop a lower carboxydismutase activity per unit
soluble protein than do plants from open habitats (Bjorkman, 1967), careful

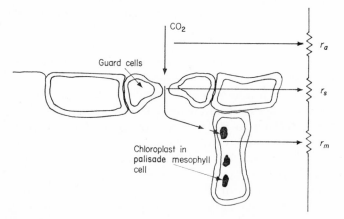

FIG. 2. Diagrammatic representation of diffusional and "chemical" resistances to CO_2
uptake in a leaf, by means of an electrical analogy

purification studies and K_m determinations are required before it can be defi-
nitely concluded, as Bjorkman has suggested, that this necessarily indicates
direct genetic differences in the actual capacity for enzyme production in the
two races. One may be involved with a situation in which epigenetic factors are
influencing the pattern of aggregation of different sub-units or modifying the
activity of the enzyme after translation as has been shown in the case of alkaline
phosphatase from *Escherischia coli* (Schlessinger and Anderson, 1968). It is
also important to note that in temperate climates where plants such as *S.
virgaurea* are found, open versus shaded habitats may also differ greatly in
patterns of temperature fluctuation as well as light intensity, both the extremes
of temperature encountered and the rates of heating and cooling of the leaves
being greater in the more exposed habitats. Thus, selection may be operating
for aspects of the thermal stability of the CO_2-fixing enzymes and the CO_2-fixing
capacity may be changing simply as a secondary consequence of this. As noted
above differences in thermal stability between two forms of an enzyme may
derive from differences of amino acid composition between the different forms
or from changes in the polymerization pattern of the sub-units; there is also,
however, the possibility of the formation of conformational isozymes which

need not involve such basic differences in the composition of the enzyme molecule.

V. Stable Conformational Variations in Enzymes

There are many aspects of what one may broadly describe as "acclimatization" responses in plants which present particularly intriguing problems at the level of enzyme structure and activity; they warrant brief consideration here since they may result in changes in properties such as thermostability and substrate affinity of enzymes which do not necessarily reflect changes in either the amino acid composition or the covalent bond formation within the enzyme molecule.

Exposure to "cold shock" treatments will in many cases cause irreversible injury leading to the death of the plant whereas a gradual lowering of the temperature results in a "hardening" process whereby the plant is able to withstand the lower temperatures. For example pea plants grown at 30°C were killed by exposure to a temperature of −3°C for 3 hours whereas plants which were grown at 30°C but exposed to a temperature of 5°C in the light for 3 hours per day for three or more days were able to withstand exposure to temperatures of −3°C for 3 hours (Kuraishi *et al.* 1968). Hesketh (1968) has provided evidence that in some species the capacity of the leaves for CO_2 assimilation may be altered by the temperature at which the plants are grown. In *Calluna vulgaris*, there is a marked decline in the rate of photosynthesis of the shoots in late summer; preliminary studies suggest that this may result from a decline in the specific activity of the carboxydismutase in the leaves although there is no measurable decrease in the amount of Fraction I protein concomitant with this decline in photosynthesis (Grace and Woolhouse, unpublished). It is not yet clear whether low temperatures or short days are the signals primarily responsible for this change.

Changes in enzyme activity resulting from conformational changes in the shape of the molecule may result from aggregation, the binding of a small molecule to the protein (the basis of the allosteric control mechanisms which have been found to regulate the activity of such enzymes as ATPases and amino acid synthesizing enzymes) or from changes in the actual folding of the polypeptide chain in the molecule. The latter process has been discussed by Levitt (1969) and proposed as a possible basis for a general theory of thermal adaptation in plants.

It is generally supposed that the tertiary structure of an enzyme in solution will tend to take that form which tends to minimize the total free energy of the system. The main temperature-sensitive bonds involved in maintaining this tertiary structure will be the hydrophobic and hydrogen bonding in the molecule (Fig. 3). Electrostatic and S—S bonds may, of course, also be involved in the structure but these will be influenced primarily by pH and redox reagents. If a protein chain is caused to unfold by appropriate treatment and is then

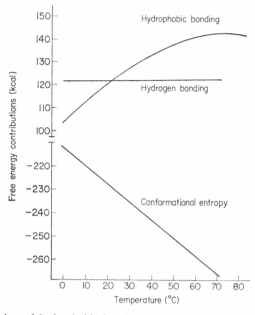

FIG. 3. The relation of hydrophobic bonding, hydrogen bonding and conformational entropy to temperature (from Levitt, 1969)

allowed to refold, a stable conformational variant of the molecule may arise if the total free energies of the two processes are both relatively large (Epstein and Schechter, 1968) (Fig. 4). A partial unfolding of an enzyme protein *in vivo* which could result in such a stable conformational change need not necessarily

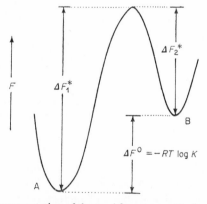

FIG. 4. Diagrammatic representation of the total free energy, F, of a protein-solvent system for a spectrum of protein conformations. Free energy minima, A and B, indicate stable and metastable conformations, respectively. The free energy of activation for the A–B transition is ΔF_1^*, and for B–A is ΔF_2^*. ΔF° is the difference in standard free energies between A and B, and is equal to $-RT \log K$, where K is the equilibrium constant of the A $\rightleftharpoons$ B conversion. Because both ΔF_1^* and ΔF_2^* are relatively large, the two species, A and B, will both appear kinetically stable (from Epstein and Schechter, 1968)

imply total loss of catalytic activity during the change; for example, RNAase has a thermolabile region which is reversibly disrupted by heat and may also be partially unfolded by treatment with 8 M urea, without complete loss of its catalytic activity in either case (Ginsburg and Carroll, 1965; Bello, 1966; Levitt, 1969). Thus, if as in some proteins, hydrophobic bonds represent the main bonds involved in determining the conformation of the molecule (Koshland and Kirtly 1966; Brandts, 1967), a lowering of temperature could permit the development of conformational variants. If, however, removal of water

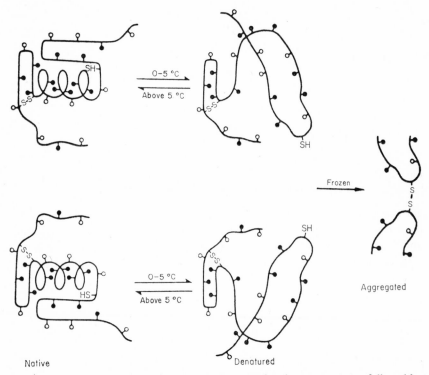

FIG. 5. Reversible denaturation of proteins at low, non-freezing temperatures, followed by extra-cellular freezing, dehydration, close approach of molecules, and intermolecular SS formation, which aggregates the proteins irreversibly (from Levitt, 1969)

from the cells accompanies the lowering of temperature, denaturation of the proteins may occur since the removal of water essential to the stabilization of the proteins in the unfolded state may lead to the closer approach of adjacent molecules with an increased probability of aggregation by intermolecular S—S bond formation (Brandts, 1967; Levitt, 1969) (Fig. 5). Levitt (1969) has pointed the way to the use of thiogel models in studies of these changes; it would clearly be of interest to extend this work to the investigation of carboxydismutase from leaves of plants grown at different temperatures and from the ecotypes or closely related species from different climatic regions. Studies of this kind are of importance not only to an understanding of the adaptive physiology of the plant but also in order that changes of this kind are considered before it is concluded that differences in the physical and catalytic properties of the enzyme from different plants must necessarily imply changes at the level of amino acid composition of the molecule.

VI. PHOSPHOHYDROLASE ENZYMES OF PLANT ROOTS AND ADAPTATION TO THE EDAPHIC ENVIRONMENT

A. INFLUENCE OF SOIL COMPOSITION ON PLANT EVOLUTION

As examples of the importance of the chemical composition of the soil in determining plant distribution, one may cite the distinctive plant communities of podzols, brown earths, sand dunes, chalk and limestone soils. Here it seems that other factors such as soil aeration, temperature, water availability or grazing regime are often of only secondary importance in the complex of selection pressures. The unique significance of the chemical composition of the soil in the selective complex arises from the fact that there is at the root surface, or at the membrane surfaces bounding the apparent free spaces of the root, a direct confrontation between structural and catalytic proteins with the chemical constituents of the soil solution. Thus when a seed germinates in a particular soil there will be an immediate contact between the surface proteins of its developing root systems and the solutes in the soil solution. A severe selective pressure is at once exerted and in consequence there is a relatively rapid selection of closely adapted individuals. This situation is particularly well illustrated by the work of Bradshaw and his colleagues showing the rapid evolution of heavy-metal tolerant races of *Agrostis tenuis* and *Festuca ovina* (Bradshaw 1952; Wilkins, 1957; Jowett, 1958, 1964; McNeilly, 1968; McNeilly and Antonovics, 1968; McNeilly and Bradshaw, 1968). The immediacy of the response of plant roots to contact with constituents of the soil solutions (for example inhibition of root elongation in *A. tenuis* by heavy metals may be demonstrated within 12 hours) suggests that the exposed sites must be important in the normal functioning of the root. *A. tenuis* is a polymorphic species which occurs widely on calcareous and acid soils, in addition to the development of races adapted to growth on soils containing heavy metals. Evidence of differences in the proteins of the plasmalemma of these ecotypes is shown in Fig. 6 in which roots of the different ecotypes have been incubated in solutions containing a range of concentrations of aluminium and the integrity of the cell membranes measured in terms of potassium leakage into the medium. It is seen that 10^{-5} M aluminium is sufficient to cause appreciable potassium leakage from the root cells of the lead-soil ecotype; 10^{-4} M aluminium is similarly effective in causing leakage from the cells of the calcareous and magnesium limestone soil ecotypes, whilst even 10^{-2} M aluminium did not immediately induce leakage from the roots of the acid-soil ecotype, i.e. the race which would normally be growing in the presence of significant concentrations of aluminium ions. In view of the technical difficulties involved in direct study of the properties of the plasmalemma, attention was turned to phosphatase enzymes associated with the root cell surfaces, which seem likely to have important functions in the mobilization of phosphate at the root surface and as constituents of the plasmalemma (Woolhouse, 1969) and are probably also involved in the metabolism of the cell wall (Elbein *et al*. 1964).

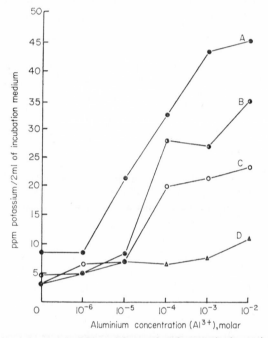

FIG. 6. Effect of aluminium on the leakage of potassium from excised root tips of four edaphic ecotypes of *Agrostis tenuis*. The data are expressed as ppm K$^+$ in 2 ml of solution after incubation for 30 min at 20°C. A: Lead soil race, B: Calcareous soil race, C: Magnesium limestone soil race, D: Acid soil race

B. ACID PHOSPHATASE ACTIVITY IN THE ROOTS OF EDAPHIC ECOTYPES OF *Agrostis tenuis*

Phosphatases associated with the surfaces of roots can be conveniently assayed by using *para*-nitrophenylphosphate as a substrate; under alkaline conditions the *p*-nitrophenol released forms the yellow-coloured phenolate ion with an absorption maximum at 410 nm. Details of the preparation of the roots and of the assay have been described previously (Woolhouse, 1969). Figs. 7 and 8 show the effects of a range of concentrations of lead and aluminium ions on the rate of hydrolysis of *p*-nitrophenylphosphate by root tips of three different edaphic ecotypes of *A. tenuis*. It is readily seen that the patterns of phosphatase inhibition by the metal ions differs for each ecotype. A peculiar feature of

Figs. 7 and 8 is that the initial decrease in enzyme activity with increasing metal ion concentration is followed by a secondary rise in activity, after which the activity again declines at yet higher concentrations of the particular metal. This apparent stimulation of the enzyme is most probably due to the fact that at the particular metal ion concentrations at which this rise occurs the semipermeability of the root cell membranes is broken, with the result that substrate

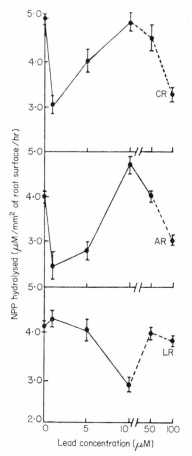

FIG. 7. Effect of lead ion concentration on the rate of hydrolysis of *p*-nitrophenylphosphate by root tips of edaphic ecotypes of *A. tenuis* (CR = calcareous race; AR = acid race; LR = lead race)

molecules can now penetrate freely into the cells and so come into contact with the internal phosphatases associated with the mitochondria and other cell components. This interpretation is supported by a comparison of Figs. 6 and 7 showing that in the case of aluminium there is a correspondence between the concentrations causing potassium leakage and the apparent stimulation of the phosphatase activity. Moreover when cell wall fractions are prepared from these roots, there is again a differential response of the wall-bound enzymes towards the metal ions (Fig. 9) but no stimulatory effect at higher concentrations is observed in these cases.

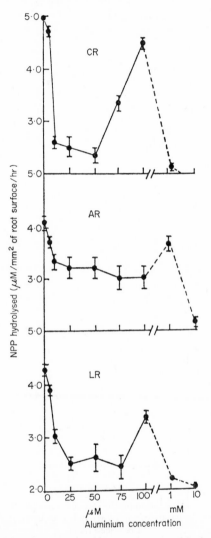

FIG. 8. Effect of aluminium ion concentration on the rate of hydrolysis of *p*-nitrophenylphosphate by root tips of edaphic ecotypes of *A. tenuis*

It is clear that there are many different phosphatase enzymes present in root cells; indeed Lamport (1965) has obtained evidence of no fewer than six peaks of *p*-nitrophenylphosphatase activity eluted from cell walls of tissue cultures of sycamore, *Acer pseudoplatanus,* by means of a sodium chloride gradient

(Fig. 10). It may be that some of these peaks represent polymeric forms of the same fundamental enzyme sub-unit but this remains to be investigated.

Treatment of cell wall preparations from the roots of the different edaphic ecotypes of *Agrostis* with KCl similarly removes a portion of the bound phosphatase activity (Table III). This enzyme component was presumably ionically bound to the cell walls, probably through the carboxyl groups of the pectin fraction. It probably represents material from inside the cells which became

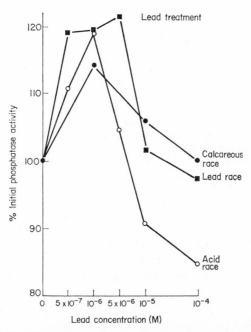

FIG. 9. Effect of Pb^{++} concentration on the rate of hydrolysis of *p*-nitrophenylphosphate by crude cell wall preparations of edaphic ecotypes of *A. tenuis*. In this case the wall preparations have not been extracted with KCl so that both ionically and covalently bound phosphatases are present. In comparing this graph with Fig. 7 for intact root tips it is seen that the patterns of activity are quite different. This is probably due to the presence of both cell wall enzyme and secondarily bound cytoplasmic phosphatases in the wall preparation, whereas at low Pb^{++} concentrations which do not destroy the semi-permeability of the cells only the cell wall enzyme is exposed to the substrate in the excised root tips

secondarily bound to the walls in the course of the extraction process. Up to 40 % of the phosphatase activity however remains firmly bound to the cell walls after KCl extraction and very little of this can be removed even by digitonin treatment. It would seem probable that this material represents a more fundamental constituent of the cell walls; further evidence that it is distinct from the ionically bound enzyme is shown by the different patterns of magnesium activation in the two fractions (Fig. 11). Figure 12 shows that there is also ecotypic differentiation in this firmly wall-bound phosphatase with respect to metal ion inhibition.

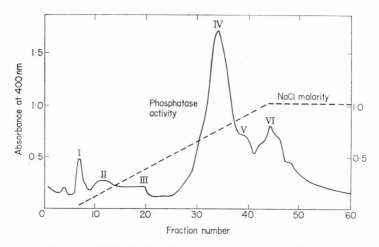

FIG. 10. From Lamport (1965). NaCl gradient elution of sycamore cell walls: acid phosphatase activity of fractions. Cell walls were isolated and washed with water 10 times. The walls were then packed as a 1 cm diameter ×5 cm long column which was then eluted with a linearly-increasing gradient of NaCl in pH 7·05 29 mM tris-HCl buffer. Two ml fractions were collected and assayed for phosphatase activity by incubating 1 ml samples for ten minutes with 22 μg p-nitrophenylphosphate in 2 ml sodium acetate buffer pH 5·8 at 20°C. The reaction was stopped by adding 3 drops of 1 M sodium carbonate. The absorbance at 400 nm was then read

In considering the nature and significance of these ecotypic modifications of the cell wall enzymes, two important aspects of the physiology of the responses of the roots to metal ions should be emphasized. Firstly there is a very high degree of specificity in the responses of particular races to particular metals and secondly if one considers effects at the threshold concentrations of the metals which affect the plants, it seems to be the cell-elongation process in the roots which is first affected.

TABLE III

The effect of 10^{-2}M Mg^{++} on the activity of acid phosphatase in cell wall fractions of the roots of a calcareous soil race of *Agrostis tenuis*

Enzyme Fraction	Activity as a percentage of activity in the absence of magnesium (%)
Enzyme extracted from wall with 0·1 M KCl for 1 hour	85
Soluble enzyme after 1 hour digitonin extraction (after KCl extraction)	115
Residual, tightly wall-bound enzyme after KCl and digitonin extraction	150

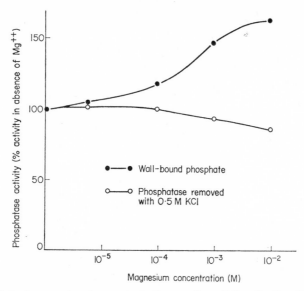

FIG. 11. Effect of magnesium ion concentration on the activity of cell wall-bound acid phosphatase and the acid phosphatase removed from the walls by extraction with 0·5 M KCl

C. SPECIFICITY OF HEAVY METAL TOLERANCE

The foregoing evidence of differential metal activation and inhibition of the phosphatases from roots of different edaphic ecotypes of *A. tenuis* must imply that there are differences in the structure of these cell wall or plasmalemma proteins between the several ecotypes. Turner (1967, 1969) has however rejected the idea that proteins may be involved in the metal tolerance mechanisms—"treatment with proteolytic enzymes of the metal-rich fractions of grass roots have revealed that proteins are not involved in heavy metal tolerance mechanisms in *A. tenuis.*" Turner proceeds to adduce evidence for differences in the carbohydrate composition of the cell walls of the various races as providing the key to the tolerance mechanism. There are however a number of serious objections to this view. Firstly, it is frequently found that the binding of proteins to cellulose or other inert matrices renders them relatively inaccessible to proteolytic attack. Thus, Lamport (1965) attempted the hydrolysis of sycamore cell wall protein with no fewer than 10 different proteolytic enzymes (chymotrypsin, pronase, papain, subtilisin, elastase, trypsin, pepsin, collagenase, bromelain, muramidase) but in no case was more than 30% of the cell wall protein degraded by this treatment and in most cases much less. Secondly, it is likely that metal-protein complexes will be much more resistant to proteolytic attack than the uncomplexed proteins. Again, it is unlikely on purely chemical grounds that the high degree of specificity found in metal ion tolerance could be based purely on carbohydrate structures.

Thus in *Agrostis*, races have been found resistant to Ni, Cu, Zn, Pb, Al and

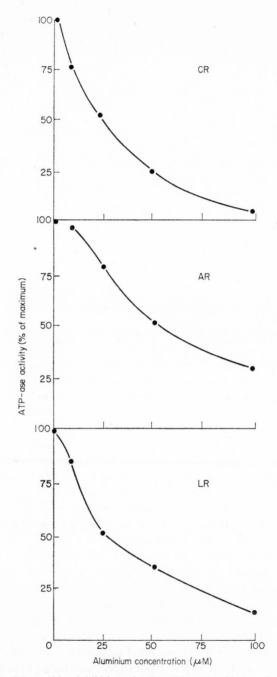

FIG. 12. Differential metal-ion inhibition of cell wall-bound acid phosphatases in edaphic ecotypes of *A. tenuis* (for key, see p. 221)

Ca and the available evidence suggests that resistances to any one of these ions does not necessarily confer resistance to any of the others (Clarkson, 1966; Gregory and Bradshaw, 1965; Gregory, 1964; Turner, 1969; Woolhouse, unpublished). Similarly in yeast (*Saccharomyces cerevisiae*) Ashida (1965) has obtained strains resistant to Co, Cu, Ni, Zn, Ag and Cd but none is cross tolerant. This specificity is the more remarkable when one considers that in several of the cases the diameters of the hydrated ions are very similar, e.g. Zn^{++} 0·68 Å and Ni^{++} 0·69 Å.

In considering further the possible chemical origins of these specific tolerance mechanisms, the work of Bayer concerning the synthesis of chelating agents suggests important possibilities. Bayer (1964) has emphasized that most of the low molecular weight compounds which have been found to chelate metals show a relatively low degree of specificity. In macromolecular structures the specificity is often greater. Investigation of haemocyanin from the octopus (*Octopus vulgaris*), a protein with molecular weight of $2·7 \times 10^6$, showed that up to 75% of the copper in the protein could be reversibly removed; no other metal could replace the copper without denaturing the protein. The copper was found to be attached to the protein through the mercapto groups of cysteine and residues linked by peptide bonds in the protein and the imidazole nitrogen of histidine (Bayer, 1962; Bayer and Fiedler, 1962). Simple, low molecular weight, chelating agents containing cysteine and histidine did not however show a unique specificity for copper and the conclusion was drawn that the specificity for copper in the haemocyanin must result from the location of these residues within the tertiary structure of the native protein molecule. Similarly in the case of the cell wall proteins, variations of the amino acid sequences or covalent bonding which would alter the folding of the molecules could profoundly change the specificity of the binding sites on the molecules towards particular metals. It is important therefore to consider further the structure of these proteins in relation to the other prominent feature of metal ion toxicity, i.e. the blocking of the cell elongation mechanism.

VII. CELL WALL PROTEIN STRUCTURE AND METAL ION INHIBITION OF ROOT ELONGATION

Although our knowledge of the mechanism of cell wall elongation is far from complete and clearly involves complex hormone, nucleic acid and protein interactions (Masuda, 1959; Lin and Key, 1967; Coartney *et al.* 1967; Cleland *et al.* 1968; Nooden, 1968) there are none-the-less a number of facts available which suggest ways in which metals could specifically influence the elongation mechanism. Nickerson and Falcone (1959) found a disulphide-rich protein in cell wall preparations of yeast; the demonstration that these bonds may be enzymically reversible led to the suggestion that such changes may form the molecular basis of localized alterations in cell wall plasticity associated with budding (Nickerson, 1963). Analyses of the cell wall proteins of higher plants showed many inter-specific differences but there was always a relatively high proportion of disulphide units present (Lamport, 1965).

The cell wall proteins of all higher plants so far examined are also found to be uniquely rich in *trans*-4-L-hydroxyproline (Dougall and Shimbayashi, 1960; Lamport and Northcote, 1960). These findings led Lamport (1965) to suggest that the disulphides and hydroxyproline may be involved in cross-linked

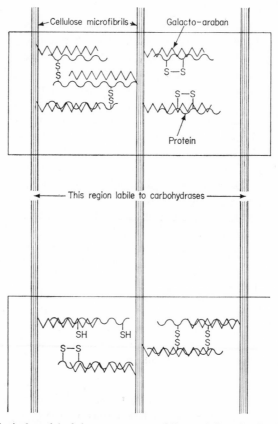

FIG. 13. Hypothetical model of the arrangement of "extensin" molecules in primary cell walls. The blocked regions would function in preserving the integrity of the wall during elongation (after Lamport, 1965). Specific heavy metal tolerance may be related to the tertiary structure of the —SH— containing proteins (see text)

structures with the wall polysaccharides. If some of these linkages were labile, they would provide a basis for cell wall plasticity; the hypothetical model is summarized in Fig. 13. Two further pieces of evidence lend support to this hypothesis; firstly, studies using partial alkaline hydrolysates of cell wall proteins from several species have shown that the hydroxyl groups of the hydroxyproline are glycosidically linked to the carbon 1 atoms of arabinose in arabinose oligosaccharides (Lamport, 1967). Secondly, there is a rapid increase of hydroxyproline-containing proteins in pea epicotyls during the transition of rapidly elongating tissue into mature, non-elongating tissue (Cleland and Karlsnes, 1967).

There are two aspects of this model which we are currently studying in relation to sites for specific metal attack. Firstly the wall elongation must depend in part upon the actual synthesis of new wall materials (Wada *et al.* 1968); thus the *p*-nitrophenylphosphatases, which as we have seen show differential metal tolerance, may prove on more specific investigation to be enzymes involved in polysaccharide synthesis. *P*-Nitrophenylphosphate is a very labile ester and can probably be degraded by almost any phosphatase enzyme; thus the phosphatase found tightly bound could prove to be an enzyme such as UDPG-pyrophosphatase involved in cellulose synthesis (Elbein *et al.* 1964). Variations in the metal ion sensitivity of this enzyme would thus influence cell wall elongation by differentially inhibiting the synthesis of cellulose. Pyrophosphatases from yeast have recently been shown to have a sub-unit structure (Araeva *et al.* 1968) and in *E. coli* the alkaline phosphatases have been shown to form isoenzymes under epigenetic or cytoplasmic control (Schlessinger and Anderson, 1968). There would thus appear to be several possibilities for structural modifications, of the types discussed earlier, in enzymes of this type which could confer differential metal ion sensitivity.

The second type of mechanism influencing the cell elongation process could involve the extensin molecule itself or the extensin-oligosaccharide complex. Here the disulphide bonds would provide ideal sites for the metal binding whilst amino acid substitutions elsewhere in the molecule leading to changes in the folding of the macromolecular complex could confer specificity as in the case of haemocyanin. Accordingly, comparative studies are currently being undertaken to make a more detailed characterization of the cell wall proteins from the roots of a range of edaphic ecotypes of *Agrostis* and also of species specifically associated with contrasting soil types. If, as already suggested, we encounter a higher degree of flexibility in the selection of amino acid sequences in these molecules one must ask again—is there really hope of finding phylogenetically definitive amino acid sequences in any particular class of proteins selected within one of the major plant groups?

ACKNOWLEDGEMENTS

The author is indebted to Mr S. J. Wainwright and Miss J. Buckhorn who have cooperated in the phosphatase studies.

REFERENCES

Araeva, S. M., Braga, E. A., and Egorov, A. M. (1968). *Biofizika*, **13**, 1126.
Ashida, J. (1965). *A. Rev. Phytopath.* 3, 153.
Bayer, E. (1962). *Chimia*, **16**, 333.
Bayer, E. (1964). *Angew. Chemie, Int. Ed.* 3, 325.
Bayer, E., and Fiedler, H. (1962). *Justus Liebigs Annln. Chem.* **653**, 149.

Bello, J. (1966). *Cryobiology*, **3**, 27.

Bjorkman, O. (1967). *Carnegie Institute Washington. Publication No. 736.*

Bjorkman, O. (1968a). *Physiologia Pl.* **21**, 1.

Bjorkman, O. (1968b). *Physiologia Pl.* **21**, 84.

Bradshaw, A. D. (1952). *Nature Lond.*, **169**, 1089.

Brandts, J. F. (1967). *In* "Thermobiology" (A. H. Rose, ed.). Academic, New York and London.

Clarkson, D. T. (1966). *J. Ecol.* **54**, 167.

Cleland, R., and Karlsnes, A. M. (1967). *Pl. Physiol. Lancaster*, **42**, 669.

Cleland, R., Thompson, M. L., Rayle, D. L., and Purves, W. K. (1968). *Nature*, **219**, 510.

Coartney, J. S., Moore, D. J., and Key, J. L. (1967). *Pl. Physiol. Lancaster*, **42**, 434.

Dayhoff, M. O., and Eck, R. V. (1968). "Atlas of Protein Sequence and Structure (1967–8)." National Biomedical Research Foundation, Silver Spring, Maryland, U.S.A.

Dorner, R. W., Kahn, A., and Wildman, S. G. (1957). *J. biol. Chem.* **229**, 945.

Dougall, D. K., and Shimbayashi, K. (1960). *Pl. Physiol. Lancaster*, **35**, 396.

Eggman, L., Singer, S. J., and Wildman, S. G. (1953). *J. biol. Chem.* **205**, 969.

Elbein, A. D., Barber, G. A., and Hassid, W. Z. (1964). *J. Am. Chem. Soc.* **86**, 309.

Epstein, C. J., and Schechter, A. N. (1968). *Ann. N.Y. Acad. Sci.* **151**, 85.

Florkin, M. (1966). "A Molecular Approach to Phylogeny." Elsevier, New York.

Gaastra, P. (1959). *Meded. LandbHoogesch. Wageningen*, **59**, 1.

Ginsburg, A., and Carroll, W. R. (1965). *Biochemistry*, **4**, 2159.

Gradmann, H. (1928). *Jb. wiss. Bot.* **69**, 1.

Gregory, R. P. G., and Bradshaw, A. D. (1965). *New Phytol.* **64**, 131.

Haldane, J. B. S. (1932). "The Causes of Evolution." Longman, Green and Co., London.

Heber, U., Pon, N. G., and Heber, M. (1963). *Pl. Physiol, Lancaster*, **38**, 355.

Hesketh, J. D. (1968). *Aust. J. Biol. Sci.* **21**, 235.

Hesselkorn, R., Fernandez-Moran, H., Kieras, F. J., and Van Bruggan, E. F. J. (1965). *Science*, **150**, 1598.

Holmgren, P., Jarvis, P. G., and Jarvis, M. S. (1965). *Physiologia Pl.* **18**, 557.

Johnson, E. J. (1966a). *Arch. Biochem. Biophys.* **114** 178.

Johnson, E. J. (1966b). *Bact. Proc.* **66**, 96.

Jowett, D. (1958). *Nature*, **182**, 816.

Jowett, D. (1964). *Evolution*, **18**, 70.

Koshland, D. E., and Kirtly, M. E. (1966). *In* "Major Problems in Developmental Biology" (M. Locke, ed.), p. 217. Academic, New York and London.

Kuraishi, S., Arai, N., Ushyima, T., and Tazaki, T. (1968). *Pl. Physiol. Lancaster*, **43**, 238.

Lamport, D. T. A. (1965). *Adv. Bot. Res.* **2**, 151.

Lamport, D. T. A. (1967). *Nature*, **216**, 1322.

Lamport, D. T. A., and Northcote, D. H. (1960). *Nature*, **188**, 665.

Levitt, J. (1969). *Symp. Soc. Exp. Biol.* **23**, 395.

Lin, C. Y., and Key, J. L. (1967). *Pl. Cell Physiol.* **9**, 553.

Lips, S. H., and Ben-Zioni, A. (1968). *Israel J. Bot.* **17**, 130.

MacElroy, R. D., Johnson, E. J., and Johnson, M. K. (1968). *Arch. Biochem. Biophys.* **127**, 310.

MacElroy, R. D., Johnson, E. J., and Johnson, M. K. (1968). *Biochem. Biophys. Res. Commun.* 678.

McNeilly, T. (1968). *Heredity*, **23**, 99.

McNeilly, T., and Antonovics, J. (1968). *Heredity*, **23**, 205.

McNeilly, T., and Bradshaw, A. D. (1968). *Evolution*, **22**, 108.

Margoliash, E., and Fitch, W. M. (1968). *Ann. N.Y. Acad. Sci.* **151**, 359.

Markert, C. L. (1963). *Science*, **140**, 1329.

Markert, C. L. (1968). *Annals New York Acad. Sci.* **151**, 14.

Masuda, Y. (1959). *Physiologia Pl.* **12**, 324.

Nickerson, W. J. (1963). *Bact. Rev.* **27**, 305.

Nickerson, W. J., and Falcone, G. (1959). *In* "Sulphur in Proteins" (R. E. Benesch *et al.* eds.), p. 409. Academic, New York and London.

Nooden, L. D. (1968). *Pl. Physiol. Lancaster*, **43**, 140.

Park, R. B., and Pon, N. G. (1961). *J. Mol. Biol.* **3**, 1.

Penman, H. L., and Schofield, R. K. (1951). *Symp. Soc. Exp. Biol.* **5**, 115.

Rabin, B. P., and Trown, P. W. (1964). *Nature*, **202**, 1290.

Rackham, O. (1966). *In* "Light as an Ecological Factor" (R. Bainbridge *et al.* eds.), p. 167. British Ecological Society Symposium No. 6.

Schlessinger, M. J., and Anderson, L. (1968). *Ann. N.Y. Acad. Sci.* **151**, 159.

Shaw, C., and Barto, E. (1963). *Proc. natn. Acad. Sci.* **50**, 211.

Slack, C. R. (1968). *Biochem. Biophys. Res. Commun.* **30**, 5.

Sugiyama, T., and Azakawa, T. (1967). *J. Biol. Chem.* **62**, 474.

Sugiyama, T., Nakayama, N., and Akazawa, T. (1968). *Arch. Biochem. Biophys.* **126**, 3.

Trown, P. W. (1965). *Biochemistry*, **4**, 908.

Turner, R. G. (1967). "Experimental Studies on Heavy Metal Tolerance." Ph.D. Thesis. University of Wales.

Turner, R. G. (1969). *In* "Ecological Aspects of the Mineral Nutrition of Plants" (I. H. Rorison *et al.* eds.), p. 399. British Ecological Society Symposium No. 9.

Wada, S., Tanimoto, E., and Masuda, Y. (1968). *Pl. cell Physiol.* **9**, 369.

Wareing, P. F., Khalifa, M. M., and Treharne, K. (1968). *Nature*, **220**, 453.

Wildman, S. G., and Bonner, J. (1947). *Arch. Biochem.* **14**, 301.

Wilkins, D. A. (1957). *Nature*, **180**, 37.

Woolhouse, H. W. (1967). *Symp. Soc. Exp. Biol.* **21**, 179.

Woolhouse, H. W. (1968). *Hilger Journal*, **11**, 15.

Woolhouse, H. W. (1969). *In* "Ecological Aspects of the Mineral Nutrition of Plants" (I. H. Rorison *et al.* eds.), p. 357. British Ecological Society Symposium No. 9.

CHAPTER 12

Secondary Constituents of Aquatic Angiosperms

JERRY W. MCCLURE

Department of Botany, Miami University, Oxford, Ohio, USA

I. INTRODUCTION

Aquatic angiosperms are considered to be descendants of terrestrial plants which have reverted to the aquatic habits of their remote ancestors. While some entire families are exclusively aquatic—fully committed to life in the water—this habitat has also been adopted to varying degrees by taxa from otherwise terrestrial groups. For example, in some families whole genera are more or less aquatic, or only certain species of a genus may be aquatics. It has been estimated that a small minority—not more than 1% (Sculthorpe, 1967)—of the angiosperms fall into the general category of aquatics; still, these forms have held considerable interest for plant scientists. For example, when it is considered that the aquatic habitat has almost surely been acquired by many unrelated angiosperms at many different times during the history of the group, a number of morphological and phytochemical modifications occur with surprising regularity. Some of these may have adaptive significance. For the most part aquatics do not appear to produce the bewildering array of secondary constituents usually found in terrestrial plants and a number of these characteristics, especially the apparent loss of the ability either to synthesize or to accumulate particular classes of compounds, draw them into a phytochemically-interesting group.

Aquatic plants are usually morphologically simple and their shapes are altered by the environment. This limited and variable morphology has created acute problems for the Systematists. Phytochemical studies have shown that the patterns of distribution of secondary constituents are frequently unambiguous in their ability to cut through the taxonomic confusion presented by insufficient, or environmentally plastic, morphological characteristics. For these reasons the application of phytochemical data to the systematic problems of aquatics should be unusually rewarding.

In determining the scope of this chapter, since it is likely that neither secondary constituents nor aquatic angiosperms can be precisely defined, Harborne's (1968) chemical categorization and Sculthorpe's (1967) taxonomic delineation served as guidelines. It is especially difficult to circumscribe this group of plants. Arber (1920) clearly defined the problem when she wrote, "At one end of the series we have plants which are normally terrestrial, but which are able to endure occasional submergence, while at the other end we have those wholly aquatic species whose organization is so closely related to water life that they have lost all capacity for a terrestrial existence. Between these extremes there is an assemblage of forms, bewildering in number and variety."

Most of our understanding of the biosynthesis and biochemical systematics of secondary metabolites in this group of plants comes from work on the phenolics of a single family, the Lemnaceae. Thus the last two sections of this treatment are oriented primarily toward these constituents in the duckweeds.

II. Patterns of Distribution

A. SOME PHYTOCHEMICAL IMPLICATIONS OF AN AQUATIC ENVIRONMENT

The question of why two species differ qualitatively and quantitatively in their secondary products is not fundamentally different from the question of why their leaves differ in size and shape. In aquatics the phenomenon of heterophily (i.e. expression of different leaf forms when grown under either emerged or submerged conditions) has attracted much attention. Allsopp (1965) has summarized much of the work in this area. There seems little doubt that the principal morphogenetic factor is the high level of hydration of the developing submerged tissues; this follows from the absence of any restriction on water entry apart from a slight resistance of the expanding cell walls. This high water intake would account more or less directly for the great extension of leaf segments, internodes, etc., encountered in a majority of hydrophytes. It is also reasonable to assume that the low level of tissue differentiation characteristic of hydrophytes is related to the high water intake; in any event, a remarkable increase in differentiation can be achieved by altering the osmotic concentration of the external solution leading to the appearance of "land" characteristics in submerged plants. Allsopp concludes that the morphological responses of hydrophytes to environmental changes differ mainly in degree and not in kind from those of typical land plants.

Equally compelling evidence about the secondary chemistry of aquatic plants is difficult to muster. Yet there is no doubt that emergence from the water is accompanied by significant alterations in plant metabolism. For example, Rubin and Loginova (1965) studied the induction of enzyme synthesis in young leaves of *Nuphar luteum* and *Cyperus angustifolia*. Comparison of young submerged leaves with those recently emerged shows that during the transition one group of iron-containing enzymes has been replaced by another and the ability to oxidize ribulose 5-phosphate and lactate has been developed. These changes are blocked by agents which antagonize protein synthesis. This appears to be a shift from an essentially anaerobic to an aerobic type of metabolism in response to a transition from an aquatic to an emergent habitat. This fundamental metabolic change would surely be expected to influence the qualitative and quantitative production of secondary constituents.

A great number of aquatic monocotyledons are without vessels and they are often devoid of normal xylem with its familiar tracheary elements (Cheadle, 1942). This has some implications in the secondary chemistry of the plants. In *Elodea densa* annular or spiral tracheids are reported to occur in the nodes of the stems, yet *E. densa* is non-lignified by several criteria (Stafford, 1964; Blazey and McClure, 1968) and thus these must be quite atypical tracheary elements. Despite its lack of lignin, *E. densa* contains flavonoids which arise from the phenylpropanoid pathway which also leads to lignin. Perhaps it is significant that when *Elodea* is supplied with peroxide and eugenol or various cinnamic acid derivatives, it will form a lignin-like product (Siegel, 1957; Stafford, 1964). Siegel (1962) suggests that under natural conditions there is a competition between metabolic pathways in these plants, the potential lignin precursors being converted into other products more rapidly than they can be transformed into lignins.

There may be still other fundamental differences between the normal phenolic biochemistry of terrestrial and aquatic plants. Pridham (1964) reports that when he investigated the ability of various plant tissues to convert quinol and resorcinol to the corresponding mono-β-D-glucopyranosides *in vivo*, the activity was prominent in the angiosperms with the striking exception of the submerged aquatics *Elodea canadensis* and *Utricularia vulgaris*, and the free-floating aquatic *Lemna minor*. Interestingly, the emergent shoot of the rooted aquatic *Nymphaea alba* formed high concentrations of both glucosides. Pridham speculates that if glucosylation serves as one method for the detoxification of harmful phenolic compounds which could either arise from normal plant metabolism or from the environment, then there may be fundamental differences in submerged and free-floating plants. The observations by Jurd *et al.* (1957) that caffeic acid esters appear in the medium in which the free-floating aquatic *Spirodela oligorhiza* is growing may relate to this phenomenon. Wallace *et al.* (1969; Wallace, personal communication) also noted that when other members of the Lemnaceae are grown on defined media under axenic conditions, cinnamic acids are quickly detected in the medium and, after several days, flavonoids are found. Fluck (1963) considers it likely that many terrestrial plants lose considerable amounts of water-soluble secondary constituents through the epidermis by contact with water (e.g. rain, dew). This may apply to aquatics as well. A final example of the unusual phenolic metabolism of some aquatic angiosperms is the report of Seidel (1963) that a few plants, such as *Scirpus lacustris* and *Juncus* spp., will grow well in water with concentrations of phenol above 250 mg litre⁻¹.

Although little convincing data are available, it may be that some secondary constituents are characteristics of the modification to an aquatic environment. It has been suggested by Tatsuta and Ochii (1956) that persicarin, an iso-rhamnetin ester of potassium bisulphate, occurs only in marsh plants and that it is in some way involved in hygrophilic adaptation. As this flavonoid is known only from species of *Polygonum* and *Oenanthe* this generalization seems to be based on insufficient evidence. Other facts for conjecture are the perplexingly atypical leucoanthocyanins of many aquatic monocotyledons (Bate-Smith and Swain, 1965), and the almost complete absence of alkaloids from the floating or submerged aquatics (Li and Willaman, 1968). Hegnauer (1967) considers that since all plants have made morphological and anatomical adaptations to various environments, the accumulation of many of the highly curious secondary plant metabolites are most probably the result of selection by the environment. He also concludes that we are still unaware of the true contributions of the secondary plant metabolites to the overall fitness for life of the plants.

For precisely this reason the concept of a secondary constituent is becoming increasingly difficult to define as more and more compounds which previously fitted into this category are shown to have biological roles. On the basis of structural class almost all of the plant growth hormones would be considered as perfectly good secondary constituents. Goodwin (1967) has reviewed the biological significance of one such class of constituents, the terpenoids. He points out that higher plants as we know them could not have evolved without the parallel development of the terpenoids. Similarly, strengthening of the vascular tissues, which are characteristic of angiosperms, could not have taken place without elaboration of the phenylpropanoid pathway which terminates in such diverse secondary constituents as the anthocyanins, cinnamic acids and coumarins.

If the secondary compounds are to be understood as they relate to the life of the plant in which they are produced, a bridge needs to be constructed between the patterns of distribution and their possible *in vivo* functions. This approach would eventually permit an assessment of the biological value of secondary constituents and thus their importance within the plant. Bate-Smith and Swain (1965) and Alston (1966) have stressed the desirability of evaluating the taxonomic value of any given group of compounds on the basis of their biological value to the species.

There are numerous examples in which a single species in an otherwise terrestrial genus is aquatic, or where one genus of a terrestrial family is adapted to life in the water. An understanding of the distribution of secondary constituents within these forms should point to phytochemical changes imposed on terrestrial forms as they revert to their ancestral environment.

Some examples of secondary constituents which may be of systematic significance are shown in Fig. 1. Ellagic acid is biosynthetically interesting as systematically equivalent to the unknown 3,4,5-trihydroxycinnamic acid

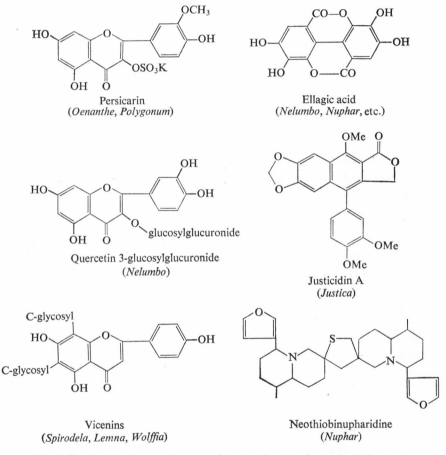

Persicarin
(*Oenanthe, Polygonum*)

Ellagic acid
(*Nelumbo, Nuphar*, etc.)

Quercetin 3-glucosylglucuronide
(*Nelumbo*)

Justicidin A
(*Justica*)

Vicenins
(*Spirodela, Lemna, Wolffia*)

Neothiobinupharidine
(*Nuphar*)

FIG. 1. Some naturally occurring secondary constituents of aquatic angiosperms

(Bate-Smith, 1968b). Its distribution has recently been used in reevaluating the positioning of the aquatic family Nymphaeaceae in the Ranales (Bate-Smith, 1968a).

The rare 4-arylnaphthalene derivatives justicidin A and B are extremely potent fish poisons known only from *Justica* (Munakata *et al.* 1965).

The genus *Nuphar* has a remarkable series of sulphur-containing alkaloids such as neothiobinupharidine (Achmatowicz and Bellen, 1962), and it would be interesting to know how the sulphur is incorporated into this class of compound which seems to be of isoprenoid origin (Ettlinger and Kjaer, 1968).

Unusual flavonoids have been found in several aquatics. For example, the 3-glucosylglucuronide of quercetin (Harborne, 1967), is quite rare and there are a series of di-C-glycosylflavones (vicenins and lucenins) which are of value in understanding the systematics of the Lemnaceae (McClure and Alston, 1966).

Table I which follows gives a synopsis of secondary constituents in aquatic angiosperms.

III. BIOSYNTHETIC STUDIES

A. INTRODUCTION

Most of the work on the biosynthesis of secondary constituents has been done with terrestrial plants and an excellent review of the practices and limitations of various techniques of studying biosynthesis has been published (Swain, 1965). Studies on secondary metabolites in aquatic plants have been limited essentially to the flavonoids and, until recently, primarily with a single clone of the duckweed *Spirodela oligorhiza*. Thimann and his associates carried out a detailed series of investigations on the biosynthesis of the anthocyanin petunidin 3,5-diglucoside in this clone between 1949 (Thimann and Edmondson) and 1964 (Furuya and Thimann). They applied the techniques of adding inhibitors, purines or pyrimidines, sugars, plant growth regulators and enzymes. They also measured the responses to light of different quality and quantity. As a result, an impressive amount of data is available on the factors which control the synthesis of a single anthocyanin in this tiny plant. Much of their work has been reviewed in perspective with other studies on the physiology of phenolic biosynthesis (Siegelman, 1964).

In the last paper in their series, Furuya and Thimann (1964) interpreted the major features of the control mechanisms essentially as follows. The synthesis of anthocyanin in *S. oligorhiza* is enhanced by sugars and light, but inhibited by the presence of copper-complexing agents or sulphur-containing compounds as well as substituted purines and pyrimidines. Thiouracil and ethionine are inhibitory only in the light, and the plant growth regulators have little direct influence on this process. When one attempts to draw a unified scheme from all of these experimental modifications, it becomes quickly apparent that the control of anthocyanin synthesis operates through a complex and perplexing series of control mechanisms. It is probable that these results will become more understandable as more is learned about the biosynthesis of flavonoids through applications of other techniques.

B. INCORPORATION STUDIES USING RADIOACTIVE PRECURSORS

Aquatic plants are well-suited for studies involving the incorporation of exogenously supplied water-soluble precursors. Growing in intimate contact

TABLE I

Synopsis of secondary constituents of aquatic angiosperms

Species	Alkaloids	Terpenoids	Simple phenolics	Flavonoids	Others	References
			Compounds[a]			
			AQUATIC FAMILIES[b]			
I. DICOTYLEDONS						
Nymphaeaceae: 8 (c60) Fl, Em, S[c]						
Nymphaea alba	nymphaeine	—	—	My, Qu, Lu, Ap	—	Willaman and Schubert, 1961; Kubitzki and Reznik, 1966
N. capensis	—	—	E, caff, pC, S, F	—	tannins	Bate-Smith, 1962; Howes, 1953
N. capensis var. Harrastick	—	—	—	Km, Qu, Cy	—	Bate-Smith, 1962
N. tetragona	—	—	E	—	—	Naya and Kotake, 1965
Nymphaea cv.,	—	—	—	Dp 3-glu	—	Harborne, 1967
Brasenia schreberi	unk.	—	—	—	—	Willaman and Schubert, 1961
Cabomba furcata	—	—	—	(none)	—	Kubitzki and Reznik, 1966
Euryale ferrox	drummine	—	—	Qu, Km	—	Kubitzki and Reznik, 1966
Nelumbo nucifera	roemerine	—	—	Km 3-glucurglu	—	Bernauer, 1964;
	nuciferine			Qu 3-glucurglu		Tomita and Furukawa, 1962;
	nornuciferine			Km 3-rham-4'-arab		Furukawa, 1965a,b;
	armepavine			Lu 7-glu		Tomita and Furukawa, 1966;
	neferine			Qu 7-glu		Bernauer, 1963;
	lotusine			Qu 3-glu		Willaman and Schubert, 1961;
	liensinine			Cy (leuco ?)		Kupchan et al. 1963;
	isoliensinine			Dp (leuco ?)		Harborne, 1967;
	pronuciferin			Km 3-galactrham		Nagarajan et al. 1966;
	anonaine			Km 3-diglu		Rahman et al. 1962;
	5-methyl-6-hydroxyl-			Qu 3-galact		Subramanian et al. 1969
	aporphine			Qu 3-glucur		

TABLE I—*continued*

Species	Compounds[a]					References
	Alkaloids	Terpenoids	Simple phenolics	Flavonoids	Others	
Nelumbo nucifera	nelumbine norarmepavine			Km 3-glucur rutin		Arata, 1965; Naya and Kotake, 1965
Nuphar japonicum	dehydrodeoxy-nupharidine	—	E	—	—	Achmatowicz and Wrobel, 1964; Achmatowicz and Bellen, 1962; Arata *et al.* 1967; Il'inskaya *et al.* 1967; Willaman and Schubert, 1961; Bate-Smith, 1968
N. luteum	Neothiobi-nupharidine thiobinupharidine allothiobinuphraidine pseudothiobi-nupharidine thiobideoxy-nupharidine, anhydronupharamine nuphleine nupharidine nupharamine	—	E, caff, pC, S, F	Lu	—	
N. variegatum	nuphenine		—	—	—	Barchet and Forrest, 1965
Victoria amazonica	—	—	—	My, Qu, Cy (leuco?)	—	Bate-Smith, 1962; Kubitzki and Reznik, 1966
Ceratophyllaceae: 1 (c6) S, Ff						
Ceratophyllum demersum	—	—	cg, caff, S, F	Cy (leuco?) Mv (leuco?) Dp (leuco?)	—	Bate-Smith, 1962; Reznik and Neuhausel, 1959
Elatinaceae: 2 (c30) Em, S						
Elatine gratioloides	(none)	(no saponins)	E	(no leuco)	—	Hegnauer, 1964; Bate-Smith, 1962
Trapaceae: 1 (c4) Ff						
Trapa natans	—	—		unk. antho.	—	Neubarger, 1968

Taxon	Saponins	Phenolics	Flavonoids/Anthocyanins	Other	References
Haloragaceae: 6 (c100) Em, S	—				
Haloragis alata	—	E, caff pC	Qu, Km Cy (leuco?)	—	Bate-Smith, 1962
H. tetragyna	unk.				
Myriophyllum proserpinacoides	unk. saponin	E	Qu, Cy (leuco?)	cyanogenetic	Willaman and Schubert, 1961; Hegnauer, 1964; Bate-Smith, 1962; Reznik and Neuhausel, 1959; Hegnauer, 1958
Hippuridaceae: 1 (1) Em, S					
Hippuris vulgaris	(no saponins)	E, caff, F scopoletin (?)	Km	tannins aucubin catalpol	Bate-Smith, 1962; Dekker, 1913; Hegnauer, 1967
Callitrichaceae: 1 (c25) Em Fl, S					
Callitriche palustris	—	caff	Ap, Lu	aucubin catalpol	Hegnauer, 1967
Menyanthaceae: 5 (c35) Em, Fl	—				
Menyanthes trifoliata	betulinic acid	caff, S(?)	Km 3-rham Km 3-galact Qu 3-rutinoside	loganin	Karrer, 1958; Bate-Smith, 1962; Harborne, 1967
Podostemaceae: c25 (c120) S	—	—	—	—	—
Hydrostachyaceae: 1 (c10) S	—	—	—	—	—
II. MONOCOTYLEDONS					
Butomaceae: 5 (c10) Em, Fl					
Butomus umbellatus (none)	—	pC, caff S	unk. antho. (no leuco)	—	Hultin and Torssell, 1965; Hegnauer, 1963
Hydrocharitaceae: 14 (c90) S, Ff					
Hydrocharis morus-ranae (none)	(no saponins)	caff	Cy, Pg rutin	(no cyanogens) (no tannins)	Tjia-Lo and Hegnauer, 1956; Hegnauer, 1963
Elodea canadensis	—	caff, cg	Cy (leuco?)	—	Reznik and Neuhausel, 1959
E. callitrichoides	—	cg	unk. antho	—	Reznik and Neuhausel, 1959
E. crispa	—	caff, cg	unk. flavone	—	Reznik and Neuhausel, 1959
E. densa	—	caff, cg	Cy digly, Qu	—	Reznik and Neuhausel, 1959

10

TABLE I—continued

Species	Compounds[a]					References
	Alkaloids	Terpenoids	Simple phenolics	Flavonoids	Others	
Limnobium stoloniferum	—	—	—	unk. antho	—	Bate-Smith, 1954
Stratiotes aloides	(none)	(no saponins)	caff	Cy (leuco?) rutin (?)	(no cyanogens)	Tjia-Lo and Hegnauer, 1956; Bate-Smith, 1968b; Hegnauer, 1963
Vallisneria spiralis	—	—	caff, pC	(no antho)	—	Bate-Smith, 1968b; Reznik and Neuhausel, 1959
Alismaceae: 12 (c70) Em, Fl, S						
Alisma plantago-aquatica	*	alisol A alisol B	—	Cy (leuco?) rutin (?)	(no cyanogens)	Hultin and Torssell, 1965; Massagetov, 1967; Murata et al. 1968; Tjia-Lo and Hegnauer, 1965; Reznik and Neuhausel, 1959
A. lanceolata	—	—	caff, pC, S, F, p-OH benzoic acid	—	—	Bate-Smith, 1968b
Sagittaria sp.	*	saponins	—	Cy (leuco?)	(no cyanogens)	Tjia-Lo and Hegnauer, 1956; Hultin and Torssell, 1965; Altman, 1954; Reznik and Neuhausel, 1959
S. platyphylla	—	—	caff, pC, F	—	—	Bate-Smith, 1968b
S. variabilis	—	—	caff, S, F	Qu, Cy, Km	—	Bate-Smith, 1968b
Scheuchzeriaceae: 1 (1) Em						
Scheuchzeria palustris	unk.	—	—	—	—	Hultin and Torssell, 1965
Juncaginaceae: 4 (c15) Em						
Tetroncium magellanicum	—	(no saponins)	—	—	—	Hegnauer, 1963

Taxon		Saponins	Phenolic acids	Flavonoids / Anthocyanins	Tannins	References
Triglochin maritimum	unk.	(no saponins)	S, pC (?)	—	—	Hultin and Torssell, 1965; Bate-Smith, 1968b; Gascoigne *et al.* 1948
T. procerum	—	—	—	My	—	—
Lilaeaceae: 1 (1) Em	—	—	—	—	—	—
Posidoniaceae: 1 (2) S	—	—	—	—	tannins	Phouphas, 1962;
Posidonia oceanica	—	—	—	—	much apiose	Van Beusekom, 1967
Aponogetonaceae: 1 (c30) Fl, S	—	—	—	—	—	—
Aponogeton distachymum	—	—	—	Cy 3,5-digly Cy 3-gly	—	Reznik and Neuhausel, 1959
A. krauseanus	—	—	caff, S, F, *p*-OH benzoic acid	Qu, Km	—	Bate-Smith, 1968
Zosteraceae: 2 (c12) S	—	—	—	—	tannins	Bate-Smith, 1968b;
Zostera nana	—	—	caff, pC	—	much apiose	Phouphas, 1962; Van Beusekom, 1967
Potamogetonaceae: 2 (c90) Fl, S	(none)	—	—	—	—	—
Potamogeton pectinatus	—	(no saponins)	—	—	(no tannins)	Hegnauer, 1963; Van Beusekom, 1967
P. perfoliatus	(none)	(no saponins)	—	—	(no tannins)	Hegnauer, 1963
P. ramosus	(none)	(no saponins)	—	—	(no tannins)	Hegnauer, 1963
P. crispus	unk.	—	—	—	—	Tjia-Lo and Hegnauer, 1956
P. natans	—	—	—	rutin	—	Hultin and Torssell, 1965
Potamogeton sp.	—	—	—	(no antho)	—	Reznik and Neuhausel, 1959
Ruppiaceae: 1 (3) S	—	—	—	—	—	—
Ruppia fillifolia	—	(no saponins)	—	—	—	Ricardi, 1958
R. maritima	—	(no saponins)	—	—	—	Ricardi, 1958
Zannichelliaceae: 6 (c25) S	—	—	—	—	—	—
Zannichellia palustris	—	(no saponins)	—	—	—	Ricardi, 1958
Najadaceae: 1(c35) S	—	—	—	—	—	—
Najas marina	—	—	—	Cy	—	Bate-Smith, 1968b
Mayacaceae: 1 (c10) Em, S	—	—	—	—	—	—
Mayaca sellowiana	—	—	caff, pC, S, F	Qu	—	Bate-Smith, 1968b
Pontederiaceae: 7 (c30) Em, S, Ff	—	—	—	—	—	—
Pontederia cordata	(none)	(no saponins)	—	—	—	Hegnauer, 1963

TABLE I—continued

Species	Compounds[a]					
	Alkaloids	Terpenoids	Simple phenolics	Flavonoids	Others	References
P. lanceolata	—	—	caff, pC, S, F	Cy, Dp	—	Bate-Smith, 1968b
Eichornia speciosa	*	triterpenes*	caff, F	Dp 3-diglu Cy (leuco?)	—	Hegnauer, 1963; Willaman and Schubert, 1961; Bate-Smith, 1968b; Shibata, 1965; Bate-Smith and Swain, 1965
Lemnaceae: 4 (c28) Ff						
Lemna minor	(none)	—	S, cg	orientin vitexin isoorientin isovitexin lutonarin lucenin vicenin saponaretin	much apiose	Bate-Smith, 1968; Van Beusekom, 1967; McClure and Alston, 1966
L. gibba	—	—	—	orientin vitexin isoorientin saponaretin	much apiose	Van Beusekom, 1967; McClure and Alston, 1966
L. obscura	—	—	—	Cy 3-glu isoorientin lucenin, saponaretin vicenin Cy 3-glu acylated vicenin acylated lucenin	—	McClure and Alston, 1966
L. trisulca	—	—	—	orientin vitexin isoorientin	—	McClure and Alston, 1966

Species				Constituents		Reference
L. perpusilla	—	—	—	lucenin, vicenin, acylated vicenin, acylated lucenin, Cy 3-glu	—	McClure and Alston, 1966
L. trinervis	—	—	—	vicenin, acylated vicenin, Ap 7-glu	—	McClure and Alston, 1966
L. valdiviana	—	—	—	lucenin, vicenin, acylated vicenin, Lu 7-glu	—	McClure and Alston, 1966
L. minima	—	—	—	lucenin, vicenin, acylated vicenin, acylated lucenin	—	McClure and Alston, 1966
Spirodela intermedia	—	—	—	orientin, isoorientin, vicenin, plus 8 unk. flavones (?)	—	McClure and Alston, 1966
S. polyrhiza	—	—	—	orientin, vitexin, isosaponarin, Cy 3-glu, Qu 3-gly, Km 3-gly, orientin, vitexin, isoorientin, lutonarin, Cy 3-glu, Lu 7-glu, Ap 7-glu	much apiose	Van Beusekom, 1967; McClure and Alston, 1966
S. biperforata	—	—	—	orientin, vitexin, isoorientin	—	McClure and Alston, 1966

TABLE I—continued

Species	Compounds[a]					References
	Alkaloids	Terpenoids	Simple phenolics	Flavonoids	Others	
S. biperforata (cont.)				lutonarin Cy 3-glu Lu 7-glu Ap 7-glu		
S. oligorhiza	—	—	—	orientin isosaponarin isoorientin lutonarin lucenin vicenin saponaretin acylated saponaretin Pt 3,5-diglu Qu 3,7-diglu Lu 7-digly	—	Jurd *et al.* 1957; Ng *et al.* 1964; McClure and Alston, 1966
Wolffiella lingulata	—	—	—	Qu 3-gly Qu 3,7-digly Qu 3,7-trigly	—	McClure and Alston, 1966
W. oblonga	—	—	—	Qu 3-gly Qu 3,7-digly Qu 3,7-trigly	—	McClure and Alston, 1966
W. gladiata	—	—	—	Qu 3-gly Qu 3,7-trigly	—	McClure and Alston, 1966
W. floridana	— —	— —	— —	Qu 3,7-trigly Qu 3-gly	— —	McClure and Alston, 1966
Wolffia punctata				Km 3-gly Qu 3-digly Qu 3,7-trigly Km 3,7-trigly		McClure and Alston, 1966

	Saponins	Phenolics	Flavonoids	Other	References
W. microscopica	—	—	Qu 3-gly Km 3-gly Qu 3,7-trigly Km 3,7-trigly	—	McClure and Alston, 1966
W. papulifera	—	—	Qu 3-gly Km 3-gly Qu 3,7-trigly Km 3,7-trigly	—	McClure and Alston, 1966
W. columbiana	—	—	orientin vitexin isoorientin vicenin saponaretin Lu 7-digly	—	McClure and Alston, 1966
W. arrhiza	—	—	orientin isoorientin vicenin saponaretin Lu 7-digly	much apiose	Duff, 1965; McClure and Alston, 1966
Sparganiaceae: 1 (c15) Em, Fl					
Sparganium americanum *	(no saponins)	—	—	(no tannins)	Willaman and Schubert, 1961; Hegnauer, 1963
S. erectum	—	caff, pC S, F	Qu, Cy	—	Bate-Smith, 1968b
Typhaceae: 1 (c10) Em					
Typha angustifolia *	—	catechins (?)	isorhamnetin rutin (?)	(no cyanogens)	Willaman and Schubert, 1961; Hegnauer, 1963; Hultin and Torssell, 1965; Tjia-Lo and Hegnauer, 1956; Fukuda, 1928
T. latifolia	—	pC, S, F	Qu, Cy, Km	—	Bate-Smith, 1968b

AQUATIC GENERA OF OTHERWISE TERRESTRIAL FAMILIES

I. DICOTYLEDONS
Lythraceae
Decodon Em, Ff

TABLE I—*continued*

Species	Compounds[a]						References
	Alkaloids	Terpenoids	Simple phenolics	Flavonoids	Others		
D. verticillatus	decalin vertalin decinin decamin vertin decodin verticallatin	—	—	—	—		Ferris, 1963
Onagraceae *Jussiaea* Em *J. repens*	—	—	pC, E	My, Dp, Qu Cy (?), Km	—		Bate-Smith, 1962
Primulaceae *Hottonia* Em *H. palustris*	—	—	caff, E	Qu, Cy, Km	—		Bate-Smith, 1962
Scrophulariaceae *Bacopa* Em *B. monniera*	nicotine 2 unk.	bacoside A bacoside B monnierin betulinic acid	—	—	—		Das *et al.* 1961; Chatterji *et al.* 1963; Dutta and Basu, 1963; Rastogi and Dhar, 1960
II. MONOCOTYLEDONS Araceae *Acorus* Em *A. calmus*	—	acoron calmene calamenol	pC, S, caff asaraldehyde asarone calamol	Cy 3-gly Qu	—		Karrer, 1958; Hegnauer, 1963; Guenther, 1952; Bate-Smith, 1968b; Price and Sturgess, 1938

	calamendiol, camphene, cinieol, geraniol, geranyl acetate, neoacoron, pinene	methyl eugenol, eugenol			References
A. gramineus	—	caff, pC S, F	—	—	Bate-Smith, 1968b
Calla Em					
C. palustris	*	caff, pC S, F	Qu, Km Cy	—	Hultin and Torssell, 1965; Massagetov, 1967; Bate-Smith, 1968b
Orontium Em					
O. aquaticum	—	S (?), scopoletin (?)	Qu, Cy, Km	—	Bate-Smith, 1968b
Pistia Ff					
P. stratiotes	—	caff (?)	Cy	—	Bate-Smith, 1968b
Gramineae					
Glyceria Em					
G. fluitans	—	pC, F	Qu, Km	—	Bate-Smith, 1968b
Zizania Em					
Z. latifolia	—	pC, S, F	—	—	Bate-Smith, 1968b

AQUATIC SPECIES OF OTHERWISE TERRESTRIAL GENERA

I. DICOTYLEDONS					
Polygonaceae					
Polygonum Em					
P. amphibium	unk.	—	—	—	Willaman and Schubert, 1961 Grippenberg, 1962;
P. hydropiper	—	—	persicarin persicarin 7-O-methyl ether rutin Cy 3-galact	—	Valentine and Wagner, 1962; Sugano and Hayashi, 1960

TABLE I—continued

Species	Compounds[a]					References
	Alkaloids	Terpenoids	Simple phenolics	Flavonoids	Others	
Lythraceae						
Lythrum Em						
L. salicaria	unk.	—	E, pC	Mv 3,5-diglu Cy 3-glu flavone-C-glycosides	—	Hultin and Torssell, 1965; Bate-Smith, 1962; Harborne, 1967; Paris and Paris, 1964; Paris, 1968
Umbelliferae						
Oenanthe Em						
O. stolonifera	—	—	—	persicarin	—	Matsuchita and Iseda, 1965
O. aquaticum	—	aldrol camphene crypton pinene phellandral phellandrene sabinene	—	rutin	—	Guenther, 1952
Hydrocotyle Em						
H. asiatica	hydrocotyline	asiaticoside thankuniside brahmoside brahminoside	—	—	—	Willaman and Schubert, 1961; Boiteau et al. 1949; Dutta and Basu, 1962; Rastogi et al. 1960
H. wilfordi	—	—	—	Qu 3-galact	—	Nakaoki and Morita, 1960
Lobeliaceae						
Lobelia Em						
L. dortmanna	Lobeline	—	—	—	—	Willaman and Schubert, 1961
Compositae						
Bidens Em						
B. laevis	unk.	—	—	Lu 7-glu	coreopsin	Willaman and Schubert, 1961;

Taxon	Alkaloids	Terpenoids	Phenolic acids	Flavonoids (Qu 3-glu, aurones, chalcones)	Misc. (sulphuretin, etc.)	Reference
Cotula Em						
C. coronopifolia	—			Qu 3-glu aurones chalcones; —	sulphuretin; dehydro-falcarinon	Haag-Berrurier and Duquenois, 1962; Hattori *et al.* 1956; Bohlmann *et al.* 1966
Acanthaceae						
Asteracantha Em						
A. longifolia	asteracanthin, asteracanthinin	lupeol		—	—	Basu and Rakhit, 1957; Chatterjee and Srimany, 1957
Justica Em						
J. hayatai	—			—	justicidin A, justicidin B	Munakata *et al.* 1965
Lentibulariaceae						
Utricularia S						
U. vulgaris	—			Cy (leuco?)	aucubin	Reznik and Neuhausel, 1959; Hegnauer, 1966
U. minor	unk.				—	Hultin and Torssell, 1965
U. caerulea	—		pC	—	—	Bate-Smith, 1962
II. MONOCOTYLEDONS						
Juncaceae						
Juncus Em						
J. effusa	—		cg, caff / F, pC, S	Lu 7-glu hesperidin diosmin	—	Stabursvik, 1968; Bate-Smith, 1968b; Plouvier, 1962
Cyperaceae						
Carex Em						
C. acutiformis	—		—	—	—	Clifford and Harborne, 1969
C. brevicollis	brevicolline, harman, brevicarine		—	carexidin	—	Terenteva and Boroukou, 1960
C. contigua	—		caff, pC, S (?), F	—	—	Bate-Smith, 1968b
C. flava	—		caff, pC, S, F	Qu, Km	—	Bate-Smith, 1968b
C. montana	—		caff, pC, S, F	—	—	Bate-Smith, 1968b
C. pendula	—		caff, pC, S (?), F	Cy	—	Bate-Smith, 1968b
C. otrubae	—		caff, pC, S, F	—	—	Bate-Smith, 1968b

TABLE I—*continued*

| Species | Compounds | | | | | References |
	Alkaloids	Terpenoids	Simple phenolics	Flavonoids	Others	
C. riparia	—	—	—	carexidin	—	Clifford and Harborne, 1969
C. scaposa	—	—	caff, pC, S (?), F	—	—	Bate-Smith, 1968b
Cyperus E						
C. articulatus	unk. sesquiterpenoid alkaloids	unk. sesquiterpenes cyperone	—	—	—	Joly, 1937; Neville *et al.* 1968
C. esculentus	—	—	—	unk. antho. (leuco?)	tannins	Hegnauer, 1963
C. letencei	—	—	—	unk. flavones	—	Hegnauer, 1963
C. longus	—	—	caff, pC, S(?), F	Cy	—	Bate-Smith, 1968b
C. papyrus	—	—	caff(?), pC, S	Qu	—	Bate-Smith, 1968b
C. prolifera	—	—	caff(?), pC, S	Qu, Cy	—	Bate-Smith, 1968b
C. rotundus	unk.	pinene cinneole cyperon bicyclic and tricyclic sesquiterpenoids	—	aureusidin	—	Willaman and Schubert, 1961; Neville *et al.* 1968; Clifford and Harborne, 1969
C. scariosus	unk.	cyperon tricyclic sesquiterpenoids	—	—	—	Willaman and Schubert, 1961; Neville *et al.* 1968; Hegnauer, 1963
Gramineae						
Echinochloa E						
E. crusgalli	hordenine	sawamilletin	—	—	—	Willaman and Schubert, 1961; Hegnauer, 1963
Phalaris E						
P. tuberosa	dimethyltryptamine 5-methoxydimethyl-tryptamine bufotenine	—	—	—	—	Culvenor *et al.* 1964

P. arundinacea	hordenine 5-methyl-*N*-methyl-tryptamine dimethyltryptamine bufotenine	—	Cy 3-arab Pn 3-arab	—	Wilkinson, 1956; Clifford and Harborne, 1967; Culvenor *et al.* 1964
P. zizanioides	—	vetoverol triciclovetivene laejujenenol khusol bicyclovetivenol veticadinol	—	—	Hegnauer, 1963

[a] Abbreviations: —, no report; (none), reportedly absent from the species; ?, tentative identification; *, conflicting reports; Ap, apigenin; arab, arabinoside; caff, caffeic acid; cg, chlorogenic acid; Cy, cyanidin; Dp, delphinidin; E, ellagic acid; F, ferulic acid; galact, galactoside; glu, glucoside; glucur, glucuronide; gly, unknown glycoside; Km, kaempferol; leuco, leucoanthocyanin; Lu, luteolin; Mv, malvidin; My, myricetin; pC, *para*-coumaric acid; Pg, pelargonidin; Pn, peonidin; Pt, petunidin; Qu, quercetin; rham, rhamnoside; S, sinapic acid; unk., unknown.

[b] The families are arranged in the table to correspond with the relative positions they occupy in Hutchinson's (1959) system of classification. The precise circumscription of the families is modified from Sculthorpe (1967).

[c] The size of the family is indicated by the estimated number of: genera (species), e.g. 8 (c60). Life forms (Sculthorpe, 1967) are abbreviated as: Em, emergent; Ff, free-floating; Fl, floating-leaves; S, submerged.

with the liquid medium, the plants quickly take them up. In characteristically thin-leaved submerged or free-floating plants the precursors normally need to be transported for only a few millimetres to reach the site of synthesis.

Wallace *et al.* (1969) selected various species of *Spirodela* and *Lemna* for investigations of the synthesis of flavone-*O*-glycosides and -*C*-glycosides. The plants were clonally subcultured axenically under controlled conditions on media containing phenylalanine-1-^{14}C. After several days the plants were

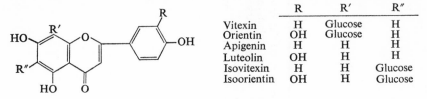

FIG. 2. ^{14}C-flavonoids which have been incorporated into *Spirodela* and *Lemna*

harvested and the ^{14}C-labelled flavones apigenin, luteolin, orientin, isoorientin and isovitexin (Fig. 2) obtained in good yield. These labelled flavones were then introduced into the media for *S. polyrhiza*, *S. oligorhiza* and *L. minor*.

The results of these investigations are summarized in Fig. 3. It may be suggested from this work that the biosynthesis of flavone-*O*-glycosides and -*C*-glycosides takes place through parallel pathways diverging at an early

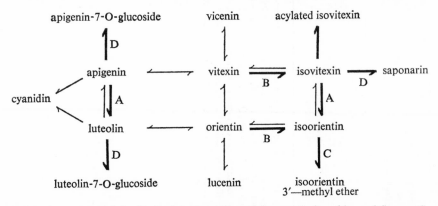

FIG. 3. A summary of the *in vivo* interconversion of flavone-*O*-glucosides and flavone-*C*-glucosides in *Spirodela* and *Lemna*. ➤, conversion; →, no interconversion; A, 3′-hydroxylation; B, isomerization; C, 3′-methylation; D, *O*-glycosylation. See Fig. 2 for the structures of the parent compounds. (Adapted from Wallace *et al.* 1969)

stage. Perhaps *O*-glycosylation is a late step taking place after ring closure to form the flavone, and *C*-glycosylation is an earlier reaction at the level of the presumed chalcone or flavanone intermediate (Wallace and Alston, 1966). Their discovery that the B-ring oxidation of apigenin to luteolin glycosides— but not vitexin to orientin—raises some interesting questions about the stage at which 3′-hydroxylation occurs.

Grisebach (1968) has summarized much of the work relating to the stage at which B-ring hydroxylation occurs. In many plants, it is best assumed that the hydroxyl group is added after the chalcone stage and that this is not a rate-limiting process. On the other hand, in *Petunia hybrida*, the substitution pattern of the anthocyanins is apparently already determined at the cinnamic acid level (Hess, 1965). It may well be, as Bu'Lock (1965) and Swain (1965) have suggested, that within a single plant there is not a single pathway leading to flavonoids with different substitutions, but that a "metabolic grid" operates in which one set of reactions may take place at one time and another set under altered conditions. Viewed as a process of natural selection, such sets of parallel transformations would ensure the production of a particular constituent under a diverse range of conditions.

C. PHOTOCONTROL

Light is probably the most important external factor controlling flavonoid synthesis. One of the most valuable experimental implications about its use as a probe for examining *in vivo* physiological processes is that it can be introduced

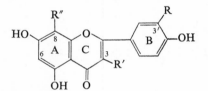

	R	R'	R"
Vitexin	H	H	glucose
Orientin	OH	H	glucose
Kaempferol	H	OH	H
Quercetin	OH	OH	H

Flavone-C-glycosides and flavonols

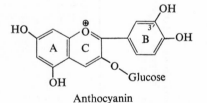

Anthocyanin

Cyanidin-3-monoglucoside

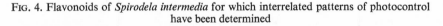

Fig. 4. Flavonoids of *Spirodela intermedia* for which interrelated patterns of photocontrol have been determined

into an intact cell with a minimal perturbation of the biological object (Siegelman, 1964). However, there are also valid criticisms to using the physiological approach in understanding the biosynthesis of secondary constituents. *In vivo* processes requiring several days for completion cannot be expected to have narrow specificity and thus the results are rarely clearcut. Furthermore, it is difficult—often almost impossible—to separate the effects of the experimental modifications on the biosynthesis of the secondary metabolite from their effects on plant growth. To balance these points of view, one must realize that

TABLE II

Control of flavonoid biosynthesis in *Spirodela intermedia*[a]

Light condition[b]	Proposed mechanism	Relative flavonoid production[c]	Growth regulator[d] interaction
Dark		vitexin + kaempferol +	Kinetin satisfies light requirements for growth
4 minutes of red light	Phytochrome-controlled 5-fold increase in growth with parallel flavonoid biosynthesis	vitexin + kaempferol +	No appreciable influence
45 minutes of white or blue light	Establishes 3′-hydroxyl, increases growth and flavonoid precursors	vitexin + kaempferol + orientin + quercetin +	Kinetin reduces vitexin and kaempferol. Gibberellic acid induces a trace of cyanidin 3-glucoside
Continuous white light	Determines oxidation level of the anthocyanin. Further promotes growth and flavonoid biosynthesis	vitexin ++ kaempferol ++ orientin ++ quercetin ++ cyanidin 3-glucoside +	Kinetin has no appreciable influence. Gibberellic acid slightly reduces both growth and cyanidin 3-glucoside synthesis

[a] Data from McClure (1968) and Norman (1968). All plants were cultured axenically on Hutner's medium with 1% sucrose at 26°.

[b] Illumination given each 24 hour period. Red light at 1·1 kerg cm^{-2} s^{-1}. 45 minute white (fluorescent) or blue light treatments at ca. 3,000 lux or 1·2 kerg cm^{-2} s^{-1} respectively. Continuous white (fluorescent) light at ca. 10,000 lux.

[c] In relative amounts of flavonoid per unit weight of plant material. See Fig. 4 for structures of these flavonoids. For molar concentrations per gramme of plant material, see McClure (1968) and Norman (1968).

[d] Incorporated into the medium at the concentrations of $1·6 \times 10^{-6}$ M (kinetin) or 5×10^{-5} M (gibberellic acid).

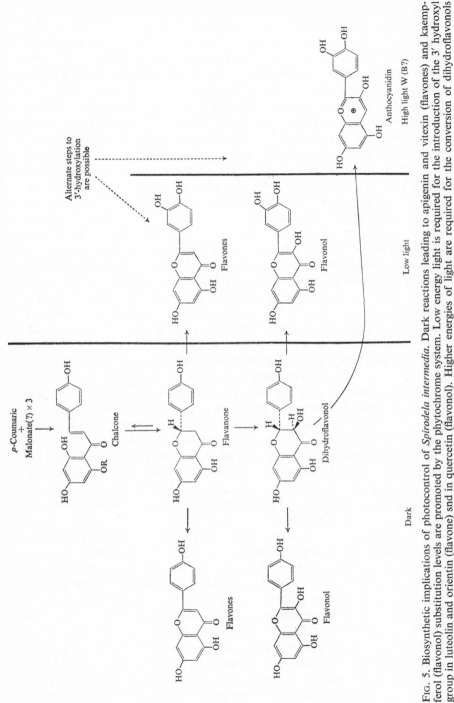

FIG. 5. Biosynthetic implications of photocontrol of *Spirodela intermedia*. Dark reactions leading to apigenin and vitexin (flavones) and kaempferol (flavonol) substitution levels are promoted by the phytochrome system. Low energy light is required for the introduction of the 3' hydroxyl group in luteolin and orientin (flavone) snd in quercetin (flavonol). Higher energies of light are required for the conversion of dihydroflavonols into anthocyanins. Photocontrols adapted from Table II. Biosynthetic pathway from Grisebach (1968)

without investigating the physiological responses, one obtains only an incomplete picture of the phytochemical *potential* of a plant.

McClure (1968) investigated the photocontrol of five flavonoids in *Spirodela intermedia*. This species produces flavonoids with both 4'-hydroxyl and 3',4'-dihydroxyl substituents in the B-ring (Fig. 4) and it makes anthocyanins, flavonols and flavone-C-glycosides. There is no evidence that any of these flavonoid classes are converted *in vivo* (Grisebach, 1968; Wallace *et al.* 1969).

Qualitative and quantitative studies of light control, coupled with time-course measurements, showed that the anthocyanin, cyanidin 3-glucoside, was present only after prolonged illumination of the plants with either white or blue, but not red, light. This wavelength dependency suggests mediation in part through the red far-red reversible phytochrome system or the less clearly defined High Energy photosystem (Siegelman, 1964). The lack of anthocyanin in prolonged red light was interpreted as an unusual response of the phytochrome system since prolonged red light is quite effective in promoting the disappearance of the active form of this photoreceptor while either blue or white light establish low, and more stable, levels of active phytochrome (Chorney and Gordon, 1966). Subsequent investigations (Norman, 1968) show that saturation of the low-energy phytochrome system alone will not induce anthocyanin formation in *S. intermedia*, even when cultured on a variety of media containing sugars or various plant growth hormones.

In sharp contrast, the 4'-hydroxylated flavonoids, vitexin and kaempferol, are produced in good yield under all photoconditions. Even when the plants were grown in complete darkness they produced approximately 25% as much of these flavonoids as they did under continuous illumination in bright light (McClure, 1968). In contrast, 3',4'-dihydroxylated orientin and quercetin require some light for synthesis in all but the most reduced amounts. This light requirement does not appear to be simply related to the phytochrome system, even under long periods of culture. For example, Norman (1968) submitted the plants to daily red light treatments (1320 kerg cm^{-2}) for 21 days and while there was a marked increase in the rate of growth over the dark-grown controls, no appreciable change was found in the flavonoid chemistry. In order to detect more than traces (< 0.05 μmole g^{-1} of fresh plant material) of either orientin or quercetin in *S. intermedia*, it was necessary to give the plants at least low-energy white or blue light for several hours each day (Norman, 1968).

Some interactions of photocontrol and the plant-growth regulators gibberellic acid and kinetin are worth mentioning. Kinetin can essentially satisfy the light requirement for growth in many of the Lemnaceae (Hillman, 1957), yet it has no appreciable influence on flavonoid synthesis in *S. intermedia* (Norman, 1968). In low light kinetin slightly reduces the production of 4'-hydroxyl flavonoids while gibberellic acid surprisingly elicits the synthesis of traces of cyanidin in 3-glucoside. However, as these growth regulators have been implicated in the control of almost every aspect of plant growth (Leopold, 1964), these results are difficult to interpret. Nevertheless, it appears that in

S. intermedia the photocontrol of flavonoid synthesis in its several aspects is essentially independent of light requirements for the production of these growth factors.

Overall aspects of the photocontrol of flavonoid synthesis in *S. intermedia* are summarized in Table II and Fig. 5. Several biosynthetic implications are apparent. Control mechanisms which elicit anthocyanin production probably operate at a late stage, since quercetin is formed under conditions where cyanidin 3-glucoside is absent and it is well established that a number of C_{15} flavonoid precursors are incorporated effectively into each of these compounds (Grisebach, 1968). The availability of some 3′,4′-dihydroxylated precursor for cyanidin 3-glucoside, quercetin and orientin probably limits the synthesis of these compounds in low light intensity.

Under all of the experimental conditions employed by McClure (1968) and Norman (1968), the relative production of flavonols and flavone-C-glycosides was related to their B-ring oxidation and not their glycosylation type. This links the biosynthesis of these two classes more closely together than would be anticipated strictly on the basis of experiments using labelled precursors. Finally, many studies have suggested that the photocontrol of flavonoid synthesis is at the level of increasing acetate or malonyl units for the flavonoid A-ring, or through providing the deamination products of aromatic amino acids for the phenylpropanoid portion of the molecule (Grisebach, 1968; Siegelman, 1964). The formation of considerable amounts of vitexin and kaempferol in the dark, coupled with the accumulation of orientin and quercetin but not cyanidin in low-light conditions, make it appear highly unlikely that the light control of anthocyanin formation works at such an early biosynthetic state or that a 3′-hydroxylated precursor is the limiting factor. These results (McClure, 1968) are quite consistent with the work of Grisebach (1968) on seedlings of buckwheat (*Fagopyrum esculentum*) in which he demonstrated that dihydroflavonols are probably the immediate precursors to both flavonols and anthocyanins. It is suggested that the specific photocontrol of anthocyanin synthesis occurs at this latter stage.

IV. BIOCHEMICAL SYSTEMATIC STUDIES

A. GENERAL CONSIDERATIONS

From the previous section two systematically relevant features emerge. First, we do not know the biological roles of secondary compounds or the evolutionary implications of environmental changes on the occurrence of these compounds. For this reason it is difficult to draw conclusions about their impact on the continued existence of a taxon. Second, and this is probably most important, biosynthetic studies have proved definite, predictable and reliable patterns of control of the biosynthesis of at least some of the secondary constituents. In fact, biosynthetic studies would not be possible if controls did not operate.

Close regulation of biosynthesis ensures the value of secondary constituents as reliable phenotypic characteristics for systematic studies.

B. STUDIES IN THE LEMNACEAE

Although representatives of the aquatic angiosperms have often been included in broadly based phytochemical or chemosystematic surveys (e.g. Bate-Smith, 1962, 1968b; Hultin and Torssell, 1965; Kubitzki and Reznik, 1966; Harborne, 1967; Li and Willaman, 1968), only in the Lemnaceae have sufficient taxa been investigated to warrant application of their chemical constituents to systematic problems at the lower taxonomic levels (Table I).

McClure and Alston (1966) investigated 186 clones of Lemnaceae representing 22 of the approximately 28 species. In this family in which morphological reduction is so extreme that the usual distinction between stem and leaf is no longer practical (Hegelmaier, 1868), there is a problem in identifying the plants.

Selecting the clone of *Spirodela oligorhiza* which had been used in studies of flavonoid biosynthesis and structure (Thimann and Edmondson, 1949; Jurd *et al*. 1957), McClure and Alston (1964) subcultured this plant under 53 regimes in which light intensity, concentration and composition of the medium, sugars and growth factors were varied. The flavonoid chemistry of this species is complex (Table I), yet the qualitative flavonoid pattern was unaltered by the treatments. A number of the compounds originally reported to be variable in McClure and Alston (1964) are now known to be either artifacts of the technique or reported as such due to the use of too-small samples of plant material (Ball *et al*. 1967; McClure, unpublished).

Many species of Lemnaceae intergrade morphologically and there are simply not enough reliable vegetative characteristics to separate them. When all 186 clones were examined after culture under conditions designed to promote flavonoid synthesis, the flavonoid chemistry was found to be species-characteristic and a highly reliable means of identifying the plants. Such a high degree of consistency was found in the flavonoid patterns of these plants that it was possible to identify the intergrading forms on the basis of their secondary chemistry. In some instances confirmation of identity was obtained by finding the plants in flower (McClure and Alston, 1966).

Intraspecific flavonoid variation is established for the Lemnaceae only in a single apigenin 7-glycoside of *L. perpusilla*. This is remarkable considering that 47 flavonoids are found in these 22 species (Table I). In *L. perpusilla* the presence or absence of this flavonoid is a clonal characteristic, which has even been used as a basis for identifying clones of *L. perpusilla* originating from different geographic areas (McClure, 1967).

The only currently practicable means of identifying the flavonoid constituents in a large number of plants appears to be through detailed investigations of one or more representatives, or clones, and then comparing the remainder by some sort of screening process. This process has the inherent limitation of depending on a small number of characteristics, such as chromatographic behaviour and colour reactions, for identifications. Since this process was applied in the work on the Lemnaceae, the possibility remains that some minor flavonoid variations exist within species, such as in *L. minor* for which more than 40 collections were examined. In fact, in only a few of these clones could anthocyanin be detected and there is evidence that other minor constituents may vary qualitatively (McClure and Alston, 1966; Wallace and Turner, personal communication). Even from this conservative viewpoint, the Lemnaceae are remarkable for the species-specificity of those flavonoids produced in sufficient amounts for identification.

This consistency may be a reflection of their lack of interbreeding. Flowering in this family is a rare phenomenon and, since these plants reproduce quite rapidly through asexual means, it is quite likely that when flowering does occur pollination will be between the numerous individuals of a single clone.

This rarity of sexual reproduction is, in fact, of general occurrence among aquatic angiosperms. Perhaps the best-documented example is the spectacular spread of the North American *Elodea canadensis* throughout the British Isles, continental Europe, Tasmania and New Zealand in the late 19th and early 20th centuries. It is probable that reproduction in these areas was exclusively vegetative and that large areas were populated with the offspring of a single clone (Sculthorpe, 1967).

For the foregoing reasons it might be predicted that other taxa of aquatic angiosperms, especially those with primarily asexual reproduction, will be found to be similarly consistent in their secondary chemistry.

In addition to serving as unambiguous criteria by which morphologically intergrading taxa can be separated, the flavonoids of the Lemnaceae also point to a phyletic reduction series paralleling the series suggested by their morphology (McClure and Alston, 1966; Turner, 1967). For example, the genus *Spirodela* has been considered to be the ancestral type of the Lemnaceae by most investigators (Hegelmaier, 1868; Brooks, 1940; Maheshwari, 1959; Daubs, 1965) on the basis of embryological, anatomical and morphological criteria. The apparent sequence is *Spirodela* → *Lemna* → *Wolffiella* → *Wolffia*. When the flavonoids are considered, *Spirodela* contains anthocyanins, flavonols, flavone-*C*-glycosides and flavone-*O*-glycosides. *Lemna* produces all of these except flavonols, *Wolffiella* forms only flavonols, and *Wolffia* quite perplexingly accumulates either flavonols or flavone-*C*-glycosides depending on the species (Table I). There appears to be a phyletic trend toward increased simplicity in both exomorphic characteristics and a concomitant loss of the ability to produce a wide range of flavonoid types. This picture is not as clear-cut as the above might indicate, however, as there is in fact very little evidence to relate closely the rooted and the rootless genera. For example, *Wolffiella* and *Wolffia* are rootless, produce new plants from a single basal reproductive pouch, and are thalloid, almost isodiametric (*Wolffia*) or linear (*Wolffiella*). *Spirodela* and *Lemna* have roots, and two lateral reproductive pouches, and the plant body is flattened and generally oval in outline.

To supplement the flavonoid work, Blazey and McClure (1968) investigated the lignification characteristics of *Spirodela* and *Lemna* to investigate further the systematic aspects of their phenolic chemistry. Since the flavonoids and lignins are both derived from the phenylpropanoid pathway, it was anticipated that several of the more complex, presumably the more primitive, members would be lignified. There is no generally accepted agreement of what constitutes lignin or which techniques are best applied in its analysis. Using a technique

TABLE III

Lignification characteristics of *Spirodela* and *Lemna*[a]

	Lignification[c]		
Species[b]	*p*-Hydroxy-benzaldehyde	Vanillin	Syringaldehyde
Spirodela intermedia	6·06	4·09	4·92
S. polyrhiza	6·58	1·74	1·07
S. oligorhiza	2·83	0·99	0·59
S. biperforata	5·34	0·06	—
Lemna perpusilla	5·46	0·96	—
L. trinervis	3·25	0·39	—
L. minor	2·99	0·18	—
L. minima	4·92	0·08	—
L. gibba	3·04	—	—
L. obscura	4·50	—	—
L. valdiviana	6·83	—	—
L. trisulca	4·33	—	—

[a] The plants were subcultured from 21 to 28 days on Hutner's medium with 1 % sucrose, homogenized, exhaustively extracted, and the insoluble residue oxidized in alkaline cupric hydroxide (Blazey and McClure, 1968).

[b] Arranged in order of decreasing lignification within the genera. If the species is not considered to be lignified, it is arranged in order of decreasing morphological complexity.

[c] Recovery of either syringaldehyde or vanillin, or both, along with *p*-hydroxybenzaldehyde is accepted as evidence for lignification. The recovery of *p*-hydroxybenzaldehyde alone is not accepted as evidence for lignification. Benzaldehyde recovery in micromoles per 300 mg sample. It should be noted that these techniques are not entirely quantitative (Towers and Mass, 1965).

modified from Towers and Mass (1965), the plants were considered to be lignified by the criterion of recovery of either syringaldehyde or vanillin, or of both, along with *p*-hydroxybenzaldehyde, from plant material which had been exhaustively extracted to remove non-lignin phenolic compounds and then subjected to rigorous alkaline cupric hydroxide oxidation (Blazey and McClure, 1968).

By these criteria the most morphologically complex members of the family are lignified. The highest yields of syringaldehyde were recovered from

Spirodela intermedia, lesser amounts from *S. polyrhiza* and *S. oligorhiza*, and none from *S. biperforata* or from any species of *Lemna*. Vanillin was present in all of the *Spirodela* species and in one group of *Lemna* (Table III).

Lignification characteristics substantiate the distinctiveness of the species *S. polyrhiza* and *S. biperforata* (Blazey and McClure, 1968). The differences between the benzaldehydes of *S. polyrhiza* and *S. biperforata* are important since Daubs (1965) has questioned the distinctiveness of these two species on morphological grounds. McClure and Alston (1966) found that the flavonoids of these two species are identical.

Applying both the lignification and flavonoid characteristics to the larger question of evolutionary relationships within *Spirodela* and *Lemna*, a pattern emerges in which many lines of evidence seem to point to the same conclusions. This evidence is derived from Brooks' (1940) investigations of their cytology

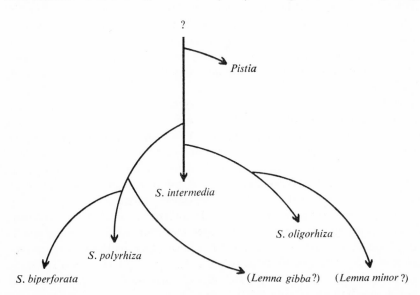

FIG. 6. Suggested origin and evolution, based on chemical and morphological evidence, of some species of *Spirodela* and *Lemna*

and morphology, Hegelmaier's (1868) and Daubs' (1965) monographic treatments, Goebel's (1921) study of their organography, and Maheshwari's (1958) work on their embryology and comparative anatomy, as well as phenolic studies by McClure and Alston (1966) and Blazey and McClure (1968).

These data suggest the hypothetical phylogenetic relationships illustrated in Fig. 6. *Spirodela* is probably the most primitive member of the Lemnaceae with a plural number of seed, a large number of roots and veins, a rich and varied flavonoid chemistry, and lignification characteristics which are comparable to many of the terrestrial monocotyledons. Although no phytochemical information seems to be available, *Pistia* is on the basis of morphological studies, a modern-day counterpart of the ancestral form of the Lemnaceae. In fact, Brooks (1940) proposed, without knowledge of *S. intermedia*, that an ancestral type for the Lemnaceae which would link *Pistia* and *Spirodela* would have the complex morphology, inflated air spaces and large number of seeds which characterize *S. intermedia*.

Two lines appear to diverge from this primitive stock. *Spirodela polyrhiza*, *S. biperforata* and *Lemna gibba* are multi-seeded, flavonol-free plants with large fronds and inflated air spaces. Plants of the other line, comprised of *S. oligorhiza* and *L. minor*, are smaller, have a single seed per fruit, produce flavonols, and they have a total of seven flavonoid glycosides in common.

When the flavonoid chemistry of *S. oligorhiza* is compared with that of all other species of *Spirodela* only a few compounds are shared; yet when *S. oligorhiza* is compared with *L. minor*, seven of the eight flavonoids of this latter species are also present in the former. Also, the anthocyanin petunidin 3,5-diglucoside is rare in the monocotyledons (Bate-Smith, 1968b), but it is known to be present in *S. oligorhiza* (Ng *et al.* 1964). It is apparently this same anthocyanin which is produced in very low concentrations in some clones of *L. minor*. All other species of Lemnaceae with anthocyanins have cyanidin 3-glucoside (McClure and Alston, 1966).

Finally, while it is not truly a secondary compound, the branched-sugar apiose and its distribution within aquatic angiosperms has been considered as a possibly significant marker compound. Van Beusekom (1967) investigated hydrolysates of 27 species of monocotyledons for the presence of apiose, and other than in the Lemnaceae, apiose in appreciable amounts was found only in certain species of the Hydrocharitaceae, Potomogetonaceae and Zannichelliaceae (Helobiae). Lawalree (1943) has presented arguments for the derivation of the Lemnaceae from the Najadaceae (Helobiae); these are based on certain similarities of flower development, vegetative homologies and endosperm development. As previously indicated, most other workers consider the Lemnaceae to be derived from the Araceae. *Pistia stratiotes* (Araceae) does not produce apiose, yet several of the Helobiae are rich sources of apiose as are the Lemnaceae (Van Beusekom, 1967).

Several points seem to be at variance if Lemnaceae are considered to be derived from the Helobiae. For example, one of Lawalree's (1943) major arguments for the similarity of Lemnaceae and Najadaceae was in helobial endosperm. Brooks (1940) and Maheshwari (1959) have both confirmed that the endosperm of the Lemnaceae is cellular (not helobial) from the beginning, as is the case in the Araceae. A further weakness in Lawaree's (1945) system is that the sequence of change within the Lemnaceae is read as a progressive one going from the thalloid *Wolffia*, through *Wolffiella* and *Lemna*, to the large and complex (by comparison) plant body of *Spirodela*.

Perhaps more importantly, while apiose is present in certain species of the Helobiae, it was not found in the Najadaceae. In fact, apiose in appreciable amounts in the Helobiae seems to be restricted to those plants which grow in the sea or in brackish water (Van Beusekom, 1967). Duff (1965) found varying amounts of apiose in each of 175 samples representing many terrestrial families from the British flora. Thus, it appears that while apiose is very widely distributed in the plant kingdom, it may be quantitatively controlled by an adaptation to certain aquatic environments.

REFERENCES

Achmatowicz, O., and Bellen, Z. (1962). *Tetrahedron Lett.* 1121.
Achmatowicz, O., and Wrobel, J. T. (1964). *Tetrahedron Lett.* 129.
Allsopp, A. (1965). *In* "Encyclopedia of Plant Physiology" (W. Ruhland, ed.), Vol. XV/1, pp. 1236–1255. Springer Verlag, Berlin.
Alston, R. E. (1966). *In* "Evolutionary Biology" (T. Dobzhansky, M. K. Hecht and W. C. Steere, eds.), Vol. I, pp. 197–305. Appleton-Century-Crofts, New York.
Altman, R. F. A. (1954). *Nature,* **173,** 1098.
Arata, Y. (1965). *Chem. Pharm. Bull (Tokyo),* **13,** 907.
Arata, Y., Ohashi, T., Yonemitsu, M., and Yasuda, S. (1967). *J. Pharm. Soc. Japan,* **87,** 1094.
Arber, A. (1920). "Water Plants: A study of Aquatic Angiosperms." University Press, Cambridge.
Ball, G. A., Beal, E. O., and Flecker, E. A. (1967). *Brittonia,* **19** 273.
Barchet, R., and Forrest, T. D. (1965). *Tetrahedron Lett.* 4229.
Basu, N., and Rakhit, S. (1957). *Indian J. Pharm.* **19,** 285.
Bate-Smith, E. C. (1954). *Biochem. J.* **58,** 122.
Bate-Smith, E. C. (1962). *J. Linn. Soc. (Bot.),* **58,** 95.
Bate-Smith, E. C. (1968a). *Phytochemistry,* **7,** 459.
Bate-Smith, E. C. (1968b). *J. Linn. Soc. (Bot.),* **60,** 325.
Bate-Smith, E. C., and Swain, T. (1965). *Lloydia,* **28,** 313.
Bernauer, K. (1963). *Helv. chim. Acta,* **46,** 1783.
Bernauer, K. (1964). *Helv. chim. Acta,* **47,** 2119.
Blazey, E. B., and McClure, J. W. (1968). *Am. J. Bot.* **55,** 1240.
Bohlmann, F., Hornowski, H., and Arndt, C. (1966). *Chem. Ber.* **99,** 2828.
Boiteau, P., Buzas, A., Lederer, E., and Polonsky, J. (1949). *Nature,* **163,** 258.
Brooks, J. S. (1940). *Dissertation,* Cornell University.
Bu'Lock, J. D. (1965). "The Biosynthesis of Natural Products." McGraw-Hill, New York.
Chatterjee, A., and Srimany, S. K. (1957). *J. Indian Chem. Soc.* **34,** 882.
Chatterji, N., Rastogi, R. P., and Dhar, M. K. (1963). *Indian J. Chem.* **1,** 212.
Cheadle, V. I. (1942). *Am. J. Bot.* **28,** 441.
Chorney, W., and Gordon, S. A. (1966). *Pl. Physiol.* **41,** 891.
Clifford, H. T., and Harborne, J. B. (1967). *J. Linn. Soc. (Bot.),* **178,** 125.
Clifford, H. T., and Harborne, J. B. (1969). *Phytochemistry,* **8,** 123.
Culvenor, C. C. J., Bon, R. D., and Smith, L. W. (1964). *Aust. J. Chem.* **17,** 1301.
Das, P. K., Malhotra, C. L., and Dhalla, N. S. (1961). *Phys. Pharmac.* **5,** 136.
Daubs, E. H. (1965). *Illinois Biol. Monogr.* 34, University of Illinois Press.
Dekker, J. (1913). "Die Gerbstoffe: botanisch–chemische Monographie der Tannide." Bornträger, Berlin.
Duff, R. B. (1965). *Biochem. J.* **94,** 768.
Dutta, T., and Basu, U. P. (1962). *J. Sci. Indian Res.* **21B,** 239.
Dutta, T., and Basu, U. P. (1963). *Indian J. Chem.* **1,** 408.

Ettlinger, M. G., and Kjaer, A. (1968). *In* "Recent Advances in Phytochemistry" (T. J. Mabry, R. E. Alston and V. C. Runeckles, eds.), Vol. I, pp. 59–144. Appleton-Century-Crofts, New York.

Ferris, J. P. (1963). *J. org. Chem.* **28**, 817.

Fluck, H. (1963). *In* "Chemical Plant Taxonomy" (T. Swain, ed.), pp. 167–186. Academic, New York and London.

Fukuda, M. (1928). *Bull. Chem. Soc. Japan*, **3**, 53.

Furukawa, H. (1965). *Yakugaku Zasshi*, **85**, 335; *Chem. Abstr.* **63**, 4351.

Furuya, M., and Thimann, K. V. (1964). *Arch. Biochem. Biophys.* **108**, 109.

Gascoigne, R. M., Ritchie, E., and White, D. R. (1948). *J. R. Soc. N.S.W.* **82**, 44.

Goebel, K. (1921). *Flora*, **114**, 278.

Goodwin, T. W. (1967). *In* "Terpenoids in Plants" (J. B. Pridham, ed.), pp. 1–23. Academic, New York and London.

Gripenberg, J. (1962). *In* "The Chemistry of Flavonoid Compounds," pp. 406–440. Macmillan, New York.

Grisebach, H. (1968). *In* "Recent Advances in Phytochemistry" (T. J. Mabry, R. E. Alston and V. C. Runeckles, eds.), Vol. I, pp. 379–406. Appleton-Century-Crofts, New York.

Guenther, E. (1952). "The Essential Oils," Vol. IV. Van Nostrand, Princeton.

Haag-Berrurier, M., and Duquenois, P. (1962). *C. r. hebd. Séanc. Acad. Sci., Paris*, **254**, 3419.

Harborne, J. B. (1967). "Comparative Biochemistry of the Flavonoids." Academic, New York and London.

Harborne, J. B. (1968). *In* "Progress in Phytochemistry" (L. Reinhold and Y. Liwschitz, eds.), Vol. I, pp. 545–588. Interscience, New York.

Hattori, S., Shimokoriyama, M., and Oka, K. (1956). *Bull. Soc. Chim. Biol.* **38**, 557; *Chem. Abstr.* **50**, 15746.

Hegelmaier, F. (1868). "Die Lemnacean. Eine monographische Untersuchung." Willhelm Engelmann, Leipzig.

Hegnauer, R. (1958). *Pharm. Weekblad*, **93**, 801.

Hegnauer, R. (1963). "Chemotaxonomie der Pflanzen," Vol. II. Birkhäuser, Basel.

Hegnauer, R. (1964). "Chemotaxonomie der Pflanzen," Vol. III. Birkhäuser, Basel.

Hegnauer, R. (1966). "Chemotaxonomie der Pflanzen," Vol. IV. Birkhäuser, Basel.

Hegnauer, R. (1967). *Pure Appl. Chem.* **14**, 173.

Hess, D. (1965). *Z. Pflphysiol.* **53**, 1.

Hillman, W. S. (1957). *Science*, **126**, 165.

Howes, F. N. (1953). "Vegetable Tanning Materials." Butterworths, London.

Hultin, E., and Torssell, T. (1965). *Phytochemistry*, **4**, 425.

Hutchinson, J. (1959). "The Families of Flowering Plants. I. Dicotyledons; II. Monocotyledons." Clarendon, Oxford.

Il'inskaya, T. N., Kuzovkov, A. D., and Monakhova, T. G. (1967). *Khim. Prir. Soedin (Tashkent)*, **3**, 178; *Biol. Abstr.* **49**, 129456.

Joly, L. (1937). *Parfum. mod.* **31**, 25.

Jurd, L., Geissman, T. A., and Seikel, M. K. (1957). *Arch. Biochem. Biophys.* **67**, 284.

Karrer, W. (1958). "Konstitution und Vorkommen der organischen pflanzenstoffe." Birkhäuser, Basel.

Kubitzki, K., and Reznik, H. (1966). *Beitr. Biol. Pfl.* **42**, 445.

Kupchan, S. M., Dasgupta, B., Fujita, E., and King, M. L. (1963). *Tetrahedron*, **19**, 227.

Lawalree, A. (1943). *La Cellule*, **49**, 335.

Leopold, C. A. (1964). "Plant Growth and Development." McGraw-Hill, New York.

Li, H. L., and Willaman, J. J. (1968). *Econ. Bot.* **22**, 239.

Maheshwari, S. C. (1959). *Nature*, **181**, 1745.

Massagetov, P. S. (1947). *Trudȳ vses. nauchno-issled. Inst. Lekarstv. Rast.* **9**, 3 (cited by Hultin and Torssell, 1965).

Matsuchita, A., and Iseda, S. (1965). *Nippon Nogei Kagaku Kaish*, **39**, 317; *Chem. Abstr.* **64**, 1012.

McClure, J. W. (1967). *Pl. Cell Physiol.* **8**, 523.

McClure, J. W. (1968). *Pl. Physiol.* **43**, 193.

McClure, J. W., and Alston, R. E. (1964). *Nature*, **201**, 311.

McClure, J. W., and Alston, R. E. (1966). *Am. J. Bot.* **53**, 849.

Munakata, K., Marumo, S., Ota, K., and Chen, Y. L. (1965). *Tetrahedron Lett.* 4167.

Murata, T., Shinohara, M., Hirata, T., and Miyamoto, M. (1968). *Tetrahedron Lett.* 849.

Nagarajan, S., Nair, A. G. R., Ramakrishnan, S., and Sankara, S. S. (1966). *Curr. Sci.* **35**, 176.

Nakaoki, T., and Morita, N. (1960). *Yakugaku Zasshi* **80**, 1473; *Chem. Abstr.* **55**, 5871.

Naya, Y., and Kotake, M. (1965). *Nippon Kagaku Zasshi*, **86**, 313; *Chem. Abstr.* **63**, 16244.

Neubauer, H. F. (1968). *Protoplasma*, **66**, 381.

Neville, G. A., Nigam, I. C., and Holmes, J. L. (1968). *Tetrahedron*, **24**, 3891.

Ng, Y. L., Thimann, K. V., and Gordon, S. A. (1964). *Arch. Biochem. Biophys.* **107**, 550.

Norman, A. M. (1968). *Thesis*, Miami University, Ohio.

Paris, M. (1968). *Pl. Med. Phytother.* **2**, 32.

Paris, R. R., and Paris, M. (1964). *C. r. hebd. Séanc. Acad. Sci., Paris*, **258**, 361.

Phouphas, C. (1962). *C. r. hebd. Séanc. Acad. Sci., Paris*, **255**, 1314.

Plouvier, V. (1962). *C. r. hebd. Séanc. Acad. Sci., Paris*, **263**, 439.

Price, J. R., and Sturgess, V. C. (1938). *Biochem. J.* **32**, 1658.

Pridham, J. B. (1964). *Phytochemistry*, **3**, 493.

Rahman, W., Ilyas, M., and Khan, A. W. (1962). *Chem. Abstr.* **57**, 10224.

Rastogi, R. P., and Dhar, M. L. (1960). *J. Sci. Indian Res.* **19B**, 455.

Rastogi, R. P., Sarkar, B., and Dhar, M. L. (1960). *J. Sci. Indian Res.* **19B**, 252.

Reznik, H., and Neuhausel, R. (1959). *Z. Bot.* **47**, 41.

Ricardi, M. *et al.* (1958). *Bul. Soc. Biol. De Concepcion (Chile)*, **33**, 29 (cited by Hegnauer, 1964).

Rubin, B. A., and Loginova, L. N. (1965). *Fiziol. Rast.* **12**, 386.

Sculthorpe, C. D. (1967). "The Biology of Aquatic Vascular Plants." Arnold, London.

Seidel, K. (1963). *Naturwissenschaften*, **50**, 452.

Shibata, M., Yamazaki, K., and Ishikura, N. (1965). *Bot. Mag. Tokyo*, **78**, 299.

Siegel, S. M. (1957). *J. Chem. Soc.* **79**, 1628.

Siegel, S. M. (1962). "The Plant Cell Wall." Pergamon, Oxford.

Siegelman, H. W. (1964). *In* "Biochemistry of Phenolic Compounds" (J. B. Harborne, ed.), pp. 437–456. Academic, New York and London.

Stabursvik, A. (1968). *Acta Chem. Scand.* **22**, 2371.

Stafford, H. A. (1964). *Pl. Physiol.* **39**, 350.

Subramanian, S. S., Joseph, K. J., and Nair, A. G. R. (1969). *Phytochemistry*, **8**, 674.

Sugano, N., and Hayashi, K. (1960). *Bot. Mag. Tokyo*, **73**, 231.

Swain, T. (1965). *In* "Biosynthetic Pathways in Higher Plants" (J. B. Pridham and T. Swain, eds.), pp. 9–36. Academic, New York and London.

Tatsuta, H., and Ochii, Y. (1956). *Sci. Rep. Tohoku Univ. ser. I*, **39**, 236.

Terenteva, I. V., and Boroukou, A. V. (1960). *Moldavsk. Filial Akad. Nauk. SSSR, Inst. Khim.* 41; *Chem. Abstr.* **58**, 2476.

Thimann, K. V., and Edmondson, Y. H. (1949). *Arch. Biochem.* **22**, 33.

Tjia-Lo, G., and Hegnauer, R. (1956). *Pharm. Weekblad.* **91**, 440.
Tomita, M., and Furukawa, H. (1962). *Yakugaku Zasshi,* **82,** 1458; *Chem. Abstr.* **58,** 11685.
Tomita, M., and Furukawa, H. (1966). *Tetrahedron Lett.* 2637.
Towers, G. H. N., and Mass, W. S. G. (1965). *Phtyochemistry,* **6,** 57.
Turner, B. L. (1967). *Pure Appl. Chem.* **14,** 189.
Valentine, J., and Wagner, C. (1962). *Pharm. Zentralhalle Dtl.* **92,** 354.
Van Beusekom, C. F. (1967). *Phytochemistry,* **6,** 573.
Wallace, J. W., and Alston, R. E. (1966). *Pl. Cell Physiol.* **7,** 699.
Wallace, J. W., Mabry, T. J., and Alston, R. E. (1969). *Phytochemistry,* **8,** 93.
Wilkinson, S. (1956). *J. Chem. Soc.* 2079.
Willaman, J. J., and Schubert, B. G. (1961). *Techn. Bull.* 1234. USDA, Washington.

CHAPTER 13

Infraspecific Variation of Sesquiterpene Lactones in *Ambrosia* (Compositae): Applications to Evolutionary Problems at the Populational Level

TOM J. MABRY*

*The Cell Research Institute and Department of Botany,
The University of Texas at Austin, Austin, Texas, USA*

I. INTRODUCTION

Chemotaxonomy (or biochemical systematics) is now well established as an interdisciplinary field of study spanning at once all of phytochemistry and plant systematics. We need not comment here on the broad chemical and taxonomic concepts involved in biochemical systematic investigations since a number of excellent reviews have recently dealt with this topic (e.g. Alston, 1967; Mabry *et al.* 1968; Swain, 1966; Turner, 1967). The main purpose of this chapter is to give some insights into the application of infraspecific variation for understanding evolutionary problems at the populational level. Although the present discussion will be restricted for the most part to the systematic significance and

* The author wishes to express his appreciation to the several Ph.D. candidates, postdoctorals and research assistants whose research contributed to this chapter, including, in particular, H. E. Miller, W. Renold, H. Yoshioka, Janet Potter, N. H. Fischer, T. H. Porter, Jeanne Lee, Akio Higo, H. B. Kagan and D. Seigler. In addition, Professors W. W. Payne and B. L. Turner provided many of the botanical interpretations presented here. The research described herein was supported by the Robert A. Welch Foundation (Grant F-130) and the National Science Foundation (Grant GB 5548X).

infraspecific variation of sesquiterpene lactones in the genus *Ambrosia* (family Compositae), the concepts developed here should be valid for many other plant and chemical systems (see also Turner, Chapter 10).

Alston (1967) has previously noted that the degree of acceptance or rejection of the methods of biochemical systematics may well depend upon the extent to which the reliability of the chemical data has been established. He emphasized that the matter of infraspecific variation is an important aspect of the problem of reliability:

> "In modern systematics, variation is often a valuable asset. Although some species are more variable than others, and in some instances there is extreme polymorphism, variation in general is accepted as a biological fact that may under certain conditions be essential to allow the effective study of populations, the study of incipient speciation, or of ecological factors and other important problems. The study of genetics is absolutely dependent upon phenotypic variation. For these and other reasons, it should not be either surprising or discouraging to find chemical variation existing even within a species. Since an important element of any scientific work is predictability, the question of variability must be evaluated according to this criterion. If chemical variation in plants is so excessive, so capricious, and so generally immune to analytical interpretation that predictability is essentially lacking, then such data hardly can be utilized scientifically at all, certainly not by systematists. But, if variation in secondary products is generally responsive to various factors similar to those factors which govern the morphology of a species, then chemical variation can be described in such a way as to introduce limits to the variation and to discern its pattern and thus, of course, to allow predictability. Since the latter alternative is nearly axiomatic, in my opinion, then the problem is merely to become familiar with chemical variation, its origin and factors affecting it, and its limits and its meaning, if possible, just as any good systematist would attempt to study morphological or cytological variation."

Since sesquiterpene lactones occur frequently in plants belonging to the Compositae tribes Heliantheae, Ambrosieae and Helenieae (see, for example, Steelink and Spitzer, 1966; Herz, 1968; Herout and Šorm, 1969), it appeared that a detailed analysis of the distribution of these compounds might prove useful in understanding the evolutionary relationships among such genera as *Ambrosia*, *Parthenium*, *Iva* and others. Therefore, we made a detailed comparative chemical, morphological and, in some instances, cytological analysis of several *Ambrosia* species in order to establish the reliability of the sesquiterpene lactone data for recognizing and understanding the structure of those taxa. The results of these studies indicate that sesquiterpene lactones may show considerable infraspecific variation, both quantitatively and qualitatively, and suggest that caution should be exercised by the "chemosystematists" who would base systematic interpretations upon the chemical analysis of one or two often poorly documented* plant populations of a given species. Nevertheless, our studies demonstrate that once the nature and extent of the infraspecific variation of the chemical data are understood, the data may be of exceptional value for resolving selected systematic problems.

The following section of this chapter presents W. W. Payne's present interpretation of the genus *Ambrosia* along with a survey of the sesquiterpene lactones known to be elaborated by the various species. The third section summarizes our detailed biochemical systematic studies for four species, *Ambrosia artemisiifolia*, *A. confertiflora*, *A. cumanensis* and *A. psilostachya*.

II. THE DISTRIBUTION OF SESQUITERPENE LACTONES IN THE GENUS *Ambrosia*

A. THE GENUS *Ambrosia*

The essentially New World genus *Ambrosia* contains those species commonly referred to as ragweeds, plants which are perhaps best known in North America for their contribution of allergenic pollens to the atmosphere.† Linnaeus established the genus in 1753, recognizing four species: *Ambrosia trifida*, *A. elatior*, *A. artemisiifolia* and *A. maritima* (Linnaeus, 1753); later, in 1793, Cavanilles erected the genus *Franseria* for a single species, *F. ambrosioides* (Cavanilles, 1793). In 1964, Payne re-interpreted the evidence and subsequently lumped into *Ambrosia* all the species which had been described by Linnaeus, Cavanilles and others as belonging to either *Ambrosia* or *Franseria* (Payne, 1964). Payne now recognizes some 40 species of *Ambrosia* (see Table I) and has suggested that the main North American species may be evolutionarily related according to the scheme shown in Fig. 1.‡

The *Ambrosia* species are characterized as shrubs or perennial or annual herbs nearly all of which "exhibit extreme morphological variability" (Payne, 1964). The basic chromosome number for the genus is $n = 18$ and a majority of the species are diploid; however, we and others have encountered populations of some species which are tetraploid, hexaploid and octoploid (Payne *et al.* 1964; Miller *et al.* 1968; Mabry and co-workers, unpublished).

We employ Payne's interpretation of the genus *Ambrosia* throughout this chapter.

B. SESQUITERPENE LACTONES IN THE GENUS *Ambrosia*

When, in 1962, we first became interested in using chemical data for resolving some of the systematic problems associated with *Ambrosia* only a few

---

* For further comments on the importance of properly vouchering the plant population under investigation, see Mabry and Mears (1970).

† The term "ambrosia" means the food, drink or perfume of the gods, and thus would seem to have little descriptive value for the ragweeds.

‡ The author wishes to thank Dr. W. W. Payne and K. M. Peterson for making these interpretations available prior to their publication elsewhere.

TABLE I

Sesquiterpene lactone data for *Ambrosia* species[a]

Species	Range	Compound	References
1. *A. acanthicarpa* Hook.	North America; from the Dakotas; Nebraska and Kansas west to New Mexico; Arizona, California; Washington and Oregon	Artemisiifolin, I Chamissonin, II Chamissonin mono- and diacetate Confertiflorin, XX; Cumambrin B, XV; Desacetylconfertiflorin, XXI; Psilostachyin C, XXXVII	Porter *et al.* (1970) Geissman and Levy (1967); Geissman *et al.* (1969); L'Homme *et al.* (1969) Mabry and co-workers (unpublished) Geissman *et al.* (1969)
2. *A. acuminata* (Brandeg.) Payne	Baja California; Mexico; known only from the type specimen	No data reported	
3. *A. ambrosioides* (Cav.) Payne	Sonoran Desert region, common through southern Arizona, extending into San Diego County, California, and to Sonora and Baja California, Mexico	Damsin, XXIII; Franserin, XXV	Romo *et al.* (1968b); (see also Herz *et al.* 1969
4. *A. arborescens* Mill.	Interandean Plains of Ecuador, north to north-central Colombia, south through Peru to western Bolivia	Coronopilin, XXII; Damsin, XXIII; Psilostachyin, XXXV; Psilostachyin C, XXXVII; possibly 11-Epidihydropsilo-stachyin	Herz *et al.* (1969)
5. *A. artemisiifolia* L. (including *A. maritima* L.)	Eastern United States	Ambrosin, XVI	Abu-Shady and Soine (1953, 1954); Bernardi and Buchi (1957); Šorm *et al.* (1959); Herz *et al.* (1962); Emerson *et al.* (1966)

Species	Distribution	Compounds	References
		Artemisiifolin, I; Coronopilin, XXII; Cumanin, XXIX; Dihydrocumanin, XXXII; Damsin, XXIII	Porter *et al.* (1970); Herz and Högenauer (1961); Porter and Mabry (1969)
		Isabelin, IV; Dihydroparthenolide, VI; Peruvin, XXXIII; Psilostachyin, XXXV	Abu-Shady and Soine (1953, 1954); Bernardi and Buchi (1957); Emerson *et al.* (1966); Porter *et al.* (1970); Bianchi *et al.* (1968); Porter and Mabry (1969); Bianchi *et al.* (1968); Mabry and co-workers (unpublished); See Herz (1968)
6. *A. artemisioides* Meyen & Walp. ex Meyen	Southern Peru (Arequipa Province) to northern Chile (Tarapaca Province)	No compounds detected	
7. *A. bidentata* Michx.	Midwestern United States	No crystalline lactones	Herz and Högenauer (1961)
8. *A. bryantii* (Curran) Payne	Central to southern Baja California, Mexico	No data reported	
9. *A. camphorata* (Greene) Payne	Throughout Baja California, sporadic in western Sonora and southern San Luis Potosi, Mexico	No data reported	
10. *A. canescens* Gray	East-central Mexico, from San Luis Potosi to Durango, south to Aguascalientes. Also reported from Chihuahua and Arizona	No data reported	
11. *A. carduacea* (Greene) Payne	Central to southern Baja California, Mexico	No data reported	
12. *A. "castanensis"*	Coahuila, Mexico	Artemisiifolin, I	
13. *A. chamissonis* (Less.) Greene	Along Pacific coast of North America from Vancouver Island,	Chamissonin, II	Mabry and co-workers (unpublished); Geissman *et al.* (1966); L'Homme *et al.* (1969)

TABLE I—*continued*

Species	Range	Compound	References
14. *A. chenopodiifolia* (Benth.) Payne	British Columbia to Baja California, Mexico; adventive along South American coast	Damsin, XXXIII	Herz *et al.* (1969)
15. *A. cheiranthifolia* Gray	Baja California, Mexico, from Comondu north into San Diego Co., California	No compounds detected	Mabry and co-workers (unpublished)
16. *A. confertiflora* DC.	South Texas	Artemisiifolin, I; Chihuahuin, III	Mabry and co-workers (unpublished)
	Southwestern United States from Texas and Colorado to California, throughout central Mexico from Nuevo Leon and Tamaulipas through Guanajuato and Jalisco to Sinaloa and Sonora; adventive in Hawaii on the islands of Oahu and Molokai, and in Puerto Rico	Confertiflorin, XX	Fischer and Mabry (1967a)
		Desacetylconfertiflorin, XXI	Fischer and Mabry (1967a); Romo *et al.* (1968b)
		Confertin, XXVIII	Romo *et al.* (1968b)
		Epoxysantamarine, XIII	Yoshioka *et al.* (1970)
		Parthenolide, V	Yoshioka *et al.* (1970)
		Peruvin, XXXIII	Yoshioka *et al.* (1970)
		Psilostachyin, XXXV	Yoshioka *et al.* (1970); Herz *et al.* (1969)
		Psilostachyin B, XXXVI	Yoshioka *et al.* (1970)
		Psilostachyin C, XXXVII	Yoshioka *et al.* (1970); Herz *et al.* (1969)
		Reynosin, XI; Santamarine, XII	Yoshioka *et al.* (1970)
		Tamaulipin-A, VII	Fischer *et al.* (1968)
		Tamaulipin-B, VIII	Fischer and Mabry (1967b)

No. / Species	Distribution	Sesquiterpene lactones	References
17. *A. cordifolia* (Gray) Payne	Southern Arizona, south and west into San Luis Potosi, Sonora, Sinaloa, and Baja California, Mexico	Psilostachyin C, XXXVII; Stereoisomer of Psilostachyin; Psilostachyin B, XXXVI	See Herz (1968); Mabry and co-workers (unpublished); Herz *et al.* (1969)
18. *A. cumanensis* Kunth	Central Mexico	Ambrosin, XVI; Coronopilin, XXII; Cumambrin A, XIV; Cumambrin B, XV; Cumanin, XXIX	Romo *et al.* (1968a); Miller *et al.* (1968); Romo *et al.* (1966a) [in connection with the latter report, see also Romo *et al.* (1968a)]
19. *A. deltoidea* (Torr.) Payne	Southern Arizona, extending into Sonora and Baja California, Mexico	Damsin, XXII; Psilostachyin A, XXXV; Psilostachyin B, XXXVI; Psilostachyin C, XXXVII; Damsin, XXIII; Psilostachyin C, XXXVII	Herz *et al.* (1969); Miller and Mabry (1967); Miller *et al.* (1968); Herz *et al.* (1969); Miller and Mabry (1967); Miller *et al.* (1968); Kagan *et al.* (1966)
20. *A. divaricata* (Brandeg.) Payne	Central Baja California, Mexico	No data reported	
21. *A. dumosa* (Gray) Payne	Arid regions of Utah, Arizona, Nevada and California, south into Sonora and Baja California, Mexico	Ambrosiol, XVIII; Apoludin, XIX; Burrodin, XXVII; Coronopilin, XXII; Parthenolide, V; Psilostachyin, XXXV; Psilostachyin C, XXXVII	Geissman and Matsueda (1968); See also Geissman and Turley (1964); Geissman and Matsueda (1968)

TABLE I—*continued*

Species	Range	Compound	References
22. *A. eriocentra* (Gray) Payne	Southern Arizona, scattered in arid regions of Nevada, Utah, and California	No compounds detected	Mabry and co-workers (unpublished)
23. *A. flexuosa* (Gray) Payne	Baja California, Mexico: known only from type collection; possibly a hybrid of perhaps *A. ilicifolia* and *A. eriocentra*	No data reported	
24. *A. grayi* (Nels.) Shinners	Low, moist areas throughout Kansas and Nebraska, sporadically in Oklahoma, Colorado, and Texas	No compounds detected	Mabry and co-workers (unpublished)
25. *A. hispida* Pursh	Florida Keys and Caribbean region	Ambrosin, XVI; Damsin, XXIII	Herz and Sumi (1964)
26. *A. ilicifolia* (Gray) Payne	Desert regions of southern Arizona and California, south into Sonora and Baja California, Mexico	Costic Acid,[b] IX; Ilicic Acid,[b] X	Herz *et al.* (1966)
27. *"A. jamaicensis"*	Jamaica and Caribbean region	No data reported	
28. *A. linearis* (Rydb.) Payne	Colorado, Lincoln and El Paso, Counties; known only from the type and one other specimen	No compounds detected	Mabry and co-workers (unpublished)
29. *A. magdalenae* (Brandeg.) Payne	Central Baja California, Mexico	No data reported	
30. *A. microcephala* DC.	Guyana	No data reported	
31. *A. nivea* (Rob. and Fern) Payne	Northwestern Chihuahua, Mexico; known only from type collection	No data reported	
32. *A. pannosa* Payne	Peruvian Andes	No data reported	
33. *A. parvifolia* Payne	Peruvian Andes	No data reported	

Species	Distribution	Compound	References
34. *A. peruviana* Willd.	Central South America	Tetrahydroambrosin, XVII; Peruvin, XXXIII	Herz *et al.* (1969); Joseph-Nathan and Romo (1966); Herz *et al.* (1969)
		Peruvinin, XXXIV	Romo *et al.* (1967); Herz *et al.* (1969); Kagan *et al.* (1966)
35. *A. polystachya* DC.	Central South America, Ecuador and Brazil	Psilostachyin C, XXXVII	No data reported
36. *A. psilostachya* DC.	Throughout central United States, from southern Canada to central Mexico	Ambrosin, XVI; Ambrosiol, XVIII	Miller *et al.* (1968); Mabry *et al.* (1966c); Miller *et al.* (1968)
		Artemisiifolin, I; Coronopilin, XXII	Porter *et al.* (1970); Herz and Högenauer (1961); Geissman and Turley (1964); Mabry *et al.* (1966c); Miller and Mabry (1967); Miller *et al.* (1968); Geissman *et al.* (1969)
		Cumanin, XXIX	Geissman *et al.* (1969); Miller *et al.* (1968); Geissman *et al.* (1969)
		Cumanin 3-acetate, XXX; Cumanin diacetate, XXXI; Damsin, XXIII; 3-Hydroxydamsin, XXIV	Miller *et al.* (1968); Miller and Mabry (1967); Miller *et al.* (1968)
		Isabelin, IV	Yoshioka *et al.* (1968); Yoshioka and Mabry (1969)
		Parthenin, XXVI	Geissman *et al.* (1969); Mabry *et al.* (1966c); Miller and Mabry (1967); Miller *et al.* (1968)

TABLE I —continued

Species	Range	Compound	References
		Psilostachyin, XXXV	Miller et al. (1965); Mabry et al. (1966b); Miller et al. (1968)
		Psilostachyin B, XXXVI	Miller et al. (1965); Mabry et al. (1966a, 1966b); Miller et al. (1968)
		Psilostachyin C, XXXVII	Miller et al. (1965); Kagan et al. (1966); Mabry et al. (1966a, 1966b); Miller et al. (1968)
		Desacetylconfertiflorin, XXI	Mabry and co-workers (unpublished)
37. A. pumila (Nutt.) Gray	San Diego County; California and northern Baja California, Mexico	Psilostachyin, XXXV; Psilostachyin C, XXXVII	Herz et al. (1969)
38. A. tenuifolia Spreng.[c]	South America (Argentina)	Psilostachyin, XXXV	Herz et al. (1969)
39. A. tomentosa Nutt.	Prairie and semi-arid regions of Wyoming, Nebraska, Colorado, New Mexico, and Arizona; sporadic in Idaho, Iowa, and South Dakota	No compounds detected	Mabry and co-workers (unpublished)
40. A. trifida L.	Throughout the midwestern region of the United States	No sesquiterpene lactones detected	Herz and Högenauer (1961)
41. A. velutina Willd.	Caribbean region; Haiti	No data reported	
42. A. senegalensis	Egypt	No data reported	

[a] The author thanks W. W. Payne for the interpretation of the genus Ambrosia which is presented here.
[b] Not a lactone but listed here to indicate the type of sesquiterpene constituent which occurs in this species.
[c] According to Payne (private communication) the citation by Romo et al. (1968b) of finding confertin in "A. tenuifolia Harv. and Gray" from northern Mexico almost certainly refers to a form of A. confertiflora DC. with attenuated leaf and lobe tips.

scattered reports regarding the occurrence of sesquiterpene lactones in this genus were available. Today, as a result of our own studies as well as those in the laboratories of W. Herz, T. A. Geissman, J. Romo and A. Romo de Vivar, considerable data have now been accumulated for a number of species (Tables I and II; Fig. 2). The genus is known to elaborate more than 30 different sesquiterpene lactones which belong to five different skeletal types (Fig. 2)

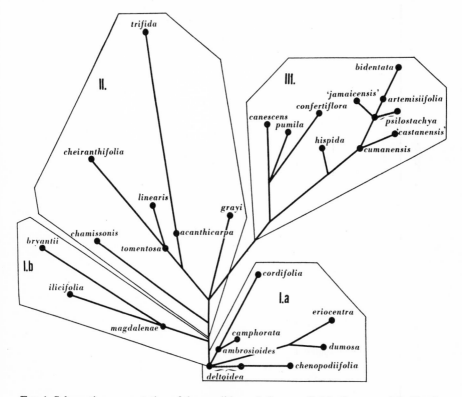

FIG. 1. Schematic representation of the possible evolutionary affinities for most of the North American *Ambrosia* species. They are arranged into three suspected phyletic groups, I, II and III, the first being further subdivided. The two taxa *castanensis* and *jamaicensis* are probable species that require additional work. The elongated mark for *A. psilostachya* reflects its character progression from fruiting involucres with a few, short spines to spineless. The elongated mark for *A. deltoidea* reflects the fact that some specimens have fruiting involucres with long, woolly pubescence, a character more properly of *A. chenopodiifolia* with which *A. deltoidea* seems to intergrade. The author thanks W. W. Payne for all these data

whose probable biogenetic relationships are shown in Fig. 3; however, as yet no direct carbon-14 labelling evidence is available to support this scheme.

It is the intent of the following section of this chapter to demonstrate that tabulations such as those presented in Table I may give little or no insight into the populational structure of a given species. It is our view that detailed populational studies can provide the sort of data which permits one to discover and

TABLE II

Sesquiterpene lactones for *Ambrosia* species

Species of *Ambrosia*

Compound	A. acanthicarpa	A. ambrosioides	A. arborescens	A. artemisiifolia	A. "castanensis"	A. chamissonis	A. chenopodiifolia	A. confertiflora	A. cordifolia	A. cumanensis	A. deltoidea	A. dumosa	A. hispida	A. ilicifolia	A. peruviana	A. psilostachya	A. pumila	A. tenuifolia
Germacranolides																		
Artemisiifolin (I)	*			*	*			*								*		
Chamissonin (II)	*					*												
Chamissonin mono- and diacetates	*																	
Chihuahuin (III)				*				*								*		
Isabelin (IV)				*								*						
Parthenolide (V)								*										
Dihydroparthenolide (VI)																		
Tamaulipin A (VII)								*										
Tamaulipin B (VIII)								*										
Eudesmanolides																		
Costic acid[a] (IX)								*						*				
Ilicic acid[a] (X)								*						*				
Reynosin (XI)								*										
Santamarine (XII)																		
Epoxysantamarine (XIII)																		
Guaianolides																		
Cumambrin A (XIV)	*									*								
Cumambrin B (XV)										*								

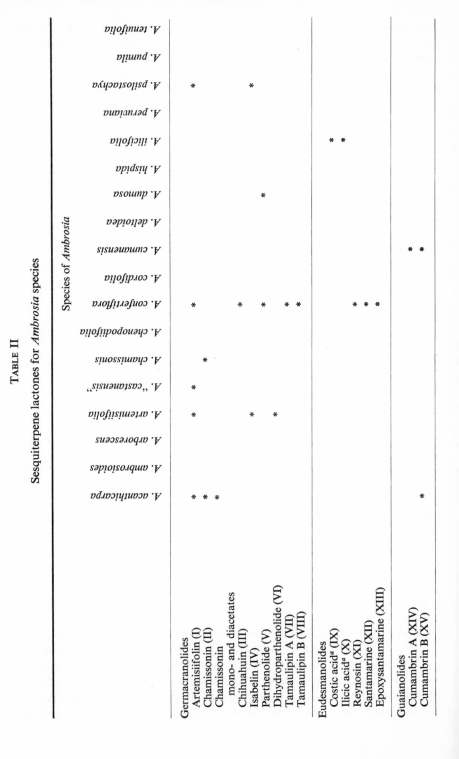

Compound	acanthicarpa	ambrosioides	arborescens	artemisiifolia	"castanensis"	chamissonis	chenopodiifolia	confertiflora	cordifolia	cumanensis	deltoidea	dumosa	hispida	ilicifolia	peruviana	psilostachya	pumila	tenuifolia
Pseudoguaianolides lactonized to C-6																		
Ambrosin (XVI)				*						*			*			*		
Tetrahydroambrosin (XVII)															*			
Ambrosiol (XVIII)												*				*		
Apoludin (XIX)	*											*						
Confertiflorin (XX)	*							*									*	
Desacetylconfertiflorin (XXI)								*										
Coronopilin (XXII)			*	*								*				*		
Damsin (XXIII)		*	*	*			*			*	*		*			*		
3-Hydroxydamsin (XXIV)										*						*		
Franserin (XXV)		*																
Parthenin (XXVI)																*		
Pseudoguaianolides lactonized to C-8																		
Burrodin (XXVII)																		
Confertin (XXVIII)								*				*						
Cumanin (XXIX)				*						*						*		
Cumanin 3-acetate (XXX)																*		
Cumanin diacetate (XXXI)																*		
Dihydrocumanin (XXXII)																		
Peruvin (XXXIII)				*				*							*			
Peruvinin (XXXIV)				*											*			
Pseudoguaianolides dilactones																		
Psilostachyin (XXXV)	*		*	*				*		*		*				*		
Psilostachyin B (XXXVI)			*					*		*						*	*	*
Psilostachyin C (XXXVII)								*	*	*	*	*			*	*		
11-Epidihydropsilostachyin (tentative)[a]																	*	

[a] Not a lactone but included for completeness.

Germacranolides

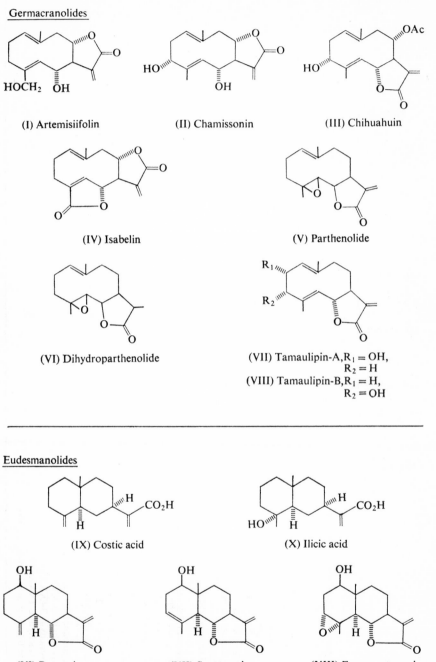

(I) Artemisiifolin   (II) Chamissonin   (III) Chihuahuin

(IV) Isabelin   (V) Parthenolide

(VI) Dihydroparthenolide   (VII) Tamaulipin-A, R_1 = OH,
                                                         R_2 = H
                                  (VIII) Tamaulipin-B, R_1 = H,
                                                         R_2 = OH

Eudesmanolides

(IX) Costic acid   (X) Ilicic acid

(XI) Reynosin   (XII) Santamarine   (XIII) Epoxysantamarine

Guaianolides

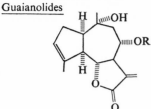

(XIV) Cumambrin A, R = Ac (see Irwin and Geissman, 1969, for stereochemistry)
(XV) Cumambrin B, R = H

Pseudoguaianolides lactonized to C-6

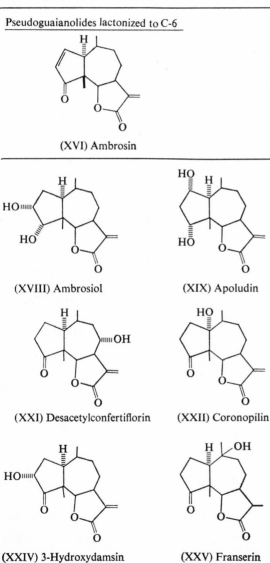

(XVI) Ambrosin

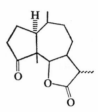

(XVII) Tetrahydroambrosin

(XVIII) Ambrosiol

(XIX) Apoludin

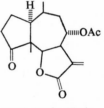

(XX) Confertiflorin

(XXI) Desacetylconfertiflorin

(XXII) Coronopilin

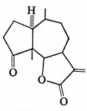

(XXIII) Damsin

(XXIV) 3-Hydroxydamsin

(XXV) Franserin

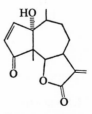

(XXVI) Parthenin

Pseudoguaianolides lactonized to C-8

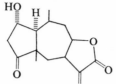

(XXVII) Burrodin

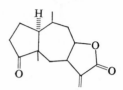

(XXVIII) Confertin

(XXIX) Cumanin

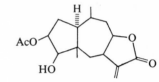

(XXX) Cumanin 3-acetate

(XXXI) Cumanin diacetate

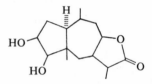

(XXXII) Dihydrocumanin

(XXXIII) Peruvin

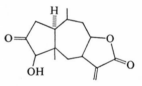

(XXXIV) Peruvinin

Dilactone Pseudoguaianolides

(XXXV) Psilostachyin   (XXXVI) Psilostachyin B   (XXXVII) Psilostachyin C

FIG. 2. Sesquiterpene lactones isolated from *Ambrosia* species; costic and ilicic acids, which are not lactones, are included for completeness

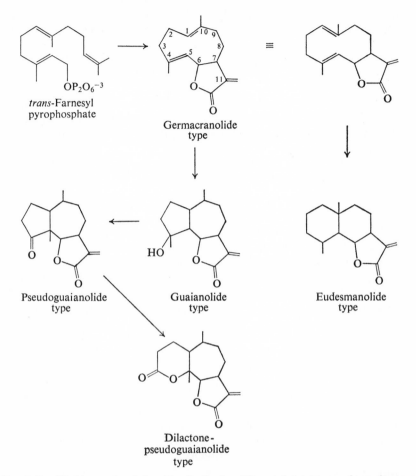

FIG. 3. Possible biogenetic relationships for the five different skeletal types of sesquiterpene lactones known to be elaborated by the genus *Ambrosia*

interpret the often complex evolutionary processes which are actually at work within a taxon.

C. ISOLATION AND DETECTION PROCEDURES

All the sesquiterpene lactones mentioned in this chapter have been obtained crystalline and were fully characterized by chromatographic and spectroscopic techniques. The crystalline samples were usually obtained after column chromatography of the crude syrup (see step 3 below) over silica gel, eluting first with chloroform, and then with mixtures of chloroform and ether or ethyl acetate. For most of the populational analyses, we found that the components present in the crude syrup could be identified by NMR and TLC analyses.

We presently employ the following standardized procedure for the isolation of sesquiterpene lactones from "small amounts" of dried plant material.

1. Twenty grammes of dried leaves and stems are ground briefly in a Waring blendor; then 100 ml of chloroform are added and the blending is continued for about 2 min.
2. The slurry is filtered and the filtrate plus chloroform washings are evaporated to dryness under water pump vacuum. The residue thus obtained is dissolved in 25 ml of ethanol (95%) and then 25 ml of 4% aqueous lead acetate are added.
3. Next, the solution is filtered and the filtrate concentrated under water pump vacuum until only water and oil remain. The oil is extracted into chloroform which is subsequently dried over anhydrous magnesium sulphate, filtered, and evaporated *in vacuo*.
4. The oily residue thus obtained is analysed by NMR and TLC. The NMR spectra of most of the compounds exhibit characteristic signals which can be used to detect them in mixtures. For the TLC analyses, we employ silica gel G plates which are developed with solvents such as: chloroform–ether (4:1), benzene–acetone (4:1), chloroform–methanol (99:1), benzene–methanol (9:1), and benzene–ether (2:3) The sesquiterpene lactones are detected by placing the plates in a chamber containing crystals of iodine.

III. INFRASPECIFIC VARIATION OF SESQUITERPENE LACTONES IN SELECTED *Ambrosia* SPECIES

A. POPULATIONAL CHEMICAL DATA FOR SESQUITERPENE LACTONES IN THE *Ambrosia psilostachya, A. cumanensis* AND *A. artemisiifolia* COMPLEX

1. Introduction

Ambrosia psilostachya DC. and *A. cumanensis* Kunth are both herbaceous perennials differing in their ranges, chromosome numbers and chemistry. The former species occurs mainly in central North America (Fig. 4) while the latter is commonly found in more tropical areas, notably central Mexico. Available chromosome numbers for *A. psilostachya* include $n = 18, 36, 54$ and 72 while to date only $n = 18$ has been recorded for *A. cumanensis* (Payne *et al.* 1964; Miller *et al.* 1968). *Ambrosia artemisiifolia* L., which also has a chromosome number of $n = 18$ (see Payne *et al.* 1964), is an annual ragweed which is widespread in the eastern part of the United States.

Although the evolutionary relationships of *A. psilostachya, A. cumanensis* and *A. artemisiifolia* are not clearly understood, Payne (in Miller *et al.* 1968) has suggested (see excerpt below) on the basis of leaf and fruit morphology that both *A. psilostachya* and *A. artemisiifolia* may be derived from *A. cumanensis* (Fig. 5).

"... the suggestion of common ancestry for *A. cumanensis* and *A. psilo-stachya* is supported by morphological and cytological evidence. They are members of a complex of closely related species that includes, in addition to themselves, *A. artemisiifolia*, *A. peruviana*, *A. velutina*, *A. tenuifolia* and *A. microcephala*, plus the more distantly related *A. hispida*. All of these are plants of the Caribbean and Gulf region, although *A. psilostachya* and *A.*

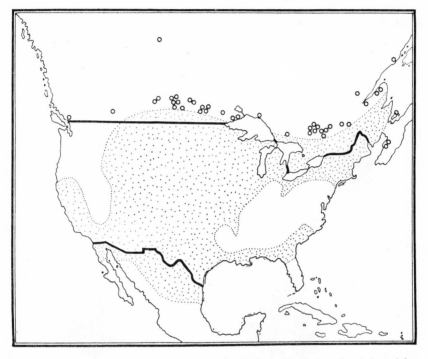

FIG. 4. The distribution of *Ambrosia psilostachya* DC. in North America (adapted from Payne, 1962). O = Canadian specimens not examined by Payne

artemisiifolia have become widespread in temperate North America, and *A. peruviana* and *A. tenuifolia* extend to central South America. All are characterized by regularly lobed and dissected leaves, and reduced capitula and fruiting involucres; the fruiting involucres have a few vestigial spines localized near the beaks and contain only a single achene. In particular, the three North American species, *A. cumanensis*, *A. psilostachya* and *A. artemisiifolia*, are very similar (Fig. 5), often being scarcely distinguishable except for usual (though not wholly constant) cytological and habit dif-ferences. On morphological and phytogeographical bases alone, it seems likely that: (a) the three species share a common ancestor; (b) diploid, perennial (root and stem proliferating) *A. cumanensis* is the central species, still probably very like the primordial form from which the three species have differentiated; (c) *A. artemisiifolia* has developed as an annual, diploid species particularly adapted to temperate, mesophytic conditions (populations of *A. artemisiifolia* from the Atlantic coastal states still closely resemble *A. cumanensis*); (d) *A. psilostachya*, of which diploid, tetraploid, hexaploid and octoploid forms are known, has differentiated as a more coriaceous-leaved, root-proliferating perennial, particularly adapted to the semi-arid regions of central Mexico and the western United States. Many Mexican populations of *A. psilostachya* are today very similar morphologically to *A. cumanensis* and are frequently collected as such. The island populations of

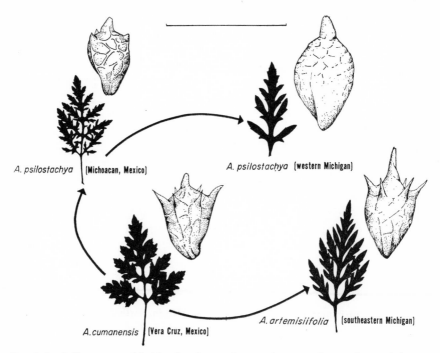

Fig. 5. Leaf silhouettes and fruiting involucres of ragweeds arranged according to their proposed evolutionary relationships (Miller *et al.* 1968)

A. psilostachya from southern Texas are chemically like *A. cumanensis* and are also morphologically more similar to *A. cumanensis* than are the mainland collections from elsewhere in the United States. Both *A. artemisiifolia* and *A. psilostachya* can be readily hybridized with *A. cumanensis*, although the former are themselves comparatively intersterile. Further, hybrid progeny involving *A. cumanensis* display good chromosome homology and are quite fertile. It appears that the chemical evidence supports the evolutionary hypotheses that may be formulated on other grounds, and is, in turn, supported by them."

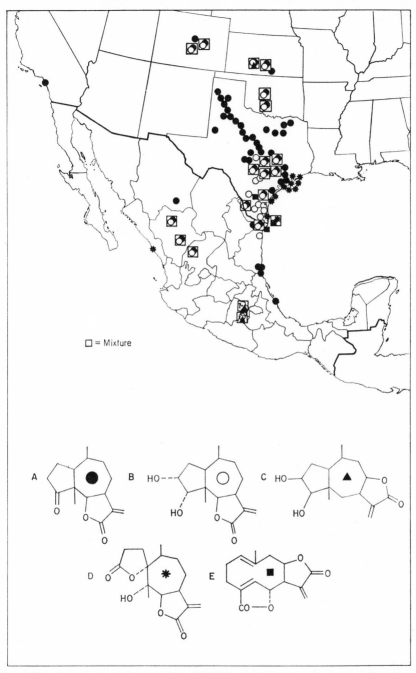

FIG. 6. Chemical races in *Ambrosia psilostachya* DC.; each structure may represent from 1 to 5 compounds: A = various combinations of ambrosin (XVI), coronopilin (XXII), damsin (XXIII), 3-hydroxydamsin (XXIV) and parthenin (XXVI); B = ambrosiol; C = cumanin (XXIX); D = psilostachyin (XXXV), psilostachyin B (XXXVI) and psilostachyin C (XXXVII); E = isabelin

2. *Chemical Races in* Ambrosia psilostachya

In 1968 we published detailed chemical data for 62 populations of *A. psilostachya*; these data and more recent results are summarized in Fig. 6. This single species produces more than a dozen different sesquiterpene lactones (see

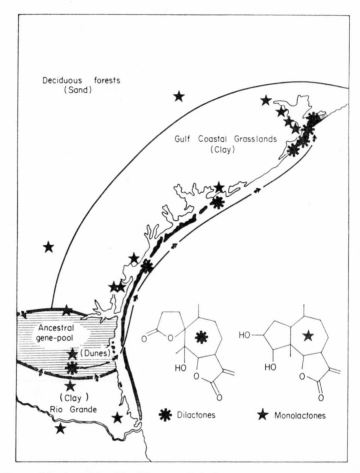

FIG. 7. Proposal for the origin of the dilactone-producing populations of *A. psilostachya* DC., which occur on islands and peninsulas lining the Texas Gulf coast, from a genetically complex population of the species which occurs north of Brownsville, Texas (Miller *et al.* 1968). The sesquiterpene lactone data for *A. cumanensis* suggest an alternative hypothesis for the origin of the dilactone races of *A. psilostachya*

Table I); however, any given population usually contains only one or two major components along with from none to as many as 3 minor ones.

With the exception of two germacranolides (Type E, Fig. 6), which are produced by some small populations of *A. psilostachya* found in Port Isabel, La Marque and Alice, Texas, the compounds isolated from all of the various populations of this species were of two types, monolactone-pseudoguaianolides (Types A, B and C, Fig. 6) and dilactone-pseudoguaianolides (Type D, Fig. 6). Plants of *A. psilostachya* collected on the several narrow islands which line the Gulf Coast of Texas (Galveston, Matagorda, Mustang and Padre Islands) were found to contain only the dilactone compounds psilostachyin, psilostachyin B and psilostachyin C. However, collections all along the mainland immediately adjacent to the islands contained only monolactones. One collection of *A. psilostachya*, obtained about 20 miles inland along the southern coast of Texas, also contained the dilactone compounds. There is, however, an intrusion of sand and dunes into the mainland in this region such that the ecology is similar to the off-shore islands. All of the more than 150 mainland populations of *A. psilostachya*, which were obtained from Canada* south to central Mexico, contained monolactones. In 1968 we suggested that the dilactone-chemical race of *A. psilostachya* found on the islands and peninsulas lining the Texas Gulf coast originated from a genetically complex population of *A. psilostachya* which occurs on the mainland north of Brownsville, Texas (Fig. 7; Miller, 1967; Miller *et al*. 1968). Although some of the chemical data for *Ambrosia cumanensis* (see next section) indicate that the island populations of *A. psilostachya* might be derived directly from the populations of *A. cumanensis* which occur near Vera Cruz, Mexico, the following excerpts from our 1968 publication (Miller *et al*. 1968) will serve to illustrate the nature of the problem.

"The distribution of the dilactone and monolactone races along the Texas coast is especially significant if one considers the origin of the offshore barrier islands. . . .

The present-day barrier islands along the Texas coast began some 5,000 years ago as small submerged bars at or near the mouths of the major rivers which feed into the Gulf of Mexico. Their emergence more-or-less coincided with post-Pleistocene sea-level changes, and by subsequent 'longshore drift' large volumes of sand were transplanted to the islands. They enlarged by beach accretion, eventually becoming linked into a very long barrier chain. These were subsequently modified by tidal action and spill-over fans, the latter mostly through the action of periodic hurricanes.

It is believed that the more northerly islands such as Galveston and Matagorda islands were the first to acquire large chain-dimensions, and at least the latter was ultimately linked with the more slowly accreting chains to the south-west. Thus while the northern barrier chain began approximately 5,000 years ago, it probably was not open for active plant colonization from the south until 3,000 years ago, or perhaps later.

As indicated above, the dune complex at or near southern-most Texas and adjacent Mexico along the Rio Grande River is quite complex, extending into mainland regions for distances of nearly 100 miles. It is this region which

* The pseudoguaianolides belonging to types A and B occur in populations of *A. psilostachya* extending into Canada; for convenience these populations are not shown in Fig. 6.

contains the most chemically variable populations of *A. psilostachya*. The recent dunes in this area, which have been dated at approximately 1,940 years, overlie earlier Pleistocene dune deposits with ^{14}C ages of 25,000 to 35,000 years. Thus this shore region has for at least 35,000 years maintained fluctuating dune habitats that might have served as the primary sites for a gene pool that could have given rise to the chemical and morphological race(s) which now occupy the area. Chance establishment of seeds into

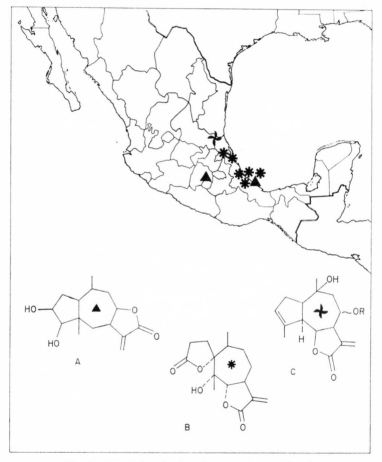

FIG. 8. Chemical races of *Ambrosia cumanensis* Kunth

newly exposed dune areas, perhaps via mechanisms postulated by Mayr (1954) in his Founder Principle, might have led to the rapid colonization of the more northern barrier islands as they became close to or in contact with the islands to the south. It should be noted that the opportunity for a more northern mainland dispersal to the barrier islands is unlikely because: (1) the islands apparently were formed de novo 3 to 10 miles off-shore with little if any mainland contact; (2) the black-waxy coastal prairie that extends along the inner mainland bays is at climax occupied by a thick coastal *Andropogon* grassland which would not permit fully developed populations of the mostly weedy *A. psilostachya* (in fact, the species occurs in these regions today probably as recent introductions along disturbed roadways leading to the gulf); (3) predominant winds and currents for seed dispersal are from the south and south-west.

Our comparative chemical studies of *Ambrosia cumanensis* offer an alternative hypothesis for the origin of the dilactone-island populations of *A. psilostachya*; these data suggest that the dilactone-island races of *A. psilostachya* are closely related to the populations of *A. cumanensis* which occur near Vera Cruz, Mexico (see next section).

3. Comparative Chemical Studies between Ambrosia cumanensis and A. psilostachya

All of the populations of *A. cumanensis* which we have examined in the vicinity of Vera Cruz, Mexico (Miller and Mabry, 1967; Miller *et al.* 1968; Mabry and co-workers, unpublished) contain the same three dilactone pseudoguaianolides that are found in the populations of *A. psilostachya* which occur on the islands lining the Texas Gulf coast (Fig. 8). Romo *et al.* (1966, 1968a) have described monolactone pseudoguaianolides and guaianolides from a few populations of *A. cumanensis* collected in central Mexico (Table I; Fig. 8). Also, Herz *et al.* (1969) have detected both dilactone and monolactone types of sesquiterpene lactones in a collection of *A. cumanensis* from Colombia. The presence of the same three structurally unique dilactone-pseudoguaianolides combined with the absence of monolactones in both *A. cumanensis* from near Vera Cruz, Mexico and the island populations of *A. psilostachya* suggest a close genetic relationship between the two taxa.

In an attempt to bring additional chemical evidence to bear upon this latter question, Mrs Janet Potter (1970) compared the volatile terpene patterns obtained for the island populations of *A. psilostachya* with the patterns observed for the mainland races of this species and the Vera Cruz, Mexico populations of *A. cumanensis*. Mrs Potter separated the volatile terpenes by preparative gas chromatography and identified the individual compounds with the aid of infrared and NMR spectroscopy. While the volatile terpene data indicate a complex relationship between *A. cumanensis* and *A. psilostachya*, they do lend support to the proposal reached on the basis of the striking sesquiterpene lactone patterns; that is, that the island populations of *A. psilostachya* are closely related to the Vera Cruz populations of *A. cumanensis*.

It should be noted that the available cytological data do not clarify the origin of the dilactone-island races of *A. psilostachya* since, in general, they exhibit chromosome number of $n = 18$, a value which has also been observed not only for *A. cumanensis* but for a few of the monolactone-mainland populations of *A. psilostachya* as well (Miller *et al.* 1968; Payne *et al.* 1964).

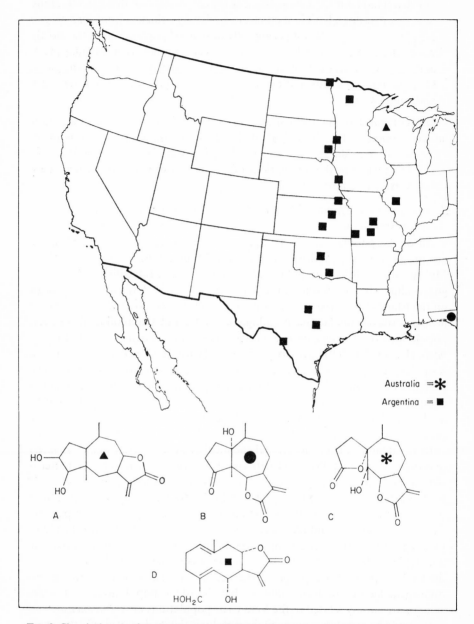

FIG. 9. Chemical races of *Ambrosia artemisiifolia* L.: Germacranolides (Type D) characterize populations of *Ambrosia artemisiifolia* from the central part of the United States while pseudo-guaianolides (Types A and B) have been detected in a few populations of the species. One population from Australia and from Texas (not shown) contained a dilactone-pseudoguaiano-lide (Type C). The Australian population also contained dihydroparthenolide

4. Chemical Races in Ambrosia artemisiifolia

The limited populational studies available for *Ambrosia artemisiifolia* suggest that some unique germacranolides (see Table I) characterize most populations of the species occurring in the central part of the United States (Fig. 9); however, additional surveys will be required to determine the detailed chemistry of this species throughout its range.

5. The Relationship of the Germacranolide-producing Populations of Ambrosia psilostachya and A. artemisiifolia

We can now turn our attention to a consideration of the origin of the two small populations of *A. psilostachya* which occur near Port Isabel and Alice, Texas. These populations produce the germacranolide isabelin which has the unusual property of existing in a 10:7 ratio of two conformational forms in solution at room temperature; the conformers can only be distinguished by NMR spectroscopy (Fig. 10) (Yoshioka *et al.* 1968; Yoshioka and Mabry, 1969).

The occurrence of isabelin along with the structurally-related germacranolide artemisiifolin (Porter *et al.* 1970) in *A. artemisiifolia* (Fig. 9) suggests that the germacranolide-producing populations of *A. psilostachya* may contain genes derived from the germacranolide-producing populations of *A. artemisiifolia*. In this connection, it should be noted that the germacranolide populations of *A. psilostachya* occur in regions where the two species, *A. psilostachya* and *A. artemisiifolia*, are sympatric.

B. CHEMICAL AND CYTOLOGICAL RACES OF Ambrosia confertiflora

I will only briefly comment on the studies of W. Renold (1970) and N. H. Fischer with regard to the heterogeneous species, *A. confertiflora*, a perennial which is widespread in the southwestern part of the United States and in northern Mexico. The sesquiterpene lactone data for more than 252 populations of this species serve to illustrate both the chemical complexity which a given species may exhibit and the way populational data can be used to detect evolution at work. Also the chemical variation, in some instances, has been correlated with the different ploidal levels of the chromosome complements in *A. confertiflora*.

Populations of *A. confertiflora* from south central Texas are characterized by a chromosome number of $n = 66$ and by the pseudoguaianolides, confertiflorin and desacetylconfertiflorin (Type A, Fig. 11), while plants obtained from northeastern Mexico yielded some new germacranolides (Type C) along with several eudesmanolides (see Tables I and II). Most of the latter populations which were cytologically examined exhibited a chromosome number of $n = 54$. A dilactone (Type B) race (with mixed chromosome numbers; $n = 54$, $n = 66$) stretches in a narrow band from just east of the Big Bend area of Texas south into Mexico along the western slopes of the Sierra Madre mountains down into

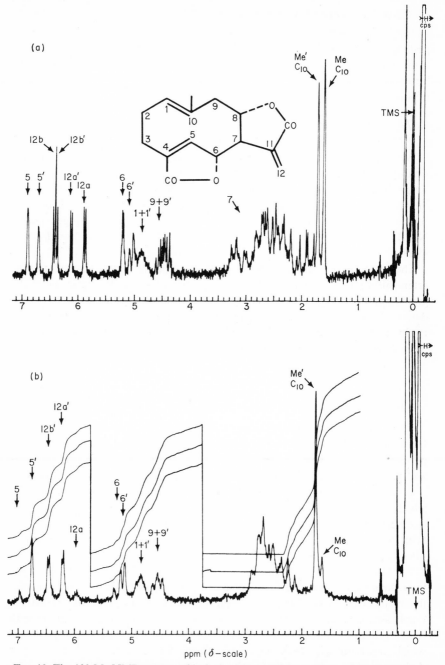

FIG. 10. The 100 Mc NMR spectra of isabelin: (A) in CDCl₃ at 25°; and (B) in CDCl₃ at
−50°. (For the latter spectrum, the crystals were dissolved in precooled CDCl₃.) In spectrum
A, the signals for the major conformer are designated with unprimed numbers; however, at
−50° (spectrum B), this conformer appears to be the minor constituent

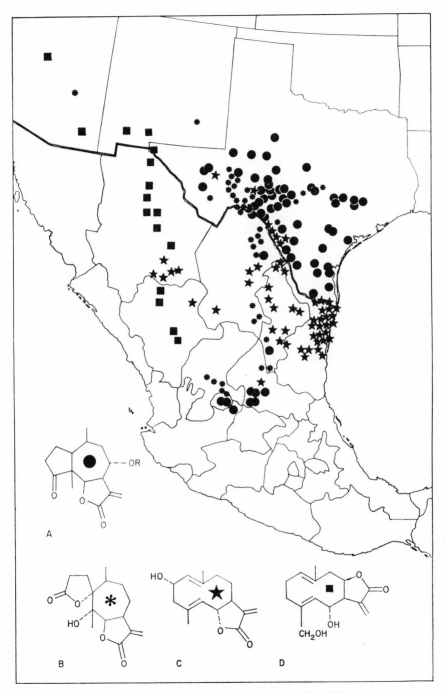

FIG. 11. Chemical races of *Ambrosia confertiflora* DC.

the region where *A. confertiflora* is sympatric with the dilactone-producing species *A. cumanensis*. A fourth major chemical race occurs in the north-western part of Mexico; it is characterized by the germacranolides (Type D) previously encountered in *A. artemisiifolia*. All the cytological data presently available for all the populations of *A. confertiflora* are reported elsewhere (Renold, 1970).

We are presently determining the volatile terpene patterns for each of the sesquiterpene lactone races detected in *A. confertiflora*; these data should per-mit further insights into the origin of the sesquiterpene lactone races and pro-vide additional information about the manner in which the species is pre-sently evolving.

IV. Concluding Statement

The terpene chemistry described in this chapter appears to be under direct genetic control since the nature and amounts of the constituents were not appreciably affected by either environmental or developmental influences; in this connection it should be mentioned that field and transplant populations which were tested at various times over the course of several years showed little variation. Therefore, the infraspecific variation of the chemical data do appear to be a valid measure of some of the dynamic evolutionary processes at work in these taxa and indicate the importance of populational studies for understand-ing the detailed structure of taxa. Some of the systematic implications of the chemical results presented in this chapter will be discussed in more detail elsewhere.

REFERENCES

Abu-Shady, H., and Soine, T. D. (1953 and 1954). *J. Am. Pharm. Ass.* **42**, 387; **43**, 365.

Alston, R. E. (1967). *In* "Evolutionary Biology" (Th. Dobzhansky, M. K. Hecht and Wm. C. Steere, eds.), pp. 197–305. Appleton-Century-Crofts, New York.

Bernardi, L., and Büchi, G. (1957). *Experientia*, **13**, 466.

Bianchi, E., Culvenor, C. C. J., and Loder, J. W. (1968). *Aust. J. Chem.* **21**, 1109.

Cavanilles, J. (1793). "Icones et Descriptiones Plantarum," Vol. 2. Matriti. (See Payne, 1964.)

Emerson, M. T., Caughlan, C. N., and Herz, W. (1966). *Tetrahedron Lett.* 3151.

Fischer, N. H., and Mabry, T. J. (1967a). *Tetrahedron*, **23**, 2529.

Fischer, N. H., and Mabry, T. J. (1967b). *Chem. Commun.* 1235.

Fischer, N. H., Mabry, T. J., and Kagan, H. B. (1968). *Tetrahedron*, **24**, 4091.

Geissman, T. A., and Levy, J. (1967). *Phytochemistry*, **6**, 899.

Geissman, T. A., and Matsueda, S. (1968). *Phytochemistry*, **7**, 1613.

Geissman, T. A., and Turley, R. J. (1964). *J. Org. Chem.* **29**, 2553.

Geissman, T. A., Turley, R. J., and Murayama, S. (1966). *J. Org. Chem.* **31**, 2269.

Geissman, T. A., Griffin, S., Waddell, T. G., and Chen, H. H. (1969). *Phytochemistry*, **8**, 145.

Herout, V., and Šorm, F. (1969). *In* "Perspectives in Phytochemistry" (J. B. Har-borne and T. Swain, eds.), pp. 139–165. Academic, London and New York.

Herz, W. (1968). *In* "Recent Advances in Phytochemistry" (T. J. Mabry, R. E. Alston and V. C. Runeckles, eds.), pp. 229–269. Appleton-Century-Crofts, New York.

Herz, W., and Högenauer, G. (1961). *J. Org. Chem.* **26**, 5011.

Herz, W., and Sumi, Y. (1964). *J. Org. Chem.* **29**, 3438.

Herz, W., Watanabe, H., Miyazaki, M., and Kishida, Y. (1962). *J. Am. Chem. Soc.* **84**, 2601.

Herz, W., Chikamatsu, H., and Tether, L. R. (1966). *J. Org. Chem.* **31**, 1632.

Herz, W., Anderson, G., Gibaja, S., and Raulais, D. (1969). *Phytochemistry*, **8**, 877.

Irwin, M. A., and Geissman, T. A. (1969). *Phytochemistry*, **8**, 305.

Joseph-Nathan, P., and Romo, J. (1966). *Tetrahedron*, **22**, 1723.

Kagan, H. B., Miller, H. E., Renold, W., Lakshmikantham, M. V., Tether, L. R., Herz, W., and Mabry, T. J. (1966). *J. Org. Chem.* **31**, 1629.

L'Homme, M. F., Geissman, T. A., Yoshioka, H., Porter, T. H., Renold, W., and Mabry, T. J. (1969). *Tetrahedron Lett.* 3161

Linnaeus, C. (1753). "Species Plantarum," Vol. 2. Holmiae. (See Payne, 1964.)

Mabry, T. J., and Mears, J. A. (1970). *In* "The Chemistry of the Alkaloids" (S. W. Pelletier, ed.) Reinhold, New York (in press)

Mabry, T. J., Kagan, H. B., and Miller, H. E. (1966a). *Tetrahedron*, **22**, 1943.

Mabry, T. J., Miller, H. E., Kagan, H. B., and Renold, W. (1966b). *Tetrahedron*, **22**, 1139.

Mabry, T. J., Renold, W., Miller, H. E., and Kagan, H. B. (1966c). *J. Org. Chem.* **31**, 681.

Mabry, T. J., Alston, R. E., and Runeckles, V. C. (eds.) (1968). "Recent Advances in Phytochemistry." Appleton-Century-Crofts, New York.

Mayr, E. (1954). *In* "Evolution as a Process" (J. Huxley, A. C. Hardy and E. B. Ford, eds.). Allen and Unwin, London.

Miller, H. E. (1967). "The Chemistry and Infraspecific Variation of Sesquiterpene Lactones in *Ambrosia psilostachya* DC. (Compositae)." Ph.D. Thesis, Univ. of Texas at Austin.

Miller, H. E., and Mabry, T. J. (1967). *J. Org. Chem.* **32**, 2929.

Miller, H. E., Kagan, H. B., Renold, W., and Mabry, T. J. (1965). *Tetrahedron Lett.* 3397.

Miller, H. E., Mabry, T. J., Turner, B. L., and Payne, W. W. (1968). *Am. J. Bot.* **55**, 316.

Payne, W. W. (1962). "Biosystematic Studies of Four Widespread Weedy Species of Ragweeds (*Ambrosia:* Compositae)". Ph.D. Thesis, The University of Michigan, Ann Arbor.

Payne, W. W. (1964). *J. of the Arnold Arboretum*, **45**, 401.

Payne, W. W., Raven, P. H., and Kyhos, D. W. (1964). *Am. J. Bot.* **51**, 419.

Porter, T. H., and Mabry, T. J. (1969). *Phytochemistry*, **8**, 793.

Porter, T. H., Mabry, T. J., Yoshioka, H., and Fischer, N. H. (1970). *Phytochemistry*, **9**, 199.

Potter, J. (1970). "Origin of the Texas Gulf Coast Island Populations of *Ambrosia Psilostachya* DC. (Compositae): A Biochemical and Numerical Systematic Investigation." Ph.D. Thesis, The University of Texas at Austin.

Renold, W. (1970). "The Chemistry and Infraspecific Variation of Sesquiterpene Lactones in *Ambrosia Confertiflora* DC. (Compositae): A Chemosystematic Study at the Population Level." Ph.D. Thesis, The University of Texas at Austin.

Romo, J., Joseph-Nathan, P., and Siade, G. (1966). *Tetrahedron*, **22**, 1499.

Romo, J., Joseph-Nathan, P., Romo de Vivar, A., and Alvarez, C. (1967). *Tetrahedron*, **23**, 529.

Romo, J., Romo de Vivar, A., and Diaz, E. D. (1968a). *Tetrahedron*, **24**, 5625.

Romo, J., Romo de Vivar, A., Vélez, A., and Urbina, E. (1968b). *Can. J. Chem.* **46**, 1535.

Steelink, C., and Spitzer, J. C. (1966). *Phytochemistry*, **5**, 357.

Swain, T., (ed.) (1966). "Comparative Phytochemistry." Academic, London and New York.

Šorm, F., Suchy, M., and Herout, V. (1959). *Coll. Czech. Chem. Commun.* **24**, 1548.

Turner, B. L. (1967). *Pure Appl. Chem.* **14**, 189.

Yoshioka, H., and Mabry, T. J. (1969). *Tetrahedron*, **25**, 4767.

Yoshioka, H., Mabry, T. J., and Miller, H. E. (1968). *Chem. Commun.* 1679.

Yoshioka, H., Renold, W., Fischer, N. H., Higo, A., and Mabry, T. J. (1970). *Phytochemistry*, in press.

Author Index

Numbers in italics are those pages on which references are listed

A

Aaronson, S., 137, *140*
Abbott, P. M., 90, *102*
Abelson, P. H., 1, *17*
Abu-Shady, H., 272, 273, *298*
Achmatowicz, O., 237, 240, *265*
Adachi, K., 65, *78*
Adams, J. M., 184, *185*
Adams, R. P., 190, 195, 196, 197, 198, *204*
Albrecht, P., 72, *77*
Alexopoulos, C. J., 88, *101*
Aliyev, K. A., 136, *142*
Allen, C. F., 107, 112, *117*, 134, *140*
Allen, M. B., 117, *118*
Allen, M. M., 121, 124, 128, *140*
Allewell, N., 158, *177*
Allsopp, A., 234, *265*
Alston, R. E., 187, 190, *204*, 235, 236, 238, 244, 245, 246, 247, 254, 258, 260, 261, 263, 264, *265*, 267, *268*, 269, 270, *298*, *299*
Altman, R. F. A., 242, *265*
Alvarez, C., 277, *299*
Anders, E., 56, *57*
Andersen, E., 52, *58*
Anderson, E., 200
Anderson, G., 272, 274, 275, 277, 278, 293, *299*
Anderson, L., 215, 229, *231*
Anderson, L. A., 137, *140*
Andrew, H. J., 14, *17*
Andrews, H. N., 24, *29*
Anfinsen, C. B., 150, *177*
Antonovics, J., 219, *230*
Aplin, R. T., 65, *77*
Appleby, R. S., 109, 110, 111, *119*
Araeva, S. M., 229, *229*
Arai, N., 216, *230*
Arata, Y., 240, *265*
Arber, A., 234, *265*
Arndt, C., 251, *265*

Aronson, J. M., 84, 85, 86, 89, 92, *101*
Ashida, J., 227, *229*
Atkin, N. B., 173, *177*
Atkinson, G. F., 98, *101*
Axelrod, D. I., 28, *29*
Azakawa, T., 211, 212, *231*

B

Bachtofen-Echt, A., 74, *77*
Bacon, J. S. D., 88, 90, *101*, *102*
Baggaley, K. H., 63, *78*
Baglioni, C., 160, *177*
Bailey, I. W., 33, *57*
Ball, G. A., 260, *265*
Ballou, C. E., 86, *102*
Banks, H. P., 14, *17*, 28, *29*
Barber, G. A., 220, 229, *230*
Barber, H. N., 202, *204*
Barchet, R., 240, *265*
Barghoorn, E. S., 5, 6, 10, 11, 12, *17*, *18*, *19*, 46, 49, 56, *57*, *58*, 81, 85, 87, *103*
Barrett, L. P., 11, *19*
Barshad, I., 85, *102*
Bartnicki-Garcia, S., 84, 86, 88, 89, 90, 91, *101*
Barto, E., 213, *231*
Basu, N., 251, *265*
Basu, U. P., 248, 250, *265*
Bate-Smith, E. C., 236, 237, 239, 240, 241, 242, 243, 244, 247, 248, 249, 250, 251, 252, 260, 264, *265*
Baxter, R. W., 24, *29*
Bayer, E., 227, *229*
Bayley, S. T., 136, *140*
Bazin, M. J., 120, *140*
Beal, E. O., 260, *265*
Beales, F. W., 10, *18*
Beck, C. W., 75, *78*
Becker, B., 91, *102*
Bellen, Z., 237, 240, *265*
Bello, J., 218, *229*

Chemical Compounds Index

Structural formulae for compounds are given on those pages indicated by a number in bold type

Genus and Species Index

Subject Index

A

Abiogenesis, 4
Acid phosphatase, inhibition of, 220–222
  roots of edaphic ecotypes, in, 220–225
  types of, in root cells, 222–224
  wall bound, 223, 224–225
Agathis resins, 76–77
Agrostis tenuis, edaphic ecotypes of, acid phosphatase in, 220–225
Albumins, seed, electrophoretic patterns of, 172
Algae (*see also* Blue-green algae)
  blue-green, fatty acids of, 114
    lipids of, 111–115
  evolutionary pattern of, 115–117
  fatty acids of, 110
  fungal ancestors, as, 99
  green, fatty acids of, 109
  lipids of, 109–115
  marine, arachidonic acid and, 109–111
Algae-like organisms, 6–7
  blue green, 11–12, 14
    complexity of, 15
  fossilization of, 22
  green, 14
Alkaloids, population variability of, 190, 191–192
Alleles, properties of, 148
Amber, Baltic, 74–76
  amount of, 74
  constituents of, 75
  succinite source in, 75
Ambrosia
  artemisiifolia, chemical races of, 294, 295
    and *psilostachya*, germacranolide-producing populations of, 295, 296
  confertiflora, races of, 295, 297–298
  cumanensis, chemical races of, 292
    and *psilostachya*, comparative chemical studies between, 293
  psilostachya, chemical races in, 289, 290–293

Ambrosia—cont.
  distribution of, 287
  Founder Principle and, 292
  sesquiterpene lactones in, infraspecific variation of, 269–298
  distribution of, 271–286
  species complex, evolutionary relationships of, 286–287
  sesquiterpene lactones in, 286–295
  species of, 278
  evolutionary affinities of, 279
Amino acid(s), code, 161
  cytochrome *c* of, 156
  hierarchy of variability in, 157, 163
  insulin, of, 154
  residues, 152–153
  sequences, analysis of, 160–167
  drift in, 163–164
  "fossil", 166
  haemoglobin, 146
  homology, 160–161
    "lining up" in, 160–161
    quantitative aspects of, 161–162
    time scales and, 163–165
    unquantitated aspects of, 162–163
    phylogenetic trees and, 165–167
    protein, 151
    variations in, 163, 166–167
  site, types of, 165
Angiosperm(s)
  aquatic, adaptive modifications of, 233
    apiose, as marker in, 264
    biochemical systematic studies in, 259–265
    biosynthetic studies in, 238, 254–259
      incorporation studies, 238, 254–255
      photocontrol in, 255, 259
    differentiation in, water and, 234
    morphology of, 233–234
    phenolic metabolism in, 235–236
    secondary constituents of, 233–265
      distribution pattern, 234–238
    sexual reproduction in, 261